Damp Indoor Spaces
AND HEALTH

Committee on Damp Indoor Spaces and Health

Board on Health Promotion and Disease Prevention

INSTITUTE OF MEDICINE
OF THE NATIONAL ACADEMIES

THE NATIONAL ACADEMIES PRESS
Washington, D.C.
www.nap.edu

THE NATIONAL ACADEMIES PRESS   500 Fifth Street, NW   Washington, DC   20001

NOTICE: The project that is the subject of this report was approved by the Governing Board of the National Research Council, whose members are drawn from the councils of the National Academy of Sciences, the National Academy of Engineering, and the Institute of Medicine. The members of the committee responsible for the report were chosen for their special competences and with regard for appropriate balance.

This study was supported by Contract No. 200-2000-0629, TO #08 between the National Academy of Sciences and Centers for Disease Control and Prevention. Any opinions, findings, conclusions, or recommendations expressed in this publication are those of the author(s) and do not necessarily reflect the view of the organizations or agencies that provided support for this project.

Library of Congress Cataloging-in-Publication Data

Institute of Medicine (U.S.). Committee on Damp Indoor Spaces and Health.
  Damp indoor spaces and health / Committee on Damp Indoor Spaces and Health, Board on Health Promotion and Disease Prevention.
      p. ; cm.
  Includes bibliographical references and index.
  ISBN 0-309-09193-4 (hardback)
  1. Indoor air pollution—Health aspects. 2. Dampness in buildings—Health aspects. 3. Air—Microbiology—Health aspects. 4. Housing and health.
  [DNLM: 1. Air Pollution, Indoor—adverse effects. 2. Air Pollution, Indoor—prevention & control. 3. Air Microbiology. 4. Bacterial Toxins—adverse effects. 5. Mycotoxins—adverse effects. 6. Respiratory Tract Diseases—epidemiology. WA 754 I5538 2004] I. Title.
  RA577.5.I565 2004
  613'.5—dc22

                                        2004014365

Additional copies of this report are available for sale from the National Academies Press, 500 Fifth Street, NW, Lockbox 285, Washington, DC 20055; call (800) 624-6242 or (202) 334-3313 (in the Washington metropolitan area); Internet, http://www.nap.edu.

For more information about the Institute of Medicine, visit the IOM home page at: **www.iom.edu.**

The serpent has been a symbol of long life, healing, and knowledge among almost all cultures and religions since the beginning of recorded history. The serpent adopted as a logotype by the Institute of Medicine is a relief carving from ancient Greece, now held by the Staatliche Museen in Berlin.

Cover: The images for the cover design were provided by Terry Brennan. The image at the center of the design is *Stachybotrys chartarum* and the border image is *Cladosporium* on paint.

"Knowing is not enough; we must apply.
Willing is not enough; we must do."
—Goethe

INSTITUTE OF MEDICINE
OF THE NATIONAL ACADEMIES

Adviser to the Nation to Improve Health

THE NATIONAL ACADEMIES
Advisers to the Nation on Science, Engineering, and Medicine

The **National Academy of Sciences** is a private, nonprofit, self-perpetuating society of distinguished scholars engaged in scientific and engineering research, dedicated to the furtherance of science and technology and to their use for the general welfare. Upon the authority of the charter granted to it by the Congress in 1863, the Academy has a mandate that requires it to advise the federal government on scientific and technical matters. Dr. Bruce M. Alberts is president of the National Academy of Sciences.

The **National Academy of Engineering** was established in 1964, under the charter of the National Academy of Sciences, as a parallel organization of outstanding engineers. It is autonomous in its administration and in the selection of its members, sharing with the National Academy of Sciences the responsibility for advising the federal government. The National Academy of Engineering also sponsors engineering programs aimed at meeting national needs, encourages education and research, and recognizes the superior achievements of engineers. Dr. Wm. A. Wulf is president of the National Academy of Engineering.

The **Institute of Medicine** was established in 1970 by the National Academy of Sciences to secure the services of eminent members of appropriate professions in the examination of policy matters pertaining to the health of the public. The Institute acts under the responsibility given to the National Academy of Sciences by its congressional charter to be an adviser to the federal government and, upon its own initiative, to identify issues of medical care, research, and education. Dr. Harvey V. Fineberg is president of the Institute of Medicine.

The **National Research Council** was organized by the National Academy of Sciences in 1916 to associate the broad community of science and technology with the Academy's purposes of furthering knowledge and advising the federal government. Functioning in accordance with general policies determined by the Academy, the Council has become the principal operating agency of both the National Academy of Sciences and the National Academy of Engineering in providing services to the government, the public, and the scientific and engineering communities. The Council is administered jointly by both Academies and the Institute of Medicine. Dr. Bruce M. Alberts and Dr. Wm. A. Wulf are chair and vice chair, respectively, of the National Research Council.

www.national-academies.org

COMMITTEE ON DAMP INDOOR SPACES AND HEALTH

Noreen M. Clark, PhD (Chair), Dean, Marshall H. Becker Professor of Public Health, and Professor of Pediatrics, University of Michigan, Ann Arbor, Michigan

Harriet M. Ammann, PhD, DABT, Senior Toxicologist, Air Quality Program, Washington State Department of Ecology, Olympia, Washington

Bert Brunekreef, PhD, Professor of Environmental Epidemiology, Institute of Risk Assessment Sciences, University of Utrecht, The Netherlands

Peyton Eggleston, MD, Professor of Pediatrics and Professor of Environment Health Sciences, Johns Hopkins University, Baltimore, Maryland

William J. Fisk, MS, PE, Senior Staff Scientist and Department Head, Indoor Environment Department, Lawrence Berkeley National Laboratory, Berkeley, California

Robert E. Fullilove, EdD, Associate Dean for Community and Minority Affairs, Columbia University School of Public Health, New York, New York

Judith Guernsey, MSc, PhD, Associate Professor, Department of Community Health and Epidemiology, Dalhousie University, Halifax, Nova Scotia, Canada

Aino Nevalainen, PhD, Head of Laboratory, Division of Environmental Health, National Public Health Institute (KTL), Kuopio, Finland

Susanna G. Von Essen, MD, Professor of Pulmonary and Critical Care Medicine, University of Nebraska Medical Center at Omaha, Nebraska

Consultants to the Committee
Terry Brennan, MS, President, Camroden Associates, Inc., Westmoreland, New York

Jeroen Douwes, PhD, Associate Director, Centre for Public Health Research, Massey University, Wellington, New Zealand

Staff
David A. Butler, PhD, Study Director
Jennifer A. Cohen, Research Associate
Joe A. Esparza, Senior Project Assistant
Elizabeth J. Albrigo, Project Assistant
Norman Grossblatt, Senior Editor
Rose Marie Martinez, ScD, Director, Board on Health Promotion and Disease Prevention

Reviewers

This report has been reviewed in draft form by persons chosen for their diverse perspectives and technical expertise, in accordance with procedures approved by the National Research Council's Report Review Committee. The purpose of this independent review is to provide candid and critical comments that will assist the institution in making its published report as sound as possible and to ensure that the report meets institutional standards of objectivity, evidence, and responsiveness to the study charge. The review comments and draft manuscript remain confidential to protect the integrity of the deliberative process. We wish to thank the following for their review of this report:

Diane R. Gold, MD, MPH, Harvard Medical School and Harvard School of Public Health

William B. Rose, MArch, School of Architecture, University of Illinois at Urbana-Champaign

Jonathan M. Samet, MD, Bloomberg School of Public Health, Johns Hopkins University

Richard J. Shaughnessy, PhD, Indoor Air Program, University of Tulsa

Linda D. Stetzenbach, PhD, Harry Reid Center for Environmental Studies, University of Nevada, Las Vegas

Mark J. Utell, MD, University of Rochester School of Medicine and Dentistry

Although the reviewers listed above have provided many constructive comments and suggestions, they were not asked to endorse the conclusions or recommendations, nor did they see the final draft of the report before its release. The review of this report was overseen by **Robert B. Wallace, MD,** University of Iowa, and **John C. Bailar III, MD, PhD,** University of Chicago. Appointed by the National Research Council and Institute of Medicine, they were responsible for making certain that an independent examination of this report was carried out in accordance with institutional procedures and that all review comments were carefully considered. Responsibility for the final content of this report rests entirely with the authoring committee and the institution.

Acknowledgments

This report could not have been prepared without the guidance and expertise of numerous persons. Although it is not possible to mention by name all those who contributed to the committee's work, the committee wants to express its gratitude to a number of them for their special contributions.

Sincere thanks go to all the participants at the workshops convened on March 26, June 17, and October 8, 2002. The intent of the workshops was to gather information regarding issues related to damp indoor spaces, health effects attributed to microbial agents found indoors, and mold- and moisture-related research. The speakers, who are listed in Appendix A, gave generously of their time and expertise to help inform and guide the committee's work.

We are deeply indebted to two hard-working people—Terry Brennan and Jeroen Douwes—who served as consultants and made major contributions to the content of this report. Special thanks are also extended to Harriet Burge, chair of the committee from its inception through October 2002, for her exceptional commitment and guidance during her tenure. The committee also thanks Ulla Haverinen-Shaughnessy and Anne Hyvärinen, who permitted excerpting of text from their doctoral dissertations. Institute of Medicine staff members Michelle Catlin, Ben Hamlin, and Michael Schneider provided valuable input and help over the course of the study. The Committee on Damp Indoor Spaces and Health, of course, takes final responsibility for all content in the report.

The committee extends special thanks to the dedicated and hard-working staff at the Institute of Medicine (IOM). The expertise and leadership of Rose Marie Martinez, director of the IOM Board on Health Promotion and Disease Prevention, helped to ensure that this report met the highest standards of quality.

Finally, the committee would like to thank the chair, Noreen Clark, for her outstanding work, leadership, and dedication to this project.

Contents

Public Health Approaches to Damp Indoor Environments, 314
Findings, Recommendations, and Research Needs, 327
References, 329

APPENDIXES

Executive Summary

A damp spot appears in a ceiling after an intense rainstorm; a hose loosens from a washing machine, spilling gallons of water onto a basement floor; weeks after a moldy odor is detected, a plumber finds a slow leak behind a wall. There are over 119 million housing units in the United States and nearly 4.7 million commercial buildings (U.S. Census Bureau, 2003), and almost all of them experience leaks, flooding, or other forms of excessive indoor dampness at some time.

Excessive indoor dampness is not by itself a cause of ill health, but it is a determinant of the presence or source strength of several potentially problematic exposures. Damp indoor environments favor house dust mites and microbial growth, standing water supports cockroach and rodent infestations, and excessive moisture may initiate chemical emissions from building materials and furnishings.

Indoor microbial growth—especially fungal growth—has recently received a great deal of attention in the mass media. It is a prominent feature of the breakdown of dampness control; its many possible causes include a breach of the building envelope, failure of a water-use device, and excessive indoor water-vapor generation. Occupants, health professionals, and others have wondered whether indoor exposure to mold and other agents might have a role in adverse health outcomes experienced by occupants of damp buildings. Prominent among these health outcomes is acute idiopathic pulmonary hemorrhage in infants, cases of which were reported in Cleveland, Ohio in the 1990s. Residence in homes with recent water damage and in homes with visible mold (including *Stachybotrys chartarum*) was among the risk factors identified in the case infants.

Against that backdrop, the Centers for Disease Control and Prevention (CDC) asked the Institute of Medicine to convene a committee of experts. CDC provided the following charge to that committee:

> The Institute of Medicine will conduct a comprehensive review of the scientific literature regarding the relationship between damp or moldy indoor environments and the manifestation of adverse health effects, particularly respiratory and allergic symptoms. The review will focus on the non-infectious health effects of fungi, including allergens, mycotoxins and other biologically active products. In addition, it will make recommendations or suggest guidelines for public health interventions and for future basic science, clinical, and public health research in these areas.

FRAMEWORK AND ORGANIZATION

Figure ES-1 describes the path by which water or moisture sources may lead to excessive indoor dampness and to exposures that may result in adverse health outcomes. The elements of this framework are reflected in the major topics addressed in the report:

- How and where buildings become wet, the signs of dampness, how dampness is measured, the risk factors for moisture problems, and what is known about their prevalence, severity, location, and duration (Chapter 2).
- How dampness influences indoor microbial growth and chemical emissions, the various agents that may be present in damp environments, and the influence of building materials on microbial growth and emissions (Chapter 2).
- The means available for assessing exposure to microorganisms and microbial agents that occur in damp indoor environments (Chapter 3).
- The experimental data on the nonallergic biologic effects of molds and bacteria, including the bioavailability of mycotoxins and toxic effects seen in cellular (in vitro) and animal (in vivo) toxicity studies of mycotoxin and bacterial toxin exposure (Chapter 4).
- The state of the scientific literature regarding health outcomes and indoor exposure to dampness and dampness-related agents (Chapter 5).
- Dampness prevention strategies, published guidelines for the removal of fungal growth (remediation), remediation protocols, and research on the effectiveness of various cleaning strategies (Chapter 6).
- The public health implications of damp indoor environments and the elements of a public health response (Chapter 7).

The committee faced a substantial challenge in conducting its review of these topics—research on fungi and other dampness-related agents is bur-

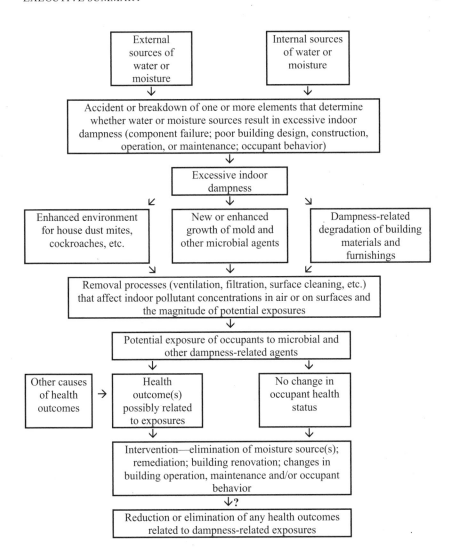

FIGURE ES-1 A framework describing the potential for water and moisture sources to lead to excessive indoor dampness and to exposures that may result in adverse health outcomes.

geoning, and important new papers are constantly being published. Although the committee did its best to paint an accurate picture of the state of the science at the time its report was completed in late 2003, it is inevitable that research advances will extend beyond the report's findings.

The sections below are a synopsis of the committee's major findings and recommendations, and the research needs they identified. Chapters 2–7 detail the reasoning underlying these and present the committee's complete findings.

THE COMMITTEE'S EVALUATION

Damp Buildings

The term *dampness* has been used to define a variety of moisture problems in buildings, including high relative humidity, condensation, water ponding, and other signs of excess moisture or microbial growth. While studies report that dampness is prevalent in residential housing in a wide array of climates, attempts to understand its scale and significance are hampered by the fact that there is no generally accepted definition of *dampness* or of what constitutes a "dampness problem."

There is no single cause of excessive indoor dampness, and the primary risk factors for it differ across climates, geographic area, and building types. Although the prevalence of dampness problems appears to increase as buildings age and deteriorate, the experience of building professionals suggests that some modern construction techniques and materials and the presence of air-conditioning also increase the risk of dampness problems. The prevalence and nature of dampness problems suggest that what is known about their causes and prevention is not consistently applied in building design, construction, maintenance, and use.

One consequence of indoor dampness is new or enhanced growth of fungi and other microbial agents. The fungi have (eukaryotic) cells like animals and plants, but are a separate kingdom. Most consist of masses of filaments, live off of dead or decaying organic matter, and reproduce by spores. Visible fungal colonies found indoors are commonly called mold or sometimes mildew. This report, following the convention of much of the literature on indoor environments, uses the terms *fungus* and *mold* interchangeably to refer to the microorganisms.

Mold spores are regularly found in indoor air and on surfaces and materials—no indoor space is free of them. There are a large number of species and genera, and those most typically found indoors vary by geographic area, climate, season, and other factors. The availability of moisture is the primary factor that controls mold growth indoors, since the nutrients and temperature range they need are usually present. While much attention is focused on mold growth indoors, it is not the only dampness-related microbial agent. Mold growth is usually accompanied by bacterial growth. Some research on fungi and bacteria focuses on specific components that may be responsible for particular health effects: spores and hyphal fragments of fungi, spores and cells of bacteria, allergens of microbial origin, structural components of fungal and bacterial cells, and such products as microbial volatile organic compounds (MVOCs) and mycotoxins. Release of these components varies, depending on many physiologic and environmental factors. Dampness can also damage building materials and furnishings, causing or exacerbating the release of chemicals and non-biologic particles from them.

Given the present state of the literature, the committee identified several kinds of research needs. Standard definitions of dampness, metrics, and associated dampness-assessment protocols need to be developed to characterize the nature, severity, and spatial extent of dampness. Precise, agreed-on definitions will allow important information to be gathered about the determinants of dampness problems in buildings and the mechanisms by which dampness and dampness-related effects and exposures affect occupant health. More than one definition may be required to meet the specific needs of health researchers (epidemiologists, physicians, and public-health practitioners) in contrast with those involved in preventing or remediating dampness (architects, engineers, builders, and those involved in building maintenance). However, definitions should be standardized to the extent possible. Any efforts to establish common definitions must be international in scope because excessive indoor dampness is a worldwide problem and research cooperation will promote the generation and dissemination of knowledge.

Research is also needed to better characterize the dampness-related emissions of fungal spores, bacteria, and other particles of biologic origin and their role in human health outcomes; the microbial ecology of buildings, that is, the link between dampness, different building materials, microbial growth, and microbial interactions; and dampness-related chemical emissions from building materials and furnishings, and their role in human health outcomes. Studies should be conducted to evaluate the effect of the duration of moisture damage of materials and its possible influence on occupant health and to evaluate the effectiveness of various changes in building designs, construction methods, operation, and maintenance in reducing dampness problems. Increased attention should be paid to heating, ventilation, and air conditioning (HVAC) systems as a potential site for the growth and dispersal of microbial contaminants that may result in adverse health effects in building occupants. And research should be performed to develop designs and construction and maintenance practices for buildings and HVAC systems that reduce moisture problems; building materials that are less prone to microbial contamination when moist; and standard, effective protocols for clean-up after flooding and other catastrophic water events that will minimize microbial growth.

Exposure Assessment

The lack of knowledge regarding the role of microorganisms in the development and exacerbation of diseases found in occupants of damp indoor environments is due largely to the lack of valid quantitative exposure-assessment methods and knowledge of which specific microbial agents may primarily account for the presumed health effects. Very few biomarkers of exposure to or dose of biologic agents have been identified, and their validity for exposure assessment in the indoor environment is often not known. The entire process of fungal-spore aerosolization, transport, deposition, resuspension, and tracking—all of which determine inhalation exposure—is poorly understood, as is the significance of exposures to fungi through dermal contact and ingestion.

There are several methods for measuring and characterizing fungal populations, but methods for assessing human exposure to fungal agents are poorly developed and are a high-priority research need. Part of the difficulty is related to the large number of fungal species that are measurable indoors and the fact that fungal allergen content and toxic potential vary among species and among morphologic forms within species. In addition, the most common methods for fungal assessment—counting cultured colonies and identifying and counting spores—have variable and uncertain relationships to allergen, toxin, and irritant content of exposures.

Based on their review of the literature, the committee recommends that existing exposure assessment methods for fungal and other microbial agents be subjected to rigorous validation and that they be further refined to make them more suitable for large-scale epidemiologic studies. This includes standardization of protocols for sample collection, transport of samples, extraction procedures, and analytical procedures and reagents. Such work should result in concise, internationally accepted protocols that will allow measurement results to be compared both within and across studies.

The committee also identified a need to develop improved exposure assessment methods, particularly methods based on nonculture techniques and techniques for measuring constituents of microorganisms—allergens, endotoxins, $\beta(1\rightarrow3)$-glucans, fungal extracellular polysaccharides, fungal spores, and other particles and emissions of microbial origin. These needs include further improvement of light and portable personal airborne exposure measurement technology, more rapid development of measurement methods for specific microorganisms that use DNA-based and other technology, and rapid and direct-reading assays for bioaerosols for the immediate evaluation of potential health risks. Application of the improved or new methods will allow more valid exposure assessment of microorganisms and their components, which should facilitate more-informed risk assessments.

Because only sparse data are available on variation in exposure to biologic agents in the home environment, it is not possible to recommend how many samples should be taken to produce an accurate assessment of risk-relevant exposure.

Toxic Effects of Fungi and Bacteria

Although a great deal of attention has focused on the effects of bacteria and fungi mediated by allergic responses, these microorganisms also cause nonallergic responses. Toxicologic studies, which examine such responses using animal and cellular models, cannot be used by themselves to draw conclusions about human health effects. However, animal studies are important in identifying hazardous substances, defining their target organs or systems and their routes of exposure, and elucidating their toxicokinetics and toxicodynamics, the mechanisms that account for the biologic effects, metabolism, and excretion of toxic substances. Animal studies are also useful for generating hypotheses that can be tested through studies of human health outcomes in controlled exposures, clinical studies, or epidemiologic investigations, and they are useful for risk assessment that informs regulatory and policy decisions.

Research reviewed in Chapter 4 shows that molds that can produce mycotoxins under the appropriate environmental and competitive conditions can and do grow indoors. Damp indoor spaces may also facilitate the growth of bacteria that can have toxic and inflammatory effects. Little information exists on the toxic potential of chemical releases resulting from dampness-related degradation of building materials, furniture, and the like.

In vitro and in vivo studies have demonstrated adverse effects—including immunotoxic, neurologic, respiratory, and dermal responses—after exposure to specific toxins, bacteria, molds, or their products. Such studies have established that exposure to microbial toxins can occur via inhalation and dermal exposure and through ingestion of contaminated food. Animal studies provide information on the potency of many toxins isolated from environmental samples and substrates from damp buildings, but the doses of such toxins required to cause adverse health effects in humans have not been determined. In vitro and in vivo research on *Stachybotrys chartarum* suggests that effects in humans may be biologically plausible, although this observation requires validation from more extensive research before conclusions can be drawn.

Among the other research needs identified in the chapter is further development of techniques for detecting and quantifying mycotoxins in tissues in order to inform questions of interactions and the determination of exposures resulting in adverse effects. The committee also recommends that animal studies be initiated to evaluate the effects of long-term (chronic) exposures to mycotoxins via inhalation. Such studies should establish dose-response, lowest-observed-adverse-effect levels, and no-observed-adverse-effect levels for identified toxicologic endpoints in order to generate information for risk assessment that is not available from presently-available studies of acute, high-level exposures.

Human Health Effects Associated with Damp Indoor Environments

The committee used a uniform set of categories to summarize its conclusions regarding the association between health outcomes and exposure to indoor dampness or the presence of mold or other agents in damp indoor environments, as listed in Box ES-1. The distinctions among categories reflect the committee's judgment of the overall strength, quality, and persuasiveness of the scientific literature evaluated. Chapter 1 details the methodologic considerations underlying the evaluation of epidemiologic evidence and details the definitions of the categories.

BOX ES-1
Summary of the Categories of Evidence Used in This Report

Sufficient Evidence of a Causal Relationship

Evidence is sufficient to conclude that a causal relationship exists between the agent and the outcome. That is, the evidence fulfills the criteria for "sufficient evidence of an association" and, in addition, satisfies the following criteria: strength of association, biologic gradient, consistency of association, biologic plausibility and coherence, and temporally correct association.

Sufficient Evidence of an Association

Evidence is sufficient to conclude that there is an association. That is, an association between the agent and the outcome has been observed in studies in which chance, bias, and confounding can be ruled out with reasonable confidence.

Limited or Suggestive Evidence of an Association

Evidence is suggestive of an association between the agent and the outcome but is limited because chance, bias, and confounding cannot be ruled out with confidence.

Inadequate or Insufficient Evidence to Determine Whether an Association Exists

The available studies are of insufficient quality, consistency, or statistical power to permit a conclusion regarding the presence of an association. Alternatively, no studies exist that examine the relationship.

Tables ES-1 and ES-2 summarize the committee's findings. The conclusions are not applicable to persons with compromised immune systems, who are at risk for fungal colonization and opportunistic infections.

Conclusions regarding exposure to agents associated with damp indoor environments are limited by the means used to assess exposure in the epidemiologic studies reviewed by the committee. For the most part, studies have relied on occupants' observations of the presence of "mold" or "moldy odor." Relatively few research efforts have used trained observers or measurements to attempt to discern which microbial agents are present, the extent of their growth, or whether there are specific common potential exposures (other than dampness). When the committee is drawing a conclusion about the association between exposure to a damp indoor environment and a health outcome, it is not imposing the assumption that the outcome is necessarily a result of exposure to a particular mold or to microbial agents in general. In some circumstances, a paper addresses the association between a particular indoor dampness-related exposure and a health outcome. However, even in those cases, it is likely that people are being exposed to multiple agents.

The committee has drawn conclusions about the state of the scientific literature regarding associations of health outcomes with two circumstances:

TABLE ES-1   Summary of Findings Regarding the Association Between Health Outcomes and Exposure to Damp Indoor Environments[a]

Sufficient Evidence of a Causal Relationship (no outcomes met this definition)	
Sufficient Evidence of an Association	
Upper respiratory (nasal and throat) tract symptoms	Wheeze
Cough	Asthma symptoms in sensitized asthmatic persons
Limited or Suggestive Evidence of an Association	
Dyspnea (shortness of breath)	Asthma development
Lower respiratory illness in otherwise-healthy children	
Inadequate or Insufficient Evidence to Determine Whether an Association Exists	
Airflow obstruction (in otherwise-healthy persons)	Skin symptoms
Mucous membrane irritation syndrome	Gastrointestinal tract problems
Chronic obstructive pulmonary disease	Fatigue
Inhalation fevers (nonoccupational exposures)	Neuropsychiatric symptoms
Lower respiratory illness in otherwise-healthy adults	Cancer
Acute idiopathic pulmonary hemorrhage in infants	Reproductive effects
	Rheumatologic and other immune diseases

[a]These conclusions are not applicable to immunocompromised persons, who are at increased risk for fungal colonization or opportunistic infections.

TABLE ES-2 Summary of Findings Regarding the Association Between Health Outcomes and the Presence of Mold or Other Agents in Damp Indoor Environments[a]

Sufficient Evidence of a Causal Relationship
(no outcomes met this definition)

Sufficient Evidence of an Association

Upper respiratory (nasal and throat) tract symptoms	Wheeze
Asthma symptoms in sensitized asthmatic persons	Cough
Hypersensitivity pneumonitis in susceptible persons[b]	

Limited or Suggestive Evidence of an Association
Lower respiratory illness in otherwise-healthy children

Inadequate or Insufficient Evidence to Determine Whether an Association Exists

Dyspnea (shortness of breath)	Skin symptoms
Airflow obstruction (in otherwise-healthy persons)	Asthma development
Mucous membrane irritation syndrome	Gastrointestinal tract problems
Chronic obstructive pulmonary disease	Fatigue
Inhalation fevers (nonoccupational exposures)	Neuropsychiatric symptoms
Lower respiratory illness in otherwise-healthy adults	Cancer
Rheumatologic and other immune diseases	Reproductive effects
Acute idiopathic pulmonary hemorrhage in infants	

[a]These conclusions are not applicable to immunocompromised persons, who are at increased risk for fungal colonization or opportunistic infections.
[b]For mold or bacteria in damp indoor environments.

exposure to a damp indoor environment, and the presence of mold or other agents in a damp indoor environment. As already noted, the term *dampness* has been applied to a variety of moisture problems in buildings. Most of the studies considered by the committee did not specify which agents were present in the buildings occupied by subjects, and this probably varied between and even within study populations.

The committee found **sufficient evidence of an association** between exposure to damp indoor environments and some respiratory health outcomes: upper respiratory tract (nasal and throat) symptoms, cough, wheeze, and asthma symptoms in sensitized asthmatic persons. Epidemiologic studies also indicate that there is sufficient evidence to conclude that the presence of mold (otherwise unspecified) indoors is associated with upper respiratory symptoms, cough, wheeze, asthma symptoms in sensitized asthmatic persons, and hypersensitivity pneumonitis (a relatively rare immune-mediated condition) in susceptible persons.

Limited or suggestive evidence was found for an association between exposure to damp indoor environments and dypsnea (the medical term for shortness of breath), lower respiratory illness in otherwise-healthy children, and the development of asthma in susceptible persons. It is not clear whether the latter association reflects exposure to fungi or bacteria or their constituents and emissions, to other exposures related to damp indoor environments, such as dust mites and cockroaches, or to some combination thereof. The responsible factors may vary among individuals. For the presence of mold (otherwise unspecified) indoors, there is limited or suggestive evidence of an association with lower respiratory illness in otherwise-healthy children.

Inadequate or insufficient information was identified to determine whether damp indoor environments or the agents associated with them are related to a variety of health outcomes listed in Tables ES-1 and -2. Included among these is acute idiopathic pulmonary hemorrhage in infants (AIPHI). The committee concluded that the available case-report information constitutes inadequate or insufficient information to determine whether an association exists between AIPHI and the presence of *Stachybotrys chartarum* or exposure to damp indoor environments in general. AIPHI is a serious health outcome, and the committee encourages the CDC to pursue surveillance and additional research on the issue to resolve outstanding questions.

The committee considered whether any of the health outcomes listed above met the definitions for the categories "sufficient evidence of a causal relationship" and "limited or suggestive evidence of no association" defined in Chapter 1, and concluded that none did.

It offers some additional observations on research needs and recommendations for action:

• Indoor environments subject occupants to multiple exposures that may interact physically or chemically with one another and with the other characteristics of the environment, such as humidity, temperature, and ventilation rate. Few studies to date have considered whether there are additive or synergistic interactions among these factors. The committee encourages researchers to collect and analyze data on a broad range of exposures and factors characterizing indoor environments in order to inform these questions and possibly point the way toward more effective and efficient intervention strategies.

• The committee encourages the CDC to pursue surveillance and additional research on acute pulmonary hemorrhage or hemosiderosis in infants to resolve questions regarding this serious health outcome. Epidemiologic and case studies should take a broad-based approach to gathering and evaluating information on exposures and other factors that would help to elucidate the etiology of acute pulmonary hemorrhage or hemosiderosis in infants, including dampness and agents associated with damp indoor environments; environmental tobacco smoke (ETS) and other potentially adverse exposures; and social and cultural circumstances, race/ethnicity, housing conditions, and other determinants of study subjects' health.

• Concentrations of organic dust consistent with the development of organic dust toxic syndrome are very unlikely to be found in homes or public buildings. However, clinicians should consider the syndrome as a possible explanation of symptoms experienced by some occupants of highly contaminated indoor environments.

• Greater research attention to the possible role of damp indoor environments and the agents associated with them in less well understood disease entities is needed to address gaps in scientific knowledge and concerns among the public.

Prevention and Remediation of Damp Indoor Environments

Homes and other buildings should be designed, operated, and maintained to prevent water intrusion and excessive moisture accumulation when possible. When water intrusion or moisture accumulation is discovered, the source should be identified and eliminated as soon as practicable to reduce the possibility of problematic microbial growth and building-material degradation. The most effective way to manage microbial contaminants, such as mold, that are the result of damp indoor environments is to eliminate or limit the conditions that foster its establishment and growth. That also restricts the dampness-related degradation of building materials and furnishings.

Information is available on the sources of excessive indoor dampness and on the remediation of damp indoor conditions and its adverse consequences. Chapter 6 summarizes several sources of guidance on how to respond to various indoor microbial contamination situations. However, as the committee observes, determining when a remediation effort is warranted or when it is successful is necessarily subjective because there are no generally accepted health-based standards for acceptable concentrations of fungal spores, hyphae, or metabolites in the air or on surfaces.

There is a great deal of uncertainty and variability in samples of mold and other microbial materials taken from indoor air and surfaces, but the information gained from a careful and complete survey may aid in the evaluation of contamination sources and remediation needs. Visible surfaces and easily accessible spaces are not the only source of microbial contaminants, however, and the potential for exposure from sources in spaces such as attics, crawl spaces, wall cavities, and other hidden or seldom-accessed areas is poorly understood.

When microbial contamination is found, it should be eliminated by means that not only limit the possibility of recurrence but also limit exposure of occupants and persons conducting the remediation. Disturbance of contaminated material during remediation activities can release microbial particles and result in contamination of clean areas and exposure of occupants and remediation workers. Containment during clean-up (through the erection of barriers, application of negative air pressure, and other means) has been shown to prevent the spread of microbial particles to noncontaminated parts of a contaminated building. The amount of containment and worker personal protection and the determination of whether occupant evacuation is appropriate depend on the magnitude of contamination.

Notwithstanding the interest in the topic, very few controlled studies have been conducted on the effectiveness of remediation actions in eliminating problematic microbial contamination in the short and long term or on the effect of remediation actions on the health of building occupants. In addition, the available literature addresses the management of microbial contamination when remediation is technically and economically feasible. There is no literature addressing situations where intervening in the moisture dynamic or cleaning or removing contaminated materials is not practicable.

Among the research needs identified by the committee are studies that better characterize the effectiveness of remediation assessment and remediation methods in different contamination circumstances, the dynamics of movements of contaminants from colonies of mold and other microorganisms in spaces such as attics, crawl spaces, exterior sheathing, and garages, and the effectiveness of various means of protection of workers and occupants during remediation activities. Standard methods should be formulated to assess the potential of new materials, designs, and construction practices to cause or exacerbate dampness problems. And research should be performed to address the other data gaps discussed above and to determine

- How free of microbial contamination a surface or building material must be to eliminate problematic exposure of occupants and in particular, how concentrations of microbial contamination left after remediation are related to those found on ordinary surfaces and materials in buildings where no problematic contamination is present.
- Whether and when microbial contamination that is not visible to the naked eye but is detectable through screening methods should be remediated.
- The best ways to open a wall or other building cavity to seek hidden contamination while controlling the release of spores, microbial fragments, and the like.
- The effectiveness of managing contamination in place by using negative air pressure, encapsulation, and other means of isolation.
- How to measure the effectiveness and health effects of a remediation effort.

The Public Health Response

On the basis of its review of the scientific papers and other information summarized above and detailed in the report, the committee concludes that excessive indoor dampness is a public-health problem. An appropriate public health goal should thus be to prevent or reduce the incidence of potentially problematic damp indoor environments, that is, environments that may be associated with undesirable health effects, particularly in vulnerable populations. However, there are serious challenges associated with achieving that goal. As the report indicates, there is insufficient information on which to base quantitative recommendations for either the appropriate level of dampness reduction or the "safe" level of exposure to dampness-related agents. The relationship between dampness or particular dampness-related agents and health effects is sometimes unclear and in many cases indirect. Questions of exposure and dose have not, by and large, been resolved. It is also not possible to objectively rank dampness-related health problems within the larger context of threats to the public's health because there is insufficient information available to confidently quantify the overall magnitude of the risk resulting from exposures in damp indoor environments.

Institutional and social barriers may hinder the widespread adoption of technical measures and practices that could prevent or reduce problematic indoor dampness. Economic factors, for example, encourage poor practice or impede remediations; they may also create incentives to forgo or limit investment in maintenance that might help to prevent moisture problems.

Given these challenges, the committee identifies seven areas of endeavor that deserve discussion in the formulation of public health mechanisms to prevent or reduce the incidence of damp indoor environments:

• Assessment and monitoring of indoor environments at risk for problematic dampness.

• Modification of regulations, building codes, and building-related contracts to promote healthy indoor environments; and enforcement of existing rules.

• Creation of incentives to construct and maintain healthy indoor environments; and financial assistance for remediation where needed.

• Development, dissemination, and implementation of guidelines for the prevention of dampness-related problems.

• Public-health-oriented research and demonstration projects to evaluate the short-term and long-term effectiveness of intervention strategies.

• Education and training of building occupants, health professionals, and people involved in the design, construction, management, and maintenance of buildings to improve efforts to avoid or reduce dampness and dampness-related health risks.

• Collaborations among stakeholders to achieve healthier indoor environments.

Among the recommendations the committee offers for implementing the actions it suggests are these:

• CDC, other public-health-related, and building-management-related funders should provide new or continuing support for research and demonstration projects that address the potential and relative benefit of various strategies for the prevention or reduction of damp indoor environments, including data acquisition through assessment and monitoring, building code modification or enhanced enforcement, contract language changes, economic and other incentives, and education and training. These projects should include assessments of the economic effects of preventing building dampness and repairing damp buildings and should evaluate the savings generated from reductions in morbidity and gains in the useful life of structures and their components associated with such interventions.

• Carefully designed and controlled longitudinal research should be undertaken to assess the effects of population-based housing interventions on dampness and to identify effective and efficient strategies. As part of such studies, attention should be paid to definitions of dampness and to measures of effect; and the extent to which interventions are associated with decreased occurrence of specific negative health conditions should be assessed when possible.

• Government agencies with housing-management responsibility should evaluate the benefit of adopting economic-incentive programs designed to reward actions that prevent or reduce building dampness. Ideally, these should be coupled with independent assessments of effectiveness.

• HUD or another appropriate government agency with responsibility for building issues should provide support for the development and dissemination of consensus guidelines on building design, construction, operation, and maintenance for prevention of dampness problems. Development of the guidelines should take place at the national level and should be under the aegis of either a government body or an independent nongovernment organization that is not affiliated with the stakeholders on the issue.

• CDC and other public-health-related funders should provide new or continuing support for research and demonstration projects that:

— Develop communication instruments to disseminate information derived from the scientific evidence base regarding indoor dampness, mold and other dampness-related exposures, and health outcomes to address public concerns about the risk from dampness-related exposures, indoor conditions, and causes of ill health.

— Foster education and training for clinicians and public-health professionals on the potential health implications of damp indoor environments.

• Government and private entities with building design, construction, and management interests should provide new or continuing support for research and demonstration projects that develop education and training for building professionals (architects, home builders, facility managers and maintenance staff, code officials, and insurers) on how and why dampness problems occur and how to prevent them.

• Those formulating the education and training programs discussed above should include means of evaluating whether their programs are reaching relevant persons and, ideally, whether they materially affect the occurrence of moisture or microbial contamination in buildings or occupant health.

REFERENCE

U.S. Census Bureau. 2003. Statistical Abstract of the United States 2002. United States Department of Commerce.

1

Background and Methodologic Considerations

INTENT AND GOALS OF THE STUDY

The Centers for Disease Control and Prevention (CDC) charged the Institute of Medicine (IOM) committee responsible for this report to conduct a comprehensive review of the scientific literature regarding the relationship between damp or moldy indoor environments and the manifestation of adverse health effects, particularly respiratory and allergic symptoms. The request came against the backdrop of escalating public and scientific community interest in the question of whether indoor exposure to mold and other agents might have a role in adverse health outcomes experienced by occupants of damp buildings. Prominent among these health outcomes is acute idiopathic pulmonary hemorrhage in infants, cases of which were reported in Cleveland, Ohio in the 1990s. Residence in homes with recent water damage and in homes with visible mold (including *Stachybotrys chartarum*) were among the risk factors identified in the case infants.

The CDC requested that the review focus on fungi and their secondary metabolites, including mycotoxins. Several issues were identified for consideration:

- The effect of damp indoor spaces on health.
- The relationship between damp indoor spaces and fungi.
- The characterization of fungal growth in homes, including the definition of the specific ecologic niches that fungi exploit in water-damaged areas.

- The conditions needed for toxin and allergen production.
- Methods of detecting fungi and secondary metabolites in indoor environments.
- Mechanisms of exposure to fungi and secondary metabolites.
- Respiratory health effects of fungal exposure, including allergic effects.
- The role of secondary metabolites—in particular, mycotoxins—in adverse health outcomes.
- Pathologies associated with mycotoxins in pulmonary tissue.
- The synergistic interaction of molds and their toxins with biologic and chemical agents in the environment.
- The evidence for and effectiveness of prevention, control, and management of exposures related to fungi.

However, CDC indicated that the committee should exercise its own judgment concerning the topics to address in its report.

The committee operationalized this charge by establishing 7 broad areas of inquiry:

- How and where buildings become wet, the signs of dampness, how dampness is measured, the risk factors for moisture problems, and what is known about their prevalence, severity, location, and duration.
- How dampness influences indoor microbial growth and chemical emissions, the various agents that may be present in damp environments, and the influence of building materials on microbial growth and emissions.
- The means available for assessing exposure to microorganisms and microbial agents that occur in damp indoor environments.
- The experimental data on the nonallergic biologic effects of molds and bacteria, including the bioavailability of mycotoxins and toxic effects seen in in vitro and animal toxicity studies of mycotoxin and bacterial toxin exposure.
- The state of the scientific literature regarding health outcomes and indoor exposure to dampness and dampness-related agents.
- Dampness prevention strategies, published guidelines for the removal of fungal growth (remediation), remediation protocols, and research on the effectiveness of various cleaning strategies.
- The public health implications of damp indoor environments and the elements of a public health response to the issue.

The ensuing chapters of the report address these topics to the extent permitted by currently available science. Because there are great differences in the amount and type of information available on specific topics, the discussions vary in their depth and focus.

The remainder of this chapter describes how the committee evaluated the evidence it reviewed. It discusses the committee's research approach, and—for the epidemiologic evidence used to assess questions about health outcomes—the methodologic considerations underlying the evaluation of information, considerations in assessing the strength of the evidence, and the categories used to summarize the committee's conclusions. Some text is derived from Chapter 2 of the report *Clearing the Air* (IOM, 2000). Institute of Medicine (IOM) reports characterizing scientific evidence regarding vaccine safety (IOM, 1991, 1993) and the health effects of herbicides used in Vietnam (IOM, 1994, 1996, 1999, 2001, 2003) have used similar approaches to summarize epidemiologic evidence.

RESEARCH APPROACH

Information Gathering

To answer the questions posed by CDC, the committee undertook a wide-ranging evaluation of the research on the determinants, characterization, and remediation of damp indoor spaces and the possible association of dampness or dampness-related agents with occupant health. While it did not review all such literature—an undertaking beyond the scope of this report—the committee attempted to cover the work it believed to be influential in shaping scientific understanding at the time it completed its task in late 2003.

The committee consulted several sources of information in the course of its work. For conclusions regarding health outcomes, the primary source was epidemiologic studies. Most of those studies examined general population exposures to dampness or indoor agents at home, reflecting the focus of researchers working in this field. Some clinical research was considered where appropriate. Animal (in vivo) and cellular (in vitro) studies were examined in the review of toxicologic literature. The literature of engineering, architecture, and the physical sciences informed the committee's discussions of building characteristics, exposure assessment and characterization, indoor dampness, pollutant transport, and related topics; and public health and behavioral sciences research was consulted for the discussion of public health implications. These disciplines have different practices regarding the publication of research results. There are, for example, relatively few papers in the peer-reviewed literature that address building construction or maintenance issues. The committee endeavored in all cases to identify, review, and consider fairly the literature most relevant to the topics it was charged to address.

Studies and reports were identified for review through extensive searches of relevant databases. The majority of these were bibliographic, providing citations to peer-reviewed scientific literature. Factual databases were also searched to provide toxicological, demographic, and other information. Committee staff examined the reference lists of major review articles, books, and reports for relevant citations. Reference lists of individual articles were also scanned for additional relevant references. Committee members independently compiled lists of potential citations, based on their areas of expertise. The input received both in written and oral form from participants at the public meetings served as a valuable source of additional information. If an initial examination revealed that the study addressed agents or means of exposure that were not relevant to the indoor environments being evaluated; details regarding the subjects, research methodology, or some other aspect of the study that lead the committee to conclude that it would not inform the review; or that the study replicated information in papers that were already being reviewed, it was not further evaluated.

Publication Bias

An important aspect of the quality of a review is the extent to which all appropriate information is considered and serious omissions or inappropriate exclusions of evidence are avoided. A primary concern in this regard is the phenomenon known as publication bias. It is well documented (Begg and Berlin, 1989; Berlin et al., 1989; Callaham et al., 1998; Dickersin, 1990; Dickersin et al., 1992; Easterbrook et al., 1991) that studies with statistically significant findings are more likely to be published than studies with nonsignificant results. Where such bias is present, evaluations of exposure–disease associations based solely on published literature could be biased in favor of showing a positive association. Other forms of bias related to reporting and publication of results have also been suggested. These include multiple publications of positive results, slower publication of nonsignificant and negative results, and publication of nonsignificant and negative results in non-English-language and low-circulation journals (Sutton et al., 1998). For example, several researchers have addressed the specific topic of possible bias in the publication of studies regarding the health effects of exposure to environmental tobacco smoke (Bero et al., 1994; Kawachi and Colditz, 1996; Lee, 1998; Misakian and Bero, 1998).

The committee did not in general consider the risk of publication bias to be high among studies of the health of people exposed to indoor dampness or dampness-related agents, because

- Numerous published studies reported no association.
- The committee was aware of the results of some unpublished research.
- The committee felt that the interest of the research community, public health professionals, government, and the general public in the issue of

mold exposure and health is so intense that any studies showing no association would be likely to be viewed as important by investigators. In short, there would also be pressure to publish negative findings.

Nonetheless, the committee was mindful of the possibility that studies showing a positive association might be over-represented in the literature.

The Role of Judgment

The examination of evidence went beyond quantitative considerations at several stages: assessing the relevance and validity of individual reports; deciding on the possible influence of such factors as error, bias, confounding, or chance on the reported results of empirical studies; integrating the overall evidence within and between diverse types of studies; and formulating the conclusions themselves. Those aspects of the committee's review required thoughtful consideration of alternative approaches at several points and could not be accomplished by adherence to a narrowly prescribed formula.

The approach to evaluating evidence therefore evolved throughout the committee process and was determined, to a large extent, by the nature of that evidence. The committee informed its expectations for the literature by the reality of the state of the science—for example, the lack of valid quantitative exposure assessment methods and a lack of knowledge of which specific microbial agents might primarily account for any presumed health effects. Although the quantitative and qualitative aspects of the process that could be made explicit were important to the overall review, ultimately the conclusions expressed in this report are based on the committee's collective judgment. The committee has endeavored to express its judgments as clearly and precisely as the data allowed.

EVALUATING THE EPIDEMIOLOGIC EVIDENCE

Methodologic Considerations

Several methodologic considerations underlie the evaluation of the epidemiologic studies reviewed in this report. Three of these—uncertainty and confidence, analytical bias, and confounding—are addressed below.

Uncertainty and Confidence

All science is characterized by uncertainty; scientific conclusions concerning the result of a particular analysis or set of analyses can range from highly uncertain to highly confident. In its review, the committee evaluated the degree of uncertainty associated with the results on which it had to base its conclusions.

In the epidemiologic studies reviewed in this report, *statistical significance* is a quantitative measure of the extent to which chance—that is, sampling variation—might be responsible for an observed association between an exposure and an adverse event. The magnitude of the probability value or the width of the confidence interval associated with an effect measure, such as the relative risk or risk difference, is generally used to estimate the role of chance in producing the observed association. Confidence intervals are a function, in part, of the sample size: all else equal, increasing the number of samples increases the precision of the estimate. This type of quantitative estimation is firmly founded in statistical theory on the basis of repeated sampling.

Empirical measures do not, however, necessarily capture all relevant considerations that should be applied when evaluating the uncertainty of conclusions about an association between an exposure and a health outcome. Therefore, to assess the appropriate level of confidence to be placed in conclusions, it is useful to also consider qualitative aspects.

Analytic Bias

Analytic bias is a systematic error in the estimate of association; for example, between an exposure and an adverse event. It can be categorized as selection bias, information bias, confounding bias, and reverse causality bias. *Selection bias* refers to the way that the sample of subjects for a study has been selected (from a source population) and retained. If the subjects in whom an exposure–adverse event association has been analyzed differ from the source population in ways linked to *both* exposure *and* development of the adverse event, the resulting estimate of association will be biased. *Information bias* is the result of a systematic error in the measurement of information on an exposure or outcome. It can result in a bias toward the null hypothesis (that is, that there is no association between the exposure and the adverse event), particularly when ascertainment of either exposure or outcome has been sloppy, or it may create a bias away from the null hypothesis through such mechanisms as recall bias or unequal surveillance of exposed and non-exposed subjects. *Confounding bias*—addressed in greater detail below—occurs when the exposure–adverse event association is biased as a result of a third factor that is both capable of causing the adverse event and is statistically associated with the exposure. Finally, *reverse causality bias* occurs when it is possible that the outcome in question influences the probability of experiencing the exposure being studied. It is not always possible to quantify the impact of such nonrandom errors in estimating the strength of an association.

Another form of bias is possible in studies that use surveys or questionnaires to obtain information about exposures or adverse events. *Response bias* is "a systematic tendency to respond to a range of questionnaire items on some basis other than . . . what the items were designed to measure" (Paulhus, 1991). In this context, it may be considered a form of information or reporting bias. It would be a factor in the studies evaluated here if, for example, those who had experienced an adverse health outcome were more likely to seek out and report instances of indoor dampness than those who had not.

Confounding

In any epidemiologic study comparing an exposed with a non-exposed group, it is likely that the two groups will differ in characteristics other than exposure. When the groups differ with respect to factors that are also associated with the risk of the outcome of interest, a simple comparison of the groups may either exaggerate or hide the true difference in disease rates that is due to the exposure of interest. For example, people with low socioeconomic status may be more likely to be exposed to a particular indoor pollutant than other people. A simple comparison of the incidence of the health outcome among the exposed and non-exposed may exaggerate an apparent difference because socioeconomic status is also thought to influence the incidence of several health problems. If exposed people were of higher socioeconomic status, the simple comparison would tend to mask any true association between exposure and outcome by spuriously increasing the risk of disease in the non-exposed group. This phenomenon, known as confounding, poses a major challenge to researchers and those evaluating their work.

Considerations in Assessing the Strength of Epidemiologic Evidence

Evaluation Criteria

A widely used set of criteria has evolved for the assessment of epidemiologic evidence (Hill, 1965; Hill and Hill, 1991; Susser, 1973; U.S. Public Health Service, 1964); they are also often used to inform public health policy recommendations and decisions (Weed, 1997). These criteria informed the committee's review of the studies addressed in Chapter 5.

1. Strength of Association: Strength of association is usually expressed in epidemiologic studies as the magnitude of the measure of effect, for example, relative risk or odds ratio. Generally, the higher the relative risk, the greater the likelihood that the exposure–disease association is "real";

that is, the less likely it is to be due to undetected error, bias, or confounding. Where the study population is small, selection bias may also play an important role. Small increases in relative risk that are consistent across a number of studies, however, may also provide evidence of an association.

2. *Biologic Gradient (Dose–Response Relationship)*: In general, a potential association is strengthened by evidence that the risk of occurrence of an outcome increases with dose or frequency of exposure. In the case of allergic diseases, this is complicated by the central roles that susceptibility and sensitization play in the disease. The same exposure may have very different effects in susceptible and nonsusceptible people and in sensitized and nonsensitized people. Thus, multiple dose-response curves may be needed to characterize a particular exposure-disease association.

3. *Consistency of Association*: Consistency of association requires that an association be found regularly in a variety of studies, for example, in more than one study population and with different study methods. Findings that are consistent among different categories of studies are supportive of an association. However, consistency does not necessarily mean that one should expect to see exactly the same magnitude of association in different populations. Rather, consistency of a positive association means that the results of most studies are positive and that the differences in measured effects are within the range expected on the basis of all types of error, including sampling, selection bias, misclassification, confounding, and differences in exposure.

4. *Biologic Plausibility and Coherence*: Biologic plausibility is based on whether a possible association fits existing biologic or medical knowledge. The existence of a possible mechanism increases the likelihood that the exposure–disease association in a particular study reflects a true association. In addition, in evaluating exposures such as those addressed in this report, one might consider such factors as evidence that an outcome was associated with documented high exposure levels outside the home.

5. *Temporally Correct Association*: If an observed association is real, exposure must precede the onset or exacerbation of the disease by at least the duration of disease induction. Temporality can be difficult to evaluate for some indoor agents because exposure to them is recurrent and pervasive. If people are exposed to an agent almost every day in an environment where they spend most of their time, it can be difficult to discern a relationship between exposure and effect. The lack of an appropriate time sequence is evidence against association, but the lack of knowledge concerning variations in exposure and exposure magnitude limits the utility of this consideration. One might also consider whether the outcome being studied occurred within a period after exposure that was consistent with current understanding of its natural history.

Other Considerations

As noted earlier, it is important also to consider whether alternative explanations—error, bias, confounding, or chance—might account for the finding of an association. If an association could be sufficiently explained by one or more of these factors, there would be no need to invoke the several considerations listed above. Because these alternative explanations can rarely be excluded sufficiently, however, assessment of the applicable considerations discussed in this chapter almost invariably remains appropriate. The final judgment is then a balance between the strength of support for the association and the degree of exclusion of alternatives.

Bornehag et al.'s (2001) review of the literature regarding building dampness and health includes an extensive discussion of the potential influences on study outcomes. These authors concluded that, while bias or confounding may have contributed to some results, there was no reason to believe that the reported associations were primarily driven by these factors.

Considerations of biologic plausibility informed the committee's decisions about how to categorize associations between relevant indoor exposures and health outcomes. However, the committee recognized that research regarding mechanisms is still in its infancy, and it did not predicate decisions on the existence of specific evidence regarding biologic plausibility.

The committee did not feel that there was sufficient evidence to support confident quantitative estimates of the risk associated with relevant indoor exposures. It is not possible to make general statements about the relative risk posed by various exposures, because this depends heavily on the characteristics of a particular environment and its occupants. Fungi are ubiquitous and can be the primary source of allergens in some arid climates. Endotoxins may be found in humidifiers in urban settings or in organic dusts that infiltrate rural homes from outdoors. Occupant choice has a role in determining indoor humidity and temperature levels as well as the specific building materials and furnishings present.

Much of the literature regarding damp indoor spaces and health outcomes focuses on some measure of indoor moisture or exposure to mold or particular dampness-related agents. Indoor environments, however, are complex. They subject occupants to multiple exposures that may interact physically or chemically with one another and with other characteristics of the environment such as temperature and ventilation levels. Synergistic effects—that is, interactions among agents that result in a combined effect greater than the sum of the individual effects—may also occur. Information on the combined effects of multiple exposures and on synergist effects among agents is cited in this report wherever possible; however, rather little information is available on this topic, and it remains one of active research interest.

SUMMARIZING CONCLUSIONS REGARDING
EPIDEMIOLOGIC EVIDENCE

Categories of Association

The committee summarized its conclusions regarding health outcomes by using a common format, described below, to categorize the strength of the scientific evidence. The five categories were adapted by the committee from those used by the International Agency for Research on Cancer (IARC, 1977) to summarize the scientific evidence of the carcinogenicity of various agents. Similar sets of categories have been used in National Academies reports characterizing scientific evidence regarding vaccine safety (IOM, 1991, 1993), the health effects of herbicides used in Vietnam (IOM, 1994, 1996, 1999, 2001, 2003), and the association between asthma and various indoor exposures (IOM, 2000). The distinctions reflect the committee's judgment that an association would be found in a large, well-designed study of the outcome in question in which exposure is sufficiently high, well characterized, and appropriately measured on an individual basis.

The categories address the association between exposure to an agent and a health outcome, not to the likelihood that any individual person's health problem is associated with or caused by the exposure.

Sufficient Evidence of a Causal Relationship

Evidence is sufficient to conclude that a causal relationship exists between the agent and the outcome. That is, the evidence fulfills the criteria for "sufficient evidence of an association" and, in addition, satisfies the evaluation criteria discussed above: strength of association, biologic gradient, consistency of association, biologic plausibility and coherence, and temporally correct association.

The finding of sufficient evidence of a causal relationship between an exposure and a health outcome does not mean that the exposure would inevitably lead to that outcome. Rather, it means that the exposure *can* cause the outcome, at least in some people under some circumstances.

Sufficient Evidence of an Association

Evidence is sufficient to conclude that there is an association. That is, an association between the agent and the outcome has been observed in studies in which chance, bias, and confounding could be ruled out with reasonable confidence. For example, if several small studies that are free from bias and confounding show an association that is consistent in magnitude and direction, there may be sufficient evidence of an association.

Limited or Suggestive Evidence of an Association

Evidence is suggestive of an association between the agent and the outcome but is limited because chance, bias, and confounding could not be ruled out with confidence. For example, at least one high-quality study shows a positive association, but the results of other studies are inconsistent.

Inadequate or Insufficient Evidence to Determine Whether or Not an Association Exists

The available studies are of insufficient quality, consistency, or statistical power to permit a conclusion regarding the presence or absence of an association. Alternatively, no studies exist that examine the relationship.

Limited or Suggestive Evidence of No Association

Several adequate studies are consistent in not showing an association between the agent and the outcome. A conclusion of "no association" is inevitably limited to the conditions, magnitude of exposure, and length of observation covered by the available studies.

REFERENCES

Begg CB, Berlin JA. 1989. Publication bias and dissemination of clinical research. Journal of the National Cancer Institute 81(2):107–115.

Berlin JA, Begg CB, Louis TA. 1989. An assessment of publication bias using a sample of published clinical trials. Journal of the American Statistical Association 84:381–392.

Bero LA, Glantz SA, Rennie D. 1994. Publication bias and public health policy on environmental tobacco smoke. Journal of the American Medical Association 272(2):133–136.

Callaham ML, Wears RL, Weber EJ, Barton C, Young G. 1998. Positive-outcome bias and other limitations in the outcome of research abstracts submitted to a scientific meeting. Journal of the American Medical Association 280(3):254–257. [Published erratum appears in JAMA 1998 280(14):1232.]

Dickersin K. 1990. The existence of publication bias and risk factors for its occurrence. Journal of the American Medical Association 263(10):1385–1389.

Dickersin K, Min YI, Meinert CL. 1992. Factors influencing publication of research results: follow-up of applications submitted to two institutional review boards. Journal of the American Medical Association 267(3):374–378.

Easterbrook PJ, Berlin JA, Gopalan R, Matthews DR. 1991. Publication bias in clinical research. Lancet 337(8746):867–872.

Hill AB. 1965. The environment and disease: association or causation. Proceedings of the Royal Society of Medicine 58:295–300.

Hill AB, Hill ID. 1991. Bradford Hill's Principles of Medical Statistics (Twelfth Edition). London: Hodder & Stoughton.

IARC (International Agency for Research on Cancer). 1977. Some Fumigants, the Herbicides 2,4-D and 2,4,5-T, Chlorinated Dibenzodioxins and Miscellaneous Industrial Chemicals. IARC Monographs on the Evaluation of the Carcinogenic Risk of Chemicals to Man, Vol. 15. Lyon: IARC.

IOM (Institute of Medicine). 1991. Adverse Effects of Pertussis and Rubella Vaccines. Washington, DC: National Academy Press.

IOM. 1993. Adverse Events Associated with Childhood Vaccines: Evidence Bearing on Causality. Stratton KR, Howe CJ, Johnston RB, eds. Washington, DC: National Academy Press.

IOM. 1994. Veterans and Agent Orange Health Effects of Herbicides Used in Vietnam. Washington, DC: National Academy Press.

IOM. 1996. Veterans and Agent Orange: Update 1996. Washington, DC: National Academy Press.

IOM. 1999. Veterans and Agent Orange: Update 1998. Washington, DC: National Academy Press.

IOM. 2000. Clearing the Air: Asthma and Indoor Air Exposures. Washington, DC: National Academy Press.

IOM. 2001. Veterans and Agent Orange: Update 2000. Washington, DC: National Academy Press.

IOM. 2003. Veterans and Agent Orange: Update 2002. Washington, DC: The National Academies Press.

Kawachi I, Colditz GA. 1996. Invited commentary: confounding, measurement error, and publication bias in studies of passive smoking. American Journal of Epidemiology 144(10):909–915.

Lee PN. 1998. Difficulties in assessing the relationship between passive smoking and lung cancer. Statistical Methods in Medical Research 7(2):137–163.

Misakian AL, Bero LA. 1998. Publication bias and research on passive smoking: comparison of published and unpublished studies. Journal of the American Medical Association 280(3):250–253.

Paulhus DL. 1991. Measurement and control of response bias. In JP Robinson, PR Shaver, LS Wrightsman, eds. Measures of personality and social psychological attitudes. Volume 1. pp. 17–59. San Diego, CA: Academic Press.

Susser M. 1973. Causal Thinking in the Health Sciences: Concepts and Strategies in Epidemiology. New York: Oxford University Press.

Sutton AJ, Abrams KR, Jones DR, Sheldon TA, Song F. 1998. Systematic reviews of trials and other studies. Health Technology Assessment 2(19):1–276.

U.S. Public Health Service, U.S. Department of Health, Education, and Welfare. 1964. Assessing Causes of Adverse Drug Reactions with Special Reference to Standardized Methods. Venulet J, ed. London: Academic Press.

Weed DL. 1997. On the use of causal criteria. International Journal of Epidemiology 26(6): 1137–1141.

2

Damp Buildings

Almost all buildings experience excessive moisture, leaks, or flooding at some point. If dampness-related problems are to be prevented, it is essential to understand their causes. From a technologic viewpoint, one must understand the sources and transport of moisture in buildings, which depend on the design, operation, maintenance, and use of buildings in relation to external environmental conditions such as climate, soil properties, and topography. From a societal viewpoint, it is necessary to understand how construction, operation, and maintenance practices may lead to dampness problems. The interactions among moisture, materials, and environmental conditions in and outside a building determine whether the building may become a source of potentially harmful dampness-related microbial and chemical exposures. Therefore, an understanding of the relationship of building moisture to microbial growth and chemical emissions is also critical.

This chapter addresses those issues to the extent that present scientific knowledge allows. It starts with a description of how and where buildings become wet; reviews the signs of dampness, how dampness is measured, and what is known about its prevalence and characteristics, such as severity, location, and duration; discusses the risk factors for moisture problems; reviews how dampness influences indoor microbial growth and chemical emissions; catalogs the various agents that may be present in damp environments; and addresses the influence of building materials on microbial growth and emissions.

The chapter does not review effects of building dampness that are unrelated to indoor air quality or health. However, dampness problems often cause building materials to decay or corrode, to become structurally weakened or lose their thermal capacity, and thus to reduce their useful life. Dampness also causes building materials and furnishings to develop an unacceptable appearance. The societal cost of such structural and visual effects of dampness may be high.

As discussed below, there is no single, generally accepted term for referring to "dampness" or "damp indoor spaces." This chapter and the remainder of the report adopts the terminology of the research being cited or uses the default term "dampness."

MOISTURE DEFINITIONS[1]

Studies use various qualitative terms to denote the presence of excess moisture in buildings. These include *dampness, condensation, building dampness, visible dampness, damp patches, damp spots, water collection, water ponding,* and *moisture problem.* Dampness—however it is expressed—is used to signify a wide array of signs of moisture damage of variable spatial extent and severity. It may represent visual observations of current or prior moisture (such as water stains or condensation on windows), observed microbial growth, measurement of high moisture content of building materials, measurement of high relative humidity in the indoor air, moldy or musty odors, and other signs that can be associated with excess moisture in a building. Some studies make separate observations of dampness and mold, and both observed dampness and visible mold have been weakly associated with measured concentrations of fungi (Verhoeff et al., 1992). Chapter 3 discusses the various signs and measurements of dampness, moisture, or mold that have been used in studies and lists several examples.

Numerous technical terms are also used to describe characteristics of moisture and moisture physics, including absorption, adsorption, desorption, diffusion, capillary action, capillary height, convection, dew point, partial pressure, and water vapor permeability. A complete discussion of all the terms is beyond the scope of this study, but some that are used in the report are defined below.

The amount of water present in a substance is expressed in relation to its volume (kg/m^3), or to its oven-dry weight (kg/kg). The former is referred to as moisture content (MC), and the latter as percentage moisture content (%-MC). MC is directly proportional to %-MC and to the density of the substance (Björkholtz, 1987).

---

[1]Material in this section and later in the chapter has been adapted or excerpted from a dissertation by Dr. Ulla Haverinen-Shaughnessy (Haverinen, 2002) that was written under the supervision of one of the committee members. It is used here with the permission of the author.

Relative humidity (RH) is the existing water vapor pressure of the air, expressed as a percentage of the saturated water vapor pressure at the same temperature. RH reflects both the amount of water vapor in air and the air temperature. For example, if the temperature of a parcel of air is decreased but no water is removed, the RH will increase. If the air is cooled sufficiently, a portion of the gaseous water vapor in the air will condense, producing liquid water. The highest temperature that will result in condensation is called the "dewpoint temperature." "Humidity ratio" is another technical term used to characterize the moisture content of air. The humidity ratio of a parcel of air equals the mass or weight of water vapor in the parcel divided by the mass or weight of dry (moisture-free) air in the parcel. Humidity ratio, unlike RH, is independent of air temperature. The indoor–outdoor humidity ratio can be used to estimate the rate of interior water vapor generation, or more qualitatively to indicate if a building has sources. Water generation rate can be computed from a moisture mass balance equation; however, the rate of outdoor air ventilation must be known. If the building has a dehumidifier or an air conditioner that dehumidifies, the rate of water removal via this device must be factored. Sorption and desorption of water and from indoor surfaces also complicates the estimation of the internal water vapor generation rate. Monthly mean water activity level has been proposed as a metric for evaluating whether mold growth will occur on surfaces of newly-designed buildings (TenWolde and Rose, 1994) but there is reason to be skeptical about its practicality because the level varies throughout a building and is not easily measured at all relevant locations (for example, in wall cavities).

The temperature of air and materials in a building varies spatially; therefore, RH also varies spatially. In the winter for example, the temperature of the interior surface of a window or wall will normally be less that the temperature of air in the center of a room. Air in contact with the window or wall will cool to below the central room temperature, increasing the local relative humidity. If the surface has a temperature below the dewpoint temperature of adjacent air, water vapor will condense on the surface, producing liquid water.

Without a source to moisten building material continuously, the MC of the material depends on temperature and the RH of the surrounding air. The RH of the atmosphere in equilibrium with a material that has a particular MC is known as the equilibrium relative humidity (ERH) (Oliver, 1997). Different materials have different distributions of pore size and degrees of hygroscopicity so materials that have the same ERH may have different MC. For example, at an ERH of 80%, the MC for mineral wool is about 0.3 kg/m^3, for concrete can be 80 kg/m^3, and for wood is about 90 kg/m^3 (Nevander and Elmarsson, 1994).

MOISTURE DYNAMICS IN BUILDINGS—
HOW BUILDINGS GET WET

Water exists in three states: solid (ice), liquid, and gas (water vapor). The molecules in liquid water and water vapor move freely; molecules in ice are bound into a crystal matrix and are unable to move except to vibrate. Liquid water is a cohesive fluid; when it interacts with other materials, it is affected by forces that originate in the new material. If a drop attaches to a surface that has a strong affinity for water, like wood, it will spread out across the surface. The attraction may be great enough that water will run along the bottom of a horizontal material—a roof truss, for example—until it comes to an air gap or a downward projection where gravity pulls it away from the surface and it falls.

Many building materials are porous, and the size of the pores affects their permeability. If the pores are small enough to keep both liquid water clusters and water vapor molecules from passing, the material is impermeable; metal foils are examples of such materials. Materials with slightly larger pores (building papers like Tyvek™ and builders felt) will shed liquid water but be relatively permeable to water vapor. If the material has pores that are large enough for tiny clusters of liquid water to enter, it will be permeable to both liquid water and water vapor. As a result of intermolecular forces, liquid water is drawn into the pores of such materials by capillary suction. Water drawn in that way is said to be *absorbed* by the porous material. Water migration through porous materials is a complex interaction of forces. Water molecules clinging to the surface of a solid material are bound to that surface by intermolecular forces. They cannot move about as freely as liquid water molecules or water vapor molecules and are in what is sometimes referred to as the *adsorbed* state. Water must accumulate on surfaces to a depth of four or five molecules before it begins to move freely as a liquid (Straube, 2001). Adsorbed water cannot be removed by drainage. In the adsorbed state, water molecules are less available for chemical and biologic purposes than they are in a nonadsorbed state.

It does not take a great deal of moisture to cause problems with sensitive materials like paper or composite wooden materials. Moisture sources in buildings include rainwater, groundwater, plumbing, construction moisture, water use, condensation, and indoor and outdoor humidity (Lstiburek, 2001; Straube, 2002). The first three are sources of liquid-water problems, construction moisture may result in both liquid-water and water-vapor problems, and condensation associated with humidity involves water vapor as well as liquid water. Moisture problems begin when materials stay wet long enough for microbial growth, physical deterioration, or chemical reactions to occur. Those may happen because of continual wetting or intermittent wetting that happens often enough to keep materials from drying. As discussed below, the important moisture-related variables in determining whether fungal growth occurs are those which affect the rate of wetting and the rate of drying (Lstiburek, 2002a).

The most damaging water leaks are those which are large enough to flood a building or small enough not to attract notice but large enough to wet or humidify a cavity space or material for a long time. Thus, the "best" leak is one that is large enough to be noticed right away but small enough that the wetting does not promote microbial growth or affect materials. Both floods and slow leaks can result in large areas of fungal growth. Condensation sometimes occurs over a large area and can also result in extensive mold growth.

Rainwater and Groundwater

Placing a building on a site does not change how much rain falls each year—it changes the path that rainwater takes on its journey through the hydrologic cycle. When building designs work properly, rainwater is collected and redirected so that it does not intrude into the buildings themselves. When collection and redirection fail, rainwater wets buildings. Buildings have been protected from rainwater for centuries by using gravity, air gaps, and moisture-insensitive materials to direct and drain water away from other materials that can be damaged by water through corrosion, microbial contamination, or chemical reaction (Lstiburek, 2001). Weakness in rainwater protection can be found in the detailing of the roof, walls, windows, doors, decks, foundation, and site. Rainwater leaks may take a long time to become noticeable because the water often leaks into cavities that are filled with porous insulation. Insulation may retain the water, keeping materials wet longer than would empty cavities.

Many roofing materials are impermeable to liquid water and can be repeatedly wetted and dried without damage. Wooden shingles and thatched roofs are exceptions. They drain the bulk of rainwater away from the interior but also absorb some of it. An air gap beneath then forms a moisture or water break and allows drying of the shingle or thatch by evaporation from inner and outer surfaces. Roof leaks typically occur at joints and penetrations; parapet walls, curbs for roof-mounted equipment and skylights, intersections between roofs and walls, and roof drains are common leakage sites. These leaks are often the result of failures in design or of installation of flashings and moisture or water breaks.

In climates that receive substantial snowfall, water can intrude through roofs as the result of melting snow. Ice dams occur when there is snow on a roof and roof temperatures reach 33°F (1°C) or higher at times when the outdoor air temperature is below freezing. Snow on the warm part of the roof melts and then follows the drainage path until it reaches roofing that is chilled below freezing by outdoor air. The water then freezes on this part of the roof and causes ice dams and icicles. Aggravating conditions for ice dams include sources of heat that warm snow-covered sheathing (air leaks and conductive heat loss from the building, recessed lighting fixtures in insulated ceilings, and uninsulated chimneys passing through attics) and valley roofs, which may collect water from a large surface area and drain it to one small location. Several design approaches are available for preventing ice dams:

- Air-seal and heavily insulate the top of the building so that escaping heat does not reach the roofing.
- Ventilate the roof sheathing from underneath with outdoor air. (In combination with the air sealing and insulation, this keeps roofing cold, so melting does not occur or is minimized to rates that do not result in ice problems).
- Avoid heat sources in the vented attic or vent bays (for example, do not use recessed lights in insulated ceilings).

Rainwater protection in walls is accomplished largely with three basic methods: massive moisture storage, drained cladding, and face-sealed cladding (Lstiburek, 2001; Straube and Burnett, 1997). Historically, walls capable of massive moisture storage have been built of thick masonry materials (such as stone in older churches). Exterior detailing channels rainwater away from entry through such walls. The walls are also able to store a large amount of water in the adsorbed state, and their storage capacity is sufficient to accommodate rainwater wetting and drying cycles without causing problems.[2] Rainwater intrusion problems occur in these walls when a pathway wicks water from the exterior to the interior, where moisture-sensitive materials are. Wooden structural members in masonry pockets, interior-finish walls made of wood or paper products, and furnishings composed of fabrics, adhesive, or composites are typical materials that may be affected by rainwater transported through walls by bridging or capillary suction.

Cladding (a protective, insulating, or decorative covering) with air gaps and a drain plane is another historical answer to rainwater intrusion. A drained-cladding wall has an exterior finish that intercepts most of the rainwater that strikes it but is backed by an air gap and water-resistant drainage material to keep any water that gets past the cladding from entering the wall beneath. Wooden clapboard, wooden shingles, board and bat, brick or block veneer, and traditional stucco are examples of cladding used in some climates in the United States that has historically been backed by an

---

[2]Condensation is not typically a problem, because, unlike many composite structures, such walls have relatively even distribution of water-vapor permeability.

air gap and drainage layer. Asphalt-impregnated felt paper, rosin paper, and high-permeability spun-plastic wraps are examples of materials that are used as the drainage layer. Foam board and foil-faced composite sheathing have also been used as drain planes beneath cladding (Lstiburek, 2000). The most frequent problems in these walls occur when moisture-sensitive sheathings—such as oriented strand board (OSB), plywood, and low-density fiberboards—are not protected by a drainage layer.

Face-sealed walls are made of materials that are impermeable to water and are sealed at the joints with caulking or gaskets (Straube, 2001). Structural glazing, metal-clad wooden or foam panels, and corrugated metal siding are examples of face-sealed cladding. The intention is to seal the joints between the panels well enough to prevent rainwater entry. Rainwater intrusion occurs when the seals fail. Seals on some face-sealed walls need to be renewed every 4–5 years.

The unavoidable weakness in rainwater protection for any wall is at the penetrations—windows, doors, light fixtures, the roofs of lower portions of the structure, decks, balconies, and porches. Rainwater leaks through poorly detailed, designed or installed flashing are most common. Common errors include failure to provide detailed instructions for flashing in construction documents, providing two-dimensional details for situations that require three-dimensional flashing, installing head flashings on top of building paper rather than installed underneath, and ignoring leaks in the window itself. Wall drain papers for windows must be installed in the same way that a raincoat is worn: over, not tucked into rain pants. Pan flashing beneath windows can prevent leaks, even of poorly installed windows, from wetting the wall below (Lstiburek, 2000).

Foundations are typically protected from moisture problems by being constructed of materials that are resistant to water problems (stone, concrete, and masonry) and having rainwater diverted away from them (Lstiburek, 2000, 2001). (In some old buildings, foundation structures could be constructed of wooden piers, which might have to be kept wet.) Excessive moisture in foundations is often the result of poorly managed rainwater, but it may also result from groundwater intrusion, plumbing leaks, ventilation with hot humid air, or water in building materials (such as concrete) or in exposed soil (for example, saturated ground in a crawl space foundation). Rainwater is diverted by sloping the finish grade away from the building; rainwater and groundwater are diverted with subsoil drainage. Drainage systems use stone pebbles, perforated drain pipe, sand and gravel, or proprietary drainage mats. Stone pebbles and perforated pipe are typically enclosed in a filter fabric to prevent clogging by fine soil particles. Below-grade foundations are coated with dampproofing to provide a capillary break. Water problems occur if rainwater collected on the roof is drained to the soil next to the foundation. This may happen if the site is inadvertently contoured to collect rainwater and drain it into the building or if paving does so. Other problematic scenarios include a drainage pipe that is missing, is installed improperly, or does not drain to daylight or a sump pump; a drainage system that fills with silt carried by water percolating through the soil and that then clogs; and a failure to install a capillary break, which would keep water from being wicked through concrete products to the interior.

Foundations may be slab on grade (or near grade), full basements, crawl spaces, piers, or a combination of these types. A slab-on-grade foundation consists of a concrete slab that constitutes the first floor of the building. The perimeter of the slab may be thickened and reinforced, or it may be bound by a perimeter wall that extends some distance into the soil. The most common water problems with slab-on-grade foundations are caused when rainwater from the roof or site wets the foundation and the water is wicked up through concrete to wall or flooring materials. If air ducts are placed in or beneath the slab, these may flood with poorly managed rainwater.

A basement is made by excavating a large, pond-like hole in the ground and constructing walls and a floor in the bottom of the hole. A basement floor slab is wholly or partially below grade. Some basement floors are at grade on one side and below grade on another. A drainage system is placed on the bottom of the hole around the perimeter of the walls, and a capillary break in the form of stone pebbles or polyethylene film is placed beneath the floor. Walls are coated with some form of dampproofing to make a capillary break. Free-draining material is placed against the walls to divert water from the foundation into footing drains. Many potential causes of dampness problems in full basements result from vagaries of weather and defects in design, construction, and maintenance. Rainwater from the roof or site can easily saturate the soil near the foundation and make it more likely for liquid water to seep or run into the basement. A more subtle problem occurs when water wicking through the walls or slab evaporates into the basement, leaving the walls dry but over-humidifying the space. Placing framing, insulation, paneling, or gypsum board against a basement wall creates a microclimate between finished wall and basement wall. In fact, if the outdoor-air dewpoint is higher than the temperature in this space, ventilating air will add moisture to the cavity, not dry it and this can result in conditions favorable for microbial growth. A solution to this problem is to insulate the foundation wall on the outside. If the foundation is insulated on the inside, a material with high insulating value and low water-vapor permeability should be used; this will keep the warm humid basement air away from earth-chilled walls. Plastic foam insulation meets this criterion. If the water vapor permeability of the insulation is low enough, it will reduce drying from the foundation wall into the basement.

Placing insulation beneath the floor slab can prevent basement floors from "sweating" during hot humid weather because it thermally isolates the concrete slab from the cool earth below.

A crawl space is constructed in the same way as a basement foundation except that it is shorter and often the floor is not covered by a concrete slab. Many crawl spaces have air vents through the walls intended to provide passive ventilation. Because crawl spaces are not intended for occupancy, drainage detailing around them is often lacking or poorly implemented. Rainwater intrusion is common. In addition, the floor is often exposed soil, which creates the potential for evaporation into the crawl space. Vents placed too close to the ground sometimes become rainwater intakes. When the outdoor-air dewpoint is higher than the temperature of the soil and foundation surfaces, ventilating air wets the crawl space rather than drying it (Kurnitski, 2000).

Pier foundations (concrete or crushed-stone footings for posts that constitute the major structural support for a building) are the most resistant to rainwater problems. Piers extend from the ground to above the surface of the soil to support the lower structure of a building. The most common water problem for this type of foundation occurs if a depression in the ground beneath the structure collects water and exposes the underside of the building to prolonged high humidity.

Plumbing and Wet Rooms

Most water intentionally brought into buildings is used for drinking, cooking, or cleaning. The bulk of this water passes harmlessly through drains to public or private treatment and is then released to the hydrologic cycle from which it was diverted. The pathway followed by such water consists of pipes, tubs, sinks, showers, dish and clothes washers, driers, and ventilating air. Most of the materials used in the pathway are moisture-insensitive—able to withstand dampness without decomposing, dissolving, corroding, hydrolyzing, or supporting microbial growth. Moisture problems occur when water leaks from pipes or from sinks, tub or shower enclosures, washing machines, ice machines, or other fixtures and appliances that have water hookups.

Pipes leak when joints are incorrectly made or fail, water freezes in them, the pipe material corrodes or decomposes, or a screw or nail is driven through them. Joints may not be correctly soldered, gasketed, cemented, or doped. Water lines lose integrity when they are exposed to acidic or caustic water or—in the case of rubber or plastic lines to washing machines—the polymers break down from oxidation or ultraviolet (UV) light exposure. Corrosive water may lead to mold growth if a large number of small leaks result. Pipes in exterior walls or unheated crawl spaces or attics may freeze and crack during subfreezing weather. A screw or nail driven through a pipe may not leak for some time, because the fastener seals the hole it made; after thermal expansion and contraction and corrosion work for some time, the pipe may begin to leak.

Drains and water traps are vulnerable to leaks. Overflows and careless installation and renovation practices also contribute to problems with fixtures and appliances that use water. The materials that surround tubs and showers—typically ceramic tiles and fiberglass panels—receive regular wettings. They must be constructed, sealed, and maintained to protect the wall and floor materials beneath them. As with rainwater protection, most problems occur at the joints. Grout between ceramic tiles often does not adequately serve as a capillary break and water wicks through to the base. In ceramic tile surrounds with paper-covered gypsum board as the base, mold growth may occur beneath the grout and on the backside of the gypsum board where water wicks through the paper facing the wall cavity. Depending on the detailing, water may also be wicked through the gaps where fiberglass panels overlap and meet tubs or shower pans. The shower pan in stand-alone showers is another weak spot. Essentially, these are basins that must hold a small depth of water. Leaks are most common at the drain penetration. Pans that are constructed on site have more joints to leak than prefabricated pans that are molded into a single piece. Poorly designed, incorrectly installed, and carelessly used shower curtains and doors are another source of problems. Tub surrounds and shower enclosures can be constructed of materials that are poor substrates for fungal growth; for example, fiber-cement board, rather than paper-covered gypsum board, can be used as the base for ceramic tile. Such steps reduce, but do not eliminate, the possibility of microbial contamination.

Construction Moisture

In newly constructed buildings, a large amount of water vapor can be released by wet building materials such as recently cast concrete, and wet wooden products (Christian, 1994). Manufactured products that were originally dry can become extensively wetted by exposure to rain during transportation, storage, and building construction. Case studies have attributed microbial contamination to the use of wet building materials or to wetting during building construction (Hung and Terra, 1996; Salo, 1999). Large areas of mold growth may occur when a floor enclosing an earth-floored crawl space is installed because the soil may be a reservoir of rainwater; the humidity in such a crawl space quickly becomes high when the floor deck is applied over moist earth. Floor decks made from OSB or plywood are vulnerable to mold growth during extended periods (23 days for OSB, 42 days for plywood) of RH greater than 95% (Doll, 2002).

Condensation and High Humidity

Condensation necessarily involves water-vapor transport. The two important variables for condensation are chilled surfaces and sources of water vapor. Materials chilled below the indoor or outdoor air temperature accumulate water molecules in the adsorbed state and are at risk for condensation; those chilled below the local dew point will begin to accumulate liquid water. Porous materials can hold more water vapor than impermeable ones before liquid water appears. The combination of high RH in indoor or outdoor air and cooled building materials increases the risk of dampness problems and microbial growth. Even without condensation, the local RH of air at the surface of cool material can be very high, leading to high moisture content in the material.

Figure 2-1 illustrates how much air needs to be cooled before the difference between the air temperature and dewpoint temperature equals zero and condensation occurs. Regardless of the initial air temperature, when the relative humidity is very high only a few degrees of cooling will result in condensation. For example, if the bulk of the air in a room has a RH of 80%, condensation will occur on a surface that is only about 7°F (4°C) cooler than the bulk room air temperature. Therefore, whenever cool

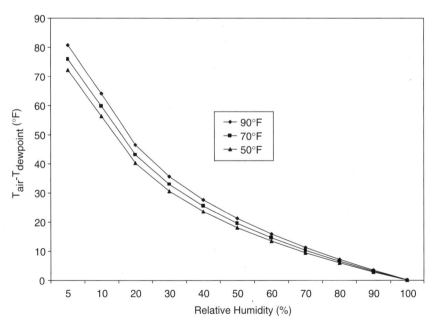

FIGURE 2-1 The difference between air and dewpoint temperatures needed for condensation to occur, expressed as a function of relative humidity, for three indoor air temperatures.

surfaces are present due to cold outdoor temperatures or air conditioning of a building, high humidity poses a condensation risk. However, at present, there is no generally accepted upper limit for indoor RH level, based on the need to prevent dampness problems. Acceptable RH levels vary with climate and building features.

During periods of cooling in air-conditioned buildings, indoor materials are colder than the outdoor air. Ventilation is then a source of indoor moisture—not a removal process—unless the incoming ventilation air is first dehumidified (Harriman et al., 2001). If nonconditioned outdoor air is accidentally drawn across a surface that is chilled sufficiently by air-conditioned indoor air, condensation will occur. In the cooling season, that is most likely to happen when an exhaust fan or the return side of an air handler lowers indoor air pressure in rooms or depressurizes wall or ceiling cavities (Brennan et al., 2002). Outdoor air is drawn in by the lower air pressure and carries water vapor with it. Water vapor in this accidental outdoor airflow may condense on the backside of gypsum board or in cabinets that have holes for wire or plumbing. The backside of interior gypsum board and the underside of vinyl wallpaper on exterior walls are common locations for mold growth resulting from this process (Lstiburek, 2001). When buildings are air-conditioned, a combination of wind-driven rain and water-vapor transport can also result in condensation and mold growth beneath vinyl wallpaper, on the backside of gypsum board on an exterior wall, or on the backside of interior foam board (Lstiburek and Carmody, 1996). Those materials act as accidental vapor retarders on the cool side of the wall. Furnishings and wall decorations—such as pictures, cabinets, mirrors, and chalkboards—can also act as accidental vapor retarders.

Materials may be chilled by outdoor air when it is cold outdoors (Brennan et al., 2002). Cold-weather condensation is often observed on the interior side of windows. Because windows usually have a lower insulating value than solid walls, the room-side surface of the glass is cooler than the surface of the surrounding walls. Indeed, if there is condensation or frost on the window, the glass temperature is necessarily below the indoor-air dew point. Condensation may also occur on cold pipes and on the bottom side of roof sheathing (the side facing the attic rather than the sky), the inside of exterior-wall sheathing, and the back side of claddings (clapboards, stone slabs, concrete panels, plywood panels, and the like).

Foundations constructed of concrete, masonry, stone, and wood are often chilled by contact with the earth. If the indoor-air dew point is higher than the temperature of the earth-chilled surfaces, water will begin to condense (Brennan et al., 2002). As water condenses on capillary materials, such as concrete or wood, it is wicked away by capillary action. Hygroscopic concrete and stone foundation materials can store moisture in a relatively harmless state until they become saturated, at which time liquid water will appear. If the materials are coated with a vapor-impermeable material, such as sheet floor covering or many paints, condensation will immediately collect under the proper conditions. Water vapor in basements or crawl spaces may come from water passing through the foundation materials as liquid or vapor or from the ventilating air when outdoor-air dew points are high, or it may be dominated by water vapor from exposed soil (Kurnitski, 2000).

Moisture-related problems become more likely when basement areas are finished (CMHC, 1996). Many of the materials used to finish basements allow water vapor to diffuse through them but are relatively good thermal insulators; thus, the materials inhibit heating of the foundation by warm indoor air but allow moisture to reach the cool surfaces. When a wooden stud wall with fiberglass insulation covered with gypsum board is placed against a concrete foundation and no vapor retarder is used, water vapor can easily pass through the wall section via the permeable materials and through gaps. However, there is very little drying potential under these conditions. Vapor pressure moves water vapor from the basement into the wall and sometimes from the outdoors into the wall. Very little air is moving behind such walls so drying by airflow cannot be achieved. The "microclimate zone" behind the stud wall stays moist throughout the cooling season, while wooden studs and paper-covered gypsum provide nutrient for mold growth. A vapor retarder in the wall will not prevent migration through the gaps and holes but will reduce the drying potential of the wall and thus increase the importance of small rainwater or plumbing leaks. Carpet systems on floor slabs produce a similar phenomenon, unless there is an insulating, low-vapor-permeability layer in the system. Finished basements may have substantial mold growth because of those phenomena. Warming the surfaces of the earth-chilled materials, by insulating them or heating them prevents condensation. Insulating material placed inside the foundation must prevent vapor in the indoor air from reaching the chilled foundation materials and present a warm surface to the indoor air. As noted above, that is best accomplished by using a material with high insulating value and low vapor permeability. Insulating material placed on the outside of the foundation must resist biologic, chemical, and physical deterioration when exposed to soil and liquid water. Condensation on earth-chilled surfaces can also be avoided by dehumidifying the indoor air to lower the dew point to below the foundation surface temperatures.

Occupants as Sources of Moisture

High humidity indoors can originate in moisture emissions from cooking, washing clothes, bathing, and keeping living plants indoors. Respiration and perspiration by building occupants contribute to humidity, as does the use of humidifiers. In improperly ventilated building spaces, those sources can account for substantial problems. In addition to plumbing leaks and flooding by water overflow, wicking along wall surfaces from poor wet-mopping practices is a problem in some indoor environments.

The practices of cooking, bathing, and drying of clothes and the density of occupation vary among cultural and economic groups. In some homes, internal moisture is high because of nearly continuous simmering of foods or extensive indoor drying of clothes. Anecdotal evidence indicates that such activities can lead to high indoor humidity and associated microbial growth. In low-rise residential buildings, a damp foundation may contribute as much water vapor as all the rest of the sources combined (Angell, 1988).

Moisture in Heating, Ventilating, and Air-Conditioning Systems

Although relatively little attention has been directed to dampness and mold growth in heating, ventilating, and air-conditioning (HVAC) systems, there is evidence of associated health effects. Pollutant emissions linked to moisture and microbial growth in HVAC systems are one of several potential explanations for the consistent association of air-conditioning systems with an increased prevalence of nonspecific health symptoms, called sick building syndrome, experienced by office workers (Seppänen and Fisk, 2002). The presence of air conditioning in homes has also been associated with statistically significant increases in wheezing and other symptoms of current asthma (Zock et al., 2002). Mendell et al. (2003) analyzed data on 80 office buildings where complaints had been made and found an increased prevalence of lower respiratory symptoms associated with poor draining of water from the drain pans beneath cooling coils of HVAC systems (OR 2.6; CI 1.3–5.2). In contrast, a preliminary analysis of data on a representative set of 100 large U.S. office buildings found that dirty cooling coils, dirty or poorly draining drain pans, and standing water near outdoor air intakes were not associated with reports of mucus membrane symptoms, lower respiratory symptoms, or neurologic symptoms (Mendell and Cozen, 2002).

Liquid water is often present at several locations in or near commercial-building HVAC systems, facilitating the growth of microorganisms that may contribute to symptoms or illnesses. Outdoor air is often drawn from the rooftop or from a below-grade "well" where water (and organic debris) may accumulate. Raindrops, snow, or fog can be drawn into HVAC systems with incoming outside air, although systems are usually designed to prevent or limit this moisture penetration.

In both commercial and residential air-conditioning units, moving the supply-air stream in the direction of airflow leads it to the cooling coil where moisture condenses (as a consequence of cooling the air or intentionally for dehumidification). Ideally, that moisture drips from the surfaces of the coil into a drain pan with a drainage pipe. Drain pipes may become clogged with the remains of microbial growth. Occasionally, drain pans contain stagnant water because they do not slope toward the drain line. In drawthrough systems, drains may also be plugged or otherwise nonfunctional because air-pressure differences prevent drainage, sometimes causing the drain pan to overflow with water. If the velocity of air passing through the cooling coils is too high, water drops on the surface of the cooling coil can become entrained in the supply-air stream and deposit in the HVAC system downstream of the cooling coil. Air leaving the cooling coils is often nearly saturated with water vapor, and the high humidity of this air increases the risk of microbial growth. HVAC systems sometimes have a humidifier that uses steam or an evaporation process to add moisture. Humidifiers, used predominantly in colder climates, may have reservoirs of water or surfaces that are frequently wetted, or they may produce water drops that do not evaporate. Thus, there are many potential sources of liquid water and high humidity in HVAC systems.

Microbial growth in HVAC systems can be limited by (Ottney, 1993)

- Using sloped drain pans with drains at the low point.
- Correctly trapping drains or using critical orifice drains that work against negative pressure in the system.
- Providing easy access to coils, drain pans, and the downstream side of cooling coils for inspection and cleaning.
- Making inner surfaces of the air-conveyance systems of materials that are impermeable to water penetration and are easy to clean.
- Protecting the system from particle buildup by using filters with greater than 25% dust spot efficiency.

Microbial contamination of HVAC systems has been reported in many case studies and investigated in a few multibuilding research efforts (Battermann and Burge, 1995; Bencko et al., 1993; Martiny et al., 1994; Morey, 1994; Morey and Williams, 1991; Shaughnessy et al., 1998). Sites of reported contamination include outside air louvers, mixing boxes (where outside air mixes with recirculated air), filters, cooling coils, cooling-coil drain pans, humidifiers, and duct surfaces. The porous insulating and sound absorbing material called duct liner that is used in some HVAC systems may be particularly prone to contamination (Morey, 1988; Morey and Williams, 1991). Bioaerosols from contaminated sites in an HVAC system may be transported to occupants and deposited on previously clean surfaces, making microbial contamination of HVAC systems a potential risk factor for adverse health effects.

A 2003 study investigated the health impact of such contamination by examining the association between ultraviolet germicidal irradiation (UVGI) of drip pans and cooling coils in buildings ventilation systems and indoor microbial concentrations and self-reported symptoms in occupants (Menzies et al., 2003). The researchers systematically turned UVGI lamps installed in the HVAC systems of three office buildings on and off over the course of a year and collected environmental and occupant data. Fungi, bacteria, and endotoxin concentrations were measured, and building occupants who were unaware of the operating condition of the UVGI lamps filled out questionnaires on their health. Other environmental data (temperature, humidity, air velocity, HVAC recirculation; and CO_2, NO_x, O_3, formaldehyde, and total volatile organic compound concentrations) and occupant data (participants' assessment of thermal, physical, and air quality; and demographic, personal, medical, and work characteristics) were also collected. Occupants reported significantly fewer work-related mucosal symptoms (adjusted OR 0.7; 95% CI 0.6–0.9) and respiratory symptoms (0.6; 0.4–0.9) when the UVGI lamps were on. Reports of musculoskeletal symptoms (0.8; 0.6–1.1) and systemic symptoms (headache, fatigue, or difficulty concentrating) (1.1; 0.9–1.3) were not significantly different. Although median concentrations of viable microorganisms and endotoxins were reduced by 99% (CI 67%–100%) on surfaces exposed to UVGI, there were no significant decreases in airborne concentration. The results suggest that limiting microbial contamination of HVAC systems may yield health benefits, and follow-up research is recommended.

PREVALENCE, SEVERITY, LOCATION, AND DURATION OF BUILDING DAMPNESS

Prevalence

Table 2-1 provides examples of published data on the prevalence of signs of dampness in buildings. The studies address a variety of locations and climates. Different dampness metrics were used; most data were collected with occupant-completed questionnaires. The reported prevalence of signs of dampness ranges from 1% to 85%. In most datasets, at least 20% of buildings have one or more signs of a dampness problem. But because moisture metrics and data-collection methods varied among studies, comparisons of prevalence data from different studies can be only qualitative.

Figure 2-2, which is based on the biennial U.S. Census American Housing Survey, plots the prevalence of water leaks in U.S. houses by year. The data indicate that the prevalence of water leaks generally decreased over the period 1985–2001 and that more leaks are from external sources of water—for example, rain—than from internal water sources—plumbing leaks

TABLE 2-1 Examples of Reported Prevalence of Signs of Building Dampness

Reference	Country	Population	Dampness Metric	Prevalence, %
Residential buildings				
Brunekreef et al., 1989	United States	homes of 6,273 school children in 6 cities	Questionnaire (city averages reported) Ever water in basement Ever water damage to building Ever mold or mildew on any surface Any of above	11–42 12–23 21–38 46–58
Dales et al., 1999	Canada	homes of 3,444 children	Questionnaire Dampness stains in last 2 years Visible mold in last 2 years Either of above	24 15 25
Engvall et al., 2001	Sweden	4,815 apartments	Questionnaire Condensation on windows High relative humidity in bathroom Water leakage in last 5 years Any of above	7 9 13 22
Evans et al., 2000	United Kingdom	8,889 homes of adults	Questionnaire Damp or condensation a serious problem Damp or condensation a minor problem	1 9
Haverinen et al., 2001a	Finland	390 homes and 240 apartments	Inspections Grade 1, no to minor moisture damage Grade 2, intermediate moisture damage Grade 3, high moisture damage	16 18 15
Jaakkola et al., 2002	Finland	932 adults who were controls in asthma case-control study	Questionnaire about homes Water damage in last year Damp stains or peeling paint in last year Visible mold in last year	2 9 3
Kilpeläinen et al., 2001	Finland	homes of 10,667 university students	Questionnaire (regarding any of the students' homes in the last year) Visible mold Visible mold or damp stains Visible mold or damp stains or water damage	5 12 15

(continued on next page)

TABLE 2-1 continued

Reference	Country	Population	Dampness Metric	Prevalence, %
Nevalainen et al., 1998	Finland	450 private houses	Inspections, surface moisture measurements, and questionnaire on previous or current damage	
			Signs of moisture damage in roof	40
			Signs of moisture damage in basement	25
			Signs of moisture damage in walls	33
			Plumbing-related moisture damage	25
			Leakage from clothes washer or dishwasher	20
			Leakage in ventilation ducts	20
			Any of above	80
Norbäck et al., 1999	Sweden	homes of 429 subjects	Questionnaire via interview	
			Water damage in last year	16
			Floor dampness in last year	5
			Visible mold in last year	9
			Moldy odor in last year	6
Zock et al., 2002[a]	14 European countries, Australia, India, New Zealand, and United States	16,687 homes	Questionnaire via interview	
			Water damage in last year	12
			Water on basement floor in last year	2
			Mold or mildew in last year	22
Nonresidential buildings				
Jaakkola et al., 2002	Finland	932 adults who were controls in asthma case-control study	Questionnaire about workplace	
			Water damage in last year	6
			Damp stains or peeling paint in last year	12
			Visible mold in last year	3
Mendell and Cozen, 2002	United States	100 U.S. office buildings	Questionnaire	
			Past water damage in building	85
			Current water damage in building	43

[a]Includes data from Norbäck et al. (1999).

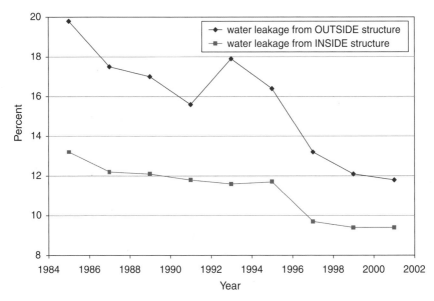

FIGURE 2-2   Prevalence of reported housing water leaks during the preceding 12 months, 1985–2001.
SOURCE: Biannual *American Housing Survey for the United States,* Bureau of the Census, U.S. Department of Commerce, http://www.census.gov/hhes/www/ahs.html.

and the like. In 2001, 11.8% and 9.4% of houses had water leakage from exterior and interior sources, respectively.

Most of the available data on dampness prevalence are related to houses and apartments, but some are related to other indoor environments. A study by Jaakkola et al. (2002) found that the self-reported prevalence of signs of dampness was similar in the workplaces and homes of subjects in the Finnish Environment and Asthma Study. In a study of 100 U.S. office buildings (Mendell and Cozen, 2002), 85% reported past water damage and 43% current water damage. Those prevalences are high relative to those typically reported in studies of homes; although localized water damage in a large office building may significantly influence exposures only of workers near the damage. Nonetheless, these data suggest that dampness in workspaces should not be ignored.

The committee did not identify any large systematic surveys of dampness in U.S. classrooms,[3] but anecdotal reports of dampness in classrooms

---

[3]Meklin et al. (2003) have assessed the occurrence of moisture damage, fungi, and airborne bacteria in schools in Finland, focusing on the impact of main building frame material (wooden vs concrete or brick). An earlier paper (Meklin et al., 2002) examined respiratory symptoms in the children attending those schools.

are common, and school data have been collected as part of broader characterizations of children's exposures. A survey of the condition of U.S. schools by the General Accounting Office (GAO) did not contain a specific question about dampness or water leaks; however, the documentation of GAO's visits to 41 schools included many references to water leaks (U.S. GAO, 1995). Daisey and Angell (1998) reviewed 49 health-hazard evaluations of educational facilities performed by the National Institute for Occupational Safety and Health in response to indoor air quality (IAQ) complaints; 28 of the evaluations reported water leaks in the building shell. Thus, the available evidence suggests that classrooms also commonly have dampness problems.

Table 2-1 includes data from two studies (Haverinen et al., 2001a; Nevalainen et al., 1998) that used inspections by trained personnel to assess the prevalence of signs of dampness. Only a few studies have analyzed differences between occupant-reported and investigator-verified prevalence of moisture and mold observations in buildings. Douwes et al. (1999) found that occupant's reports of damp spots or mold spots were better correlated with a measure of indoor mold than investigator's reports of these visible signs. Bornehag et al. (2001) concluded that, although in most studies occupants had reported more dampness than investigators had, this was due to the occupants' longer time perspective than the investigators' "snapshot" observations. A conflicting study by Williamson et al. (1997) found that occupants reported dampness less often than trained surveyors. Nevalainen et al. (1998) reported similar results, suggesting that the explanation was a result of a trained eye and of knowledge of what represents a critical problem. Dharmage et al. (1999a) examined the validity and reliability of interviewer-administered questionnaires against observations and measurements made by an independent researcher. Among 44 items examined for *validity* (defined as correspondence between occupant reports and independent observations or objective measurements), there was perfect or almost perfect agreement on 21 and substantial agreement on 19 others. Among 10 items examined for *reliability* (defined as correspondence between interviews conducted 1 year apart), there was perfect or almost perfect agreement on nine items and substantial agreement on the other one. They concluded that the data collected with questionnaires were both reliable and valid. In another study, Dales et al. (1997) concentrated on the validity and determinants of reported home dampness and molds. They established associations between occupant-reported water damage, mold and mold odors, and objectively measured concentrations of viable indoor fungi in dust. However, they found little association between questionnaire responses and an objective measure of total airborne fungal matter (ergosterol concentration) and there was evidence that—in the presence of low concentrations of viable fungi in dust—respondents reporting allergies were more likely to report visible mold growth than asymptomatic respondents. The authors therefore recommended that objective exposure measures, not questionnaires, be used in studies of the health effects of indoor fungi.

Information on the prevalence or severity of moisture damage reported by occupants is likely to be highly subjective. The validity and reliability of data gathered from questionnaires are affected by several survey factors, such as sample size, response rate, recall period, and factors related to the design of the questionnaire. Underreporting, overreporting, and systematic reporting bias that would not be corrected by increasing sample sizes are possible. But questionnaires are a relatively cost-effective method of collecting information on perceived indoor-air quality, especially if the sample is large, and questionnaire responses collected from the occupants themselves provide first-hand information; occupants' perceptions are also important in assessing the condition of a building.

Trained building inspectors have experience in observing and evaluating structures and may also be more objective than occupants, who have a personal relationship with the building. Those advantages would not, however, exclude the subjectivity of trained investigators. Thorough building investigations need both expert assessment of the building's condition and occupant knowledge of its history and current problems to complement each other.

Chapter 6 has further information on this topic, addressing the evaluation of moisture problems in the context of identifying sources and planning remediation.

Severity

Most studies have not attempted to quantify the severity of dampness or of damage associated with dampness. It is clear that the severity of dampness varies widely, from occasional minor condensation on windows to the wetting of a large portion of a building during a flood. The evaluation of the severity or magnitude of moisture problems can use several criteria, most of which are subjective. Excess moisture in a building environment may induce physical damage, but it may also manifest biologic or chemical damage. Direct, immediate impacts include structural, microbiologic, chemical, or aesthetic effects. Indirect consequences include health effects and remediation or repair costs. Because of the complexity of the evaluation, there is no agreed-on basis for determining the severity of damage from either the engineering or the health point of view.

In buildings that have moisture-induced damage, people can be exposed to a complex mixture of microorganisms, organic and inorganic dust, and volatile chemicals (Husman, 1996). It is difficult to measure and distinguish between the various agents and their effects, and exposures have often been defined indirectly and cumulatively as "damp housing" or living in a "water damaged" or "moldy" building. However, as noted in this chapter, there is no generally accepted definition of dampness or of what constitutes a dampness problem, and no generally accepted metric for characterizing dampness.

Several factors might be considered in evaluating the severity of moisture damage. Four of these are discussed below: the size of the damaged area, the presence of visible signs of moisture damage, the duration of its presence, and the building material on which the damage is observed.

The size or extent of damage is an important moisture-damage characteristic assumed to be related to source strength. It is reasonable to expect larger or more extensive damage to be associated with higher potential exposure. However, the literature does not provide much information on the estimation of damage size. Williamson et al. (1997) used a subjective grading of the extent of visible mold on a four-point scale: 0 = absent, 1 = trace, 2 = obvious but localized, and 3 = obvious and widespread.

Subjectively, the extent of visible mold contamination on surfaces in buildings has been taken into account in guidelines for cleanup procedures issued by government and professional organizations. Chapter 6 discusses those guidelines and their recommendations.

In a study intended to seek insights into the type of moisture damage that could be critical as a risk indicator for adverse health effects, a random sample of residential buildings was inspected for signs of moisture damage (Haverinen et al., 2001a). Trained building inspectors estimated the size of observed damage in square meters, and a dose-dependent association with respiratory infections and lower respiratory symptoms was observed. A later study used multivariate Poisson regression models to examine respiratory symptoms (Haverinen et al., 2003); the relative importance of a variable characterizing the size of moisture damage appeared to be high, and the authors concluded that the size of the damage is an important characteristic related to the severity of damage. It should be remembered, however, that estimation of the size of damage is difficult and that estimation accuracy varies because damage is often hidden.

Location

Intuition suggests that the location of moisture damage or mold growth might be important in evaluating exposure because it will be related to the amount of pollutants that may come into contact with a person. However, few studies have examined it in any detail. Some have concentrated on the more frequently or densely occupied locations within a home, such as bedrooms and living rooms (Dharmage et al., 1999b; Reponen et al., 1989; Su et al., 1992; Verhoeff et al., 1994; Wickman et al., 1992). Ross et al.

(2000) examined the association between asthma symptoms and indoor bioaerosols in an area where severe flooding had taken place. The study focused on locations on the basis of how they potentially influenced the exposure: bedrooms (location in relation to exposed people) and basements (location in relation to pollutant source). Little between-room variability was observed. Of the 44 homes evaluated, 26 showed no difference in concentrations between rooms; only eight of the remaining 18 had significantly higher concentrations in one room than the home average.

Duration

The period of or duration of moisture damage might also be expected to be important, but little research has investigated it. "Ongoing damage"—defined as damage resulting from either recent wetting or a lack of change in moisture conditions within 6 months of construction—has been associated with higher concentrations of culturable fungi in building materials than "dry damage"—past damage due to high moisture conditions where the materials had subsequently dried without remediation (Pasanen et al., 2000a). However, the time-frame of the damage has not been associated with health effects in a straightforward manner (Haverinen et al., 2001a).

The definition of duration of damage and examination of its possible influence on occupant health deserves more consideration. Well-designed studies could allow important data to be gathered on whether time-related characteristics of moisture-induced deterioration of materials influence the manifestation of health effects. The resulting information could be used to guide prevention or remediation strategies.

RISK FACTORS FOR MOISTURE PROBLEMS

Building Characteristics

Indoor moisture is linked with some building characteristics. Reported dampness has been associated with age of the building, lack of central heating, humidifiers, and pets (Spengler et al., 1994; Tariq et al., 1996). Low temperatures and high RH indoors can result from cold climatic conditions or from such building characteristics as the lack of thermal insulation and heating. Evans et al. (2000) found a linear association between reported indoor dampness and low temperature and adult health. Older buildings tend to be colder (Hunt and Gidman, 1982) and therefore to have higher RH. Thus, the age of the building can indirectly be associated with both indoor temperature and dampness, all else equal. Martin et al. (1987) found a relationship between damp housing and overcrowding, but not duration of occupancy, household income, use of gas heating, or occupant smoking behavior.

Microbial growth has also been associated with building characteristics. In residences, measures of microbial contamination have been found to be positively correlated with indoor temperature and humidity, age and size of buildings, use of wood stoves and fireplaces, absence of mechanical ventilation, and presence of pets and old wall-to-wall carpeting (Dharmage et al., 1999b, Lawton et al., 1998). Garrett et al. (1998) found airborne fungal concentrations and signs of moisture damage (including musty odor, water intrusion, and high RH) to be associated with smaller amount of thermal insulation, cracks in cladding, and poor ventilation. A factor analysis found an association between airborne concentrations of soil fungi and a dirt-floor, crawl space type of basement in residences (Su et al., 1992). The same study measured increased concentrations of water-requiring fungi in the air of residences where water accumulation was observed. Lawton et al. (1998) developed a "calculated internal moisture source strength" metric that was associated with high biologic contamination and age of houses but not with RH or number of occupants. Verhoeff et al. (1994) found an association between number of fungal propagules in settled dust and type of flooring; no association with other characteristics—such as ventilation and heating facilities, building materials, insulation, and observed dampness—was identified. In another study, Verhoeff et al. (1992) found that indoor viable mold propagules were weakly correlated with several risk factors for moisture problems (age of building, moisture-retaining building materials, and the presence of a crawl space) and observed dampness (damp spots, mold growth, wood rot, silverfish or sowbugs, stale odor, and wet crawl space).

Barriers to Prevention

Information on controlling moisture in residences and larger buildings has been developed and published (Lstiburek, 2001, 2002a,b; Lstiburek and Carmody, 1996; Rose, 1997), but the high incidence of indoor dampness suggests that it is not consistently applied by those designing, constructing, or maintaining buildings.

A number of institutional barriers hinder good practice. One may be that building professionals do not have the knowledge needed to design and build structures to minimize moisture problems. Systematic surveys of curricula are lacking, but generally there is minimal instruction in moisture-control principles for architects and structural engineers[4] and a lack of

---

[4]It must be noted that the broad array of subject competencies required of architects and structural engineers may leave little time for such focused training. Design teams might thus need specialists in moisture dynamics and control.

formal training in moisture-protective building techniques and materials for the construction workforce. Indeed, increased interest in dampness issues has resulted in workshops, continuing education, and new design tools that are addressing this need (Karagiozis, 2001; ORNL/IBP, 2003).

Additional barriers result from building-code requirements that inadvertently or indirectly increase the risk of moisture problems. Most codes require passive or active ventilation of crawl spaces. That requirement makes it difficult to construct a crawl space that is included as part of the conditioned space or simply inside the thermal envelope (Advanced Energy, 2001). If rainwater and groundwater are kept out, sealed insulated crawl spaces are often drier than ventilated ones. The entry of warm, humid outdoor air into ventilated crawl spaces, which are often cooler than outdoors, serves as a moisture source for the crawl spaces. Some codes, such as the 2000 International Residential Code, contain exceptions that provide a path to constructing sealed insulated crawl spaces (ICBO, 2000). Sealing crawl spaces can reduce moisture problems there, but sealing can increase concentrations of radon in a crawl space and the associated house. Because radon exposure increases the risk of lung cancer, sealed crawl spaces may be inappropriate in locations where radon concentrations tend to be high.

Building-code requirements for vapor retarders on the interior side of exterior walls and ceilings may also have an impact on building dampness. Adherence to some codes may result in condensation problems when air conditioning is used and—in combination with low-permeability exterior sheathings—reduce the drying potential of a wall section. When a building is air-conditioned, the vapor-pressure gradient is from the exterior toward the interior, where condensation on the back side of the intentional vapor retarder may occur. The situation is aggravated if the building's cladding is composed of a material, such as brick or split-face block veneer, that absorbs rainwater, because the vapor-pressure driving force is greatly increased when the sun raises the temperature of the veneer to that of liquid water. A similar situation occurs when the building interior is depressurized relative to the outdoors; depressurization causes warm air to be drawn through the building envelope, and it washes the backside of chilled surfaces with humid air from which water may condense. Such circumstances point to the need for building codes and design and construction recommendations that take climate into account. Lstiburek and Pettit (2000, 2001, 2002a,b), for example, have produced a series of books that offer design and construction advice specific to various housing types and climatic conditions found in the United States, including advice on avoiding water intrusion and excessive indoor dampness.

Finally, building codes—which guide new construction—may sometimes also apply to renovations. The advocacy group Smart Growth in America asserted that in the late 1990s there were conditions in which even a simple repair could trigger requirements to bring an entire building up to code in Maryland (Smart Growth in America, 1999). Such circumstances could make upgrades uneconomical and limit the funds available for remediation.

Chapter 7 addresses other barriers to preventing and remediating moisture problems.

FROM MOISTURE TO MICROBIAL GROWTH

Dampness and other excess moisture accumulation in buildings are closely connected to observations of mold, mildew, or other microbial growth. The behavior of moisture and air movements can be characterized with physical parameters, but the biological phenomena take place according to a complicated network of regulating factors. Several phenomena make up the microbial ecology of an indoor environment.

Buildings as Microbial Habitats

In principle, common saprophytic environmental microorganisms[5] and their spores are present everywhere and they start to grow wherever their basic needs for growth are met. They differ enormously in their needs for environmental conditions and some fungi or bacteria always do well in practically any indoor microenvironmental conditions. As previously noted, one important factor is the availability of moisture. Many environmental microorganisms easily start growing on any surface that becomes wet or moistened. The minimal moisture need for microbial growth may be characterized in terms of the water activity of the substrate, a_w, which is the ratio of the moisture content of the material in question to the moisture content of the same material when it is saturated. In a situation where the material is in equilibrium with surrounding air that has a RH of 100%, $a_w = 1$.

The lowest a_w at which the most tolerant, so-called xerophilic fungi may grow is 0.7, which corresponds to an ERH of 70%. A few species—such as *Penicillium brevicompactum*, *Eurotium* spp., *Wallemia sebi*, and *Aspergillus versicolor*—may start growing in these conditions. At higher moisture levels, such intermediate species as *Cladosporium sphaerospermum*, *C. cladosporioides*, and *Aspergillus flavus* may germinate and start their mycelial growth. Most fungi and bacteria require nearly saturated conditions; that is, a_w of at least 0.85–0.90 (Grant et al., 1989). Examples

---

[5]The term "environmental microorganism" is used here to distinguish the microorganisms that are usually found in indoor or outdoor spaces from those more typically found in humans or other living hosts.

of such fungi are *Mucor plumbeus*, *Alternaria alternata*, *Stachybotrys atra*, *Ulocladium consortiale*, and yeasts (Flannigan and Miller, 2001).

Determinants of Microbial Growth Indoors

Along the life span of a building, weather changes and other events often cause at least temporary wetting of some of its parts. Signs of microbial growth can thus be detected on many parts of a structure. Airborne spores and cells also accumulate in the parts of the structure that are in contact with soil or outdoor air, especially parts that act as sites of infiltration of intake air. Accumulated spores may or may not grow in these sites, depending primarily on moisture conditions.

Because their growth is regulated by the available resources, conditions, and competing organisms, the development of a microbial community may be slow in slowly changing conditions or fast whenever there is a sudden increase in one or more of the limiting factors. Examples of such incidents are floods, firefighting, and acute water damage (Pasanen et al., 2000b; Pearce et al., 1995; Rautiala et al., 2002).

The time it takes for fungi to grow on a particular material depends on the material's characteristics, the fungal species, and the amount of moisture (Doll, 2002). Molds are also capable of producing large quantities of spores within a short time. Rautiala et al. (2002) reported massive fungal growth within a week after firefighting efforts. According to Pasanen et al. (1992a), a fungus can grow and sporulate within a day in moist conditions and within a week on occasionally wet indoor surfaces. Viitanen (1997) modeled the time factor in the development of fungi and found that at RH above 80% for several weeks or months, mold can grow in wood when the temperature is 40–120°F (5–50°C). At RH above 95%, mold can be seen within a few days. In wetted gypsum board inoculated with spores, fungal growth started within 1–2 weeks (Murtoniemi et al., 2001). Chang et al. (1995) reported a latent period of 3 days for fungal growth on ceiling tiles, during which the germination and mold growth could be arrested.

Besides water, microorganisms need proper nutrients and temperatures to grow; some also need particular light conditions.[6] Those circumstances are usually met in buildings. Even if modern building materials do not appear to be readily biodegradable, they may support microbial action.

Microbial nutrients may be carbohydrates, proteins, lipids and other biologic molecules and complexes, or they may be nonbiologic compounds. Nutrients are provided by house dust and available moisture and by many surface and construction materials, such as wallpapers, textiles, wood,

---

[6]Light is needed for the growth of many fungi and bacteria but the lack of light does not prevent microbial action. Thus, in general, light is not a critical factor in building microbiology.

paints, and glues. Even nonbiodegradable material, such as ceramic tiles and concrete, may support microbial growth (Hyvärinen et al., 2002) by providing a surface for colonies. That explains why fungal colonies may be found on mineral fiber insulation—a material that would not seem hospitable to microbial growth (Wålinder et al., 2001; Hyvärinen et al., 2002).

Prevailing temperatures in living spaces and other sections of buildings are usually 32–130°F (0–55°C), that is, greater than freezing and less than the temperature at which the denaturalization of proteins would start. That range permits the growth of most environmental microorganisms even if the temperature is not optimal for a particular genus or species. Many environmental microorganisms are not especially strict in their temperature demands, in contrast with many pathogenic microorganisms that need the human body temperature to be able to grow.

Time is another integral element in the assessment of microbial growth in buildings. Growth may be slowed by decreasing or increasing temperatures or other limiting factors, and the time window that must be considered in building microbiology is weeks, months, or even years. It is known that microbial degradation normally consists of a chain of events, in which different groups of microorganisms follow each other (Grant et al., 1989), but present knowledge of building microbial ecology does not allow accurate estimation of the age of microbial damage on the basis of the particular fungal or bacterial flora observed.

MICROORGANISMS OCCURRING IN INDOOR SPACES AND ON BUILDING MATERIALS

Microorganism is a catch-all term that refers to any form of life of microscopic size. This section focuses on fungi and bacteria associated with damp indoor spaces. Other microorganisms that may be found in such environments—notably, house dust mites—are not addressed here, although their presence may have important effects on occupants; the health effects of exposure to them and to others more generally related to indoor environments are covered in detail in the Institute of Medicine (IOM) reports *Clearing the Air* (IOM, 2000) and *Indoor Allergens* (IOM, 1993) which discuss asthma and general allergic responses, respectively. Larger organisms, such as cockroaches, also inhabit damp spaces and may be responsible for some of the health problems attributed to these spaces; they are also addressed in the IOM reports cited above.

Fungi and Bacteria in Outdoor and Indoor Air

The fungi have (eukaryotic) cells like animals and plants, but are a separate kingdom. Most consist of masses of filaments, live off of dead or decaying organic matter, and reproduce by spores. Visible fungal colonies found indoors are commonly called mold or sometimes mildew. This report, following the convention of much of the literature on indoor environments, uses the terms fungus and mold interchangeably to refer to the microorganisms.

Filamentous fungi, yeasts, and bacteria are common in outdoor soil and vegetation, and outdoor air is an important transport route to the indoor environment for spores and other particles of microbial origin. Spores are often monitored outdoors with direct microscopic counting instead of culturable methods; when so measured, total spore counts may often reach an order of magnitude of 10^4 spores/m^3 (Mullins, 2001). Microorganisms from outdoor air often enter indoor environments through open doors and windows and through ventilation intakes.

Spores of common molds, bacteria, and other microbial particles are regularly found in indoor air and on surfaces and materials—no indoor space is free of microorganisms. They are continuously deposited and removed by various mechanisms, such as gravitational settling on surfaces, by exhaust ventilation,[7] and by diffusion to vertical surfaces and cavities. Deposited spores are also removed or released by cleaning, vibration, filtration, accidental ventilation, fan-powered outdoor air (which may pressurize the building and squeeze air out rather than exhaust it), and thermophoresis. Those mechanisms depend primarily on the size of the particle: the larger the particle, the faster the gravitational settling. Small microbial particles (<5 μm) may not settle on surfaces before they are removed by ventilation. After settling on surfaces, microbial particles integrate with other house dust, and they may be removed by cleaning. Part of the settled house dust is resuspended into the air as a result of occupants' movements and other mechanical disturbance (Buttner and Stetzenbach, 1993; Thatcher and Layton, 1995).

Common fungal genera found in outdoor air include *Cladosporium*, *Aspergillus*, *Penicillium*, *Alternaria*, and *Saccharomyces* (yeasts) (Mullins, 2001), but the overall diversity of outdoor fungi is great. The genus *Aspergillus*, for example, has over 185 known species, including *Aspergillus fumigatus*, *A. versicolor*, *A. flavus*, *A. penicilloides*, and *A. niger*. Among the other fungal genera observed in outdoor air are *Acremonium*, *Aureobasidium*, *Cunninghamella*, *Curvularia*, *Drechslera*, *Epicoccum*, *Fusarium*, *Geotrichum*, *Hyalodendron*, *Leptosphaeria*, *Neurospora*, *Paecilomyces*, *Rhinocladiella*, *Trichoderma*, *Tritirachium*, and such basidiomycete genera as *Coprinus* and *Ganoderma* (Mullins, 2001; Shelton et al., 2002).

The concentrations and diversity of outdoor-air fungi vary with the geographic area, climate, season, weather conditions, and individual

---

[7]Chapter 10 of *Clearing the Air* (IOM, 2000) provides greater detail on building ventilation and air cleaning and their effect on exposure to indoor pollutants.

sources, such as agricultural activities. In temperate climates, the concentrations are usually highest in summer and fall and lowest in winter and spring (Shelton et al., 2002). Variation is also reflected in the counts and mycoflora of the indoor environment. Indoor concentrations of fungi are usually lower than the corresponding outdoor concentrations, but they vary considerably with the same range as outdoor air: 10^0–10^4 cfu/m^3 (Shelton et al., 2002). Thus, it is difficult to give any "typical" counts of airborne fungi that would apply to more than a specific, defined set of conditions.

Fungal contamination of the indoor environment creates a source of spores, fungal fragments, and other products that may become airborne and cause changes in the microbial status of the environment outside the range of "normal" conditions. Measurements of airborne fungi are often used to detect such contamination. However, even an actively growing mold mycelium does not release spores continuously; release depends on many physiologic and environmental factors, and it is not possible to detect the presence of such a source solely from the fungal-spore content of the indoor air.

Sampling methods also cause variation in the data collected on fungal concentrations and speciation. For airborne fungi, the characteristics of the sampling device—such as its cutoff size and collection efficiency—influence the recovery of fungal particles (Reponen et al., 2001). Fungal counts are obtained either by direct microscopic counting or by culturing the spores into colonies, which are then counted and identified according to their morphologic features. Direct counting usually allows a rough genus- or group-level identification, although some species can be identified by microscopic examination of spore trap plates or tape lifts (such as *Stachybotrys chartarum*, *Cladosporium sphaerospermum*, *C. cladosporioides*, *Alternaria alternata*, and *Aspergillus niger*). Instead, the culturing results depend on the growth media and conditions selected. Ren et al. (1999a) noted that the type and concentrations of fungi measured in house-dust samples were not representative of those isolated in indoor air. No sampling and analytic technique will cover all the fungi and allow their equal detection and identification. Therefore, reported profiles also depend on the sampling, counting, and culturing methods used (ACGIH, 1999). Chapter 3 discusses sampling methods in detail.

Table 2-2 summarizes studies that have aimed at differentiating buildings with and without moisture damage by fungal counts of the indoor air. As can be seen, there is no general pattern whereby a characteristic fungal concentration is associated with either moisture-damaged or nondamaged homes, and the variation in measured quantities is large in both cases. Some studies have shown that increased airborne concentrations of fungi are associated with moisture damage in a building, and others have failed to show any such pattern. Taken together, the studies indicate that the fungal counts alone provide little information about the microbial status of an indoor environment.[8] However, information about the species found is useful in assessing whether the microbial constituents of a given indoor environment differ from what are considered typical in those particular conditions.

Indoor concentrations of fungi are usually lower than outdoor concentrations, but the indoor concentrations follow the outdoor ones (Shelton et al., 2002). The large variation in and sometimes dominating effect of outdoor-air fungal concentrations cause difficulties in interpreting measurements made in indoor environments. It is common to use the indoor:outdoor (I/O) concentration ratios to reflect the presence of indoor sources of microorganisms. Because fungal spores circulating in indoor air deposit on surfaces and are caught by air filtration, I/O ratios are typically less than 1.0. However, if there is a strong microbial source indoors, the ratio can exceed 1.0. In a compilation of data from indoor air quality investigations in the United States, Shelton et al. (2002) found I/O ratios of 0.1–200. However, the ratios in most cases were well under 1.0. Species-level identification of the fungi allows even more accurate assessment. Where the I/O ratio of an individual species is repeatedly over 1.0, it suggests the presence of an indoor source of the species. It should be remembered, though, that where the numbers of both indoor and outdoor spores are low, ratios may yield misleading values.

Most fungi found indoors come from outdoor sources, but bacteria have outdoor and indoor sources. Occupants of a building are a major source of bacteria, although the large majority of bacteria shed by people are not considered harmful to other people (Burge et al., 1999). Bacteria of human origin include gram-positive cocci, such as micrococci and staphylococci. Among typical outdoor-air bacteria are *Bacillus, Corynebacterium, Flavobacterium, Micrococcus, Pseudomonas, Streptomyces,* and other actinomycetes. Like that of fungal flora, the genus and species diversity of outdoor-air bacteria is large.

Environmental bacteria also grow in all wet spaces and are found in most cases where there is mold growth (Hyvärinen et al., 2002), but the profile of bacterial genera and species growing on moist building materials differs from that originating from humans.

Fungi and Bacteria on Building Materials

Most fungi and bacteria that grow on moistened building materials can also be found in outdoor natural habitats and air. However, the rank order

---

[8]The "Sampling Strategy" section of Chapter 3 also addresses the use of fungal counts in the assessment of indoor microbial contamination.

TABLE 2-2 Summary of Studies of Airborne Fungal Concentrations in Residences in Relation to Building Dampness Characteristics

Study	Number and Type of Sites	Study Design	Method[a]
Gallup et al., 1987	127 residences	Moisture problem Non-problem	6-stage impactor
Hunter et al., 1988	62 residences	Monitoring complaint home	6-stage impactor (MEA)
Miller et al., 1988	50 residences	Characterize concentrations of fungi and fungal metabolites in winter	RCS (rose bengal malt extract); Filter (rose bengal malt extract, MEA + sucrose)
Waegemaekers et al., 1989	36 residences	Damp (24) Reference (8) [Unspecified (4)]	6-stage impactor (MEA)
Strachan et al., 1990	88 residences	Homes of children with wheeze (34) Controls (54)	6-stage impactor (MEA)
Reynolds et al., 1990	6 residential and office environments	Monitoring complaint buildings	2-stage impactor (Sabouraud dextrose agar)
Nevalainen et al., 1991	48 residences	Mold-damaged (30) Reference (18)	6-stage impactor (Hagem)
Pasanen et al., 1992b	46 residences	Damp (25) Reference (21)	6-stage impactor (Hagem)
Pasanen, 1992	57 residences	Urban (21) Damp urban (22) Rural (13): 7 old + 6 new [Unspecified (1)]	6-stage impactor (Hagem/MEA)
Verhoeff et al., 1992	130 residences	Relation of fungal concentrations to dampness	1-stage impactor (DG18)

Levels of Airborne Fungi[b,c]		Fungal Concentrations in Relation to Building Characteristics
Problem: Nonproblem:	AM 5,950 cfu/m^3 AM 716 cfu/m^3	Concentrations higher in problem homes
Visible mold: No mold:	<12–449,800 cfu/m^3 <12–23,070 cfu/m^3	High concentrations associated with visible mold growth, construction work, and activity
RCS: Filter:	AM 345 cfu/m^3 (0–3,125) AM 111 cfu/m^3	No conclusion on effect of moisture or dampness
Damp: Reference:	GM 192 cfu/m^3 GM 102 cfu/m^3	Fungal concentrations associated with dampness
Visible mold: No mold:	<41,300 cfu/m^3 (MD 200–294) <38,600 cfu/m^3 (MD 21–283)	Median concentrations of viable fungi associated with visible mold
Indoors: Outdoors:	<18,900 cfu/m^3 <1,090 cfu/m^3	High indoor:outdoor ratio and flora indicated indoor-air sources
Damaged: Reference:	10–2,300 cfu/m^3 (GM 102) 165–850 cfu/m^3 (GM 308)	Mean concentrations of viable fungi lower in damaged than reference residences, but higher mean indoor:outdoor ratio in damaged residences (4.2/0.6) indicates indoor sources
Damp: Reference:	<2,291 cfu/m^3 (GM 80) <1,445 cfu/m^3 (GM 78)	Concentrations not higher in damp houses
Urban: Damp: New rural: Old rural:	<1,445 cfu/m^3 (GM 78) 2–1,198 cfu/m^3 (GM 69) 25–1,916 cfu/m^3 (GM 70) 98–5,730 cfu/m^3 (GM 1012)	Concentrations not higher in damp houses; concentrations higher in old rural houses
Indoors:	62–43,045 cfu/m^3 (GM 640–822)	Fungal concentrations correlated weakly with dampness

(continued on next page)

TABLE 2-2   continued

Study	Number and Type of Sites	Study Design	Method[a]
Beguin and Nolard, 1994	130 residences	Monitoring patient homes	RCS (HS medium with rose bengal)
DeKoster and Thorne, 1995	41 residences	Health-based home categories: Noncomplaint (27) Intervention (10) Complaint (4)	6-stage impactor (MEA)
Li and Kendrick, 1995	15 residences	Homes of allergic (13) Homes of nonallergic (2)	Samplair MK1/MK2
Rautiala et al., 1996	7 buildings	Monitoring of effect of mold-damage repair	6-stage impactor (MEA) filter
Dill and Niggemann, 1996	20 residences	Homes of children with allergic diseases	RCS (MEA + Czapek Dox)
Garrett et al., 1998	80 residences	Homes of Asthmatics (43) Nonasthmatics (37)	1-stage impactor (MEA)
Rautiala et al., 1998	3 buildings	Reducing microbial exposure during demolition of moldy structures	Filter cultivation (MEA, DG18)
Dharmage et al., 1999a	485 residences	Homes of Random sample (349) Asthmatics (139) [note: Σ = 488]	2-stage impactor (PDA)
Johanning et al., 1999	2 residences	Mold-damaged (1) Control (1)	1-stage impactor (MEA) Filter
Klánová, 2000	Residences and offices 68 rooms	A) no complaints + no mold (20) B) complaints + no mold (20) C) no complaints + visible mold (10) D) complaints + visible mold (18)	RCS + aeroscope (YMA)

Levels of Airborne Fungi[b,c]	Fungal Concentrations in Relation to Building Characteristics
375–3,750 cfu/m^3	No conclusion on effect of moisture or dampness
Noncomplaint: GM <1,290 cfu/m^3 Intervention: GM <1,100 cfu/m^3 Complaint: GM <6,700 cfu/m^3	Indoor:outdoor ratios higher in complaint homes; concentrations higher in basement than main floor
Damp: 2,727 spores/m^3 Nondamp: 2,051 spores/m^3	Concentrations higher in damp residences
Before repairs: GM 370 cfu/m^3 (<1,150) GM 59,000 spores/m^3 (<500,000) After repairs: GM 200 cfu/m^3 (<300)	Demolition of moldy structures increases concentrations remarkably; concentrations on baseline after 6 months
Visible mold: 64 to over 4,000 cfu/m^3 No mold: <13–1,652 cfu/m^3	Airborne concentrations not correlated with visible fungal growth
<20–54,749 cfu/m^3 (MD 812)	No association between mean concentrations and visible mold; increased concentrations of fungi associated with musty odor, moisture or humidity, poor ventilation, and failure to clean indoor mold growth
Before repairs: 860–1,300 cfu/m^3 During repairs: <8 × 10^5	Local exhaust method most effective for control; personal protection still needed
37–7,619 cfu/m^3 (MD 549)	Higher concentrations in residences with visible mold
Damaged: 1,993 to >7,069 cfu/m^3 1.8–6.6 × 10^5 spores/m^3 Control: 194–336 cfu/m^3 3.7–4.7 × 10^3 spores/m^3	Higher concentrations in residence with visible mold
A) 0–230 cfu/m^3 (AM 78) B) 0–140 cfu/m^3 (AM 58) C) 60–3,190 cfu/m^3 (AM 1033) D) 120–17,930 cfu/m^3 (AM 2476)	Concentrations of viable fungi higher in rooms with visible mold

(continued on next page)

TABLE 2-2   continued

Study	Number and Type of Sites	Study Design	Method[a]
Miller et al., 2000	58 residences	Relation of air sampling and damaged materials	RCS (rose bengal malt extract)
Pessi et al., 2002	88 residences	Fungal concentrations in relation to microbial growth in external walls	6-stage impactor (MEA)

[a]MEA = malt extract agar; RCS = Reuter centrifugal sampler; YMA = yeast and mold agar; PDA = potato dextrose agar.
[b]AM = arithmetic mean; GM = geometric mean; MD = median.

of the most prevalent species in indoor growth sites is generally different from that of species normally found in outdoor air, and otherwise unusual species may prevail indoors. Table 2-3 lists examples of fungal genera that have been isolated from "moldy" building materials or surfaces. Most fungal genera have several species, many of which occur on moldy building materials. Therefore, the species diversity is far more extensive than the genus diversity shown in the table.

Some fungi are considered "typical" or "indicators" of mold growth on building materials because they are often isolated from mold samples. However, the mere presence of a fungus at a low concentration does not necessarily indicate mold damage. Instead, the simultaneous presence of several otherwise unusual or indicator fungi at concentrations that exceed the

TABLE 2-3   Examples of Fungal Genera Found in Infested Building Materials

Acremonium	*Gliocladium*	*Scopulariopsis*
Alternaria	*Humicola*	*Sphaeropsidales*
Aspergillus	*Mucor*	*Stachybotrys*
Aureobasidium	*Oidiodendron*	*Torula*
Botrytis	*Paecilomyces*	*Trichoderma*
Chaetomium	*Penicillium*	*Tritirachium*
Cladosporium	*Phialophora*	*Ulocladium*
Doratomyces	*Phoma*	*Verticillium*
Eurotium	*Rhinocladiella*	*Wallemia*
Fusarium	*Rhizopus*	Yeasts
Geomyces	*Rhodotorula*	

SOURCES: Flannigan and Miller, 2001; Gravesen et al., 1999; Hyvärinen et al., 2002.

Levels of Airborne Fungi[b,c]	Fungal Concentrations in Relation to Building Characteristics
15 residences with lowest visible growth: AM 214 cfu/m3 15 residences with highest visible growth: AM 329 cfu/m^3	The mean levels were not associated with severity of damage. More species different from outdoor air in homes with severe damage
Low growth: 9–516 cfu/m^3 (AM 112) Growth: 2–1,784 cfu/m^3 (AM 121)	Microbial growth in insulated external wall did not affect indoor air levels.

[c]Concentrations reported in these studies cannot be used as reference values because of methodologic limitations in measurement techniques. Chapter 3 addresses this topic in greater detail.

SOURCE: Excerpted and adapted from Hyvärinen, 2002.

background concentrations in outdoor air or other reference samples can be regarded as an indication of indoor mold colonization.

Although there is no general international consensus on which species should be regarded as indicators of the presence of mold, several fungi are often isolated from moldy areas. Table 2-4 shows examples of such fungi.

Mold growth on materials is usually accompanied by bacterial growth (Hyvärinen et al., 2002). Such bacteria have been studied much less than fungi, but they are a part of the phenomenon of dampness and microbial growth on materials and therefore among the agents occupants may be exposed to, so they deserve attention. Bacteria that have been identified in samples of moldy-building materials are shown in Table 2-5.

Components of Microbial Agents

Some studies of fungi and bacteria examine specific microbial components found in damp indoor environments. Among the components characterized so far are spores and hyphal fragments of fungi, spores and cells of bacteria, allergens of microbial origin, structural components of fungal and

TABLE 2-4 Examples of Fungi and Other Microorganisms Often Associated with Dampness or Mold Growth in Buildings

Aspergillus fumigatus	*Phialophora* spp.	*Wallemia* spp.
Aspergillus versicolor	*Stachybotrys chartarum*	Actinomycetes
Aspergillus penicilloides	*Trichoderma* spp.	Gram-negative bacteria
Exophiala spp.	*Ulocladium* spp.	

SOURCES: Gravesen et al., 1994; Jarvis and Morey, 2001; Samson et al., 1994.

TABLE 2-5   Bacterial Genera Isolated from Moldy Building Materials

Acinetobacter	Dietzia	Rhodococcus
Agrobacterium	Flavobacterium	Spirillospora
Arthrobacter	Gordonia	Streptomyces
Bacillus	Methylobacterium	Streptosporangia
Brevibacterium	Microbacterium	Thermomonospora
Cellulomonas	Mycobacterium	
Clavibacter	Nocardia	
Corynebacterium	Nocardiopsis	

SOURCES: Andersson et al., 1997; Peltola et al., 2001a,b.

bacterial cells (such as $\beta(1{\rightarrow}3)$-glucans of fungi, endotoxins produced by gram-negative bacteria, and peptidoglycans of bacteria), and such products as microbial volatile organic compounds (MVOCs) and toxic products of microbial secondary metabolism. Information on those agents is briefly summarized below. Chapter 3 discuss exposures to these agents in more detail.

Spores and Fragments of Fungi

Fungi produce and release spores that are cells with well-developed resistance to environmental stresses, such as desiccation and UV radiation. They are the essential means of distribution of filamentous fungi. The particle size of most fungal spores is roughly 2–10 µm, so they are easily transported by winds and air currents, and they may enter the respiratory system (Reponen et al., 2001). Fungal types vary remarkably in their capacity to produce and release spores. *Penicillium* and *Aspergillus* typically produce large numbers of spores that are easily released into the air. *Stachybotrys* and *Chaetomium* are examples of fungi that produce fewer spores and release them only occasionally. *Penicillium* and *Aspergillus* spores are regularly found in air samples, and *Stachybotrys* and *Chaetomium* spores are rarely found in the air, even in environments where they are growing (Andersen and Nissen, 2000).

Fungi also release smaller particles (<1 µm) from the mycelium, as experimentally shown by Górny et al. (2002). The microbial origin of the small fragments was verified with antigen characterization. In the experimental study, the smaller particles were released in greater numbers than whole spores, but the concentrations of the small fragments in indoor environments have not yet been characterized. Their small size makes them capable of penetrating deeply into the alveolar region. However, their specific role—if any—in adverse health outcomes has not been studied.

Spores and Cells of Bacteria

Like fungi, spore-forming bacteria release spores from the growth site into the air. Among spore-forming bacteria are *Bacillus* spp. and actinomycetes, such as *Streptomyces*. Bacterial spores are smaller than those of fungi—about 1 μm—but bacterial growth may release fragments smaller than the spores (Górny et al., 2003).

Non-spore-forming bacteria may also enter the air as a result of various processes, but these bacteria have no specific mechanism that causes them to become aerosolized. As mentioned above, humans shed bacteria from their skin and respiratory system. Waterborne gram-negative bacteria may enter the air via aerosolization or other mechanical disturbances of standing water. Gram-negative bacteria are also common in house dust, soil, and plants, and they are probably carried indoors on pets and dust.

Allergens of Microbial Origin

Fungi produce an enormous array of potentially allergenic compounds; each fungus produces many allergens of different potencies. Table 2-6 lists the major defined allergens isolated from fungi. Others have been identified but are clinically "minor" (few patients react to them); still others remain to be identified. Fungal allergen production varies with the isolate (strain),

TABLE 2-6 · Major Defined Allergens Isolated from Fungi

Fungus	Major Allergen	Nature of Allergen(s)	Reference
Aspergillus fumigatus	Asp f 1	18 kD; mitogillin	Arruda et al., 1990
	Asp f 3	Peroxisomal membrane protein	Crameri, 1998
Aspergillus oryzae		Alkaline serine protease	Shen et al., 1998
Alternaria alternata	Alt a 1		Yunginger et al., 1980
	Alt a 2		Sanchez and Bush, 1994
Cladosporium herbarum	Cla h 1	13-kD glycoprotein	Aukrust and Borch, 1979
Penicillium chrysogenum		68-kD protein	Shen et al., 1995
Penicillium citrinum		33-kD protein	Shen et al., 1997
Psilocybe cubensis	Psi c 2	23-kD protein; cyclophilin	Horner et al., 1995
Malassezia furfur	Mal f 1	36-kD protein	Schmidt et al., 1997
Trichophyton tonsurans	Tri t 1	30-kD protein	Deuell et al., 1991

SOURCE: IOM, 2000.

species, and genus (Burge et al., 1989). Different allergen amounts and profiles are contained in spores, mycelium, and culture medium (Cruz et al., 1997; Fadel et al., 1992). In addition, the substrate strongly influences the amount and patterns of allergen production. Fungi, for example, release proteases during germination and growth, and fungal extracts contain sufficient protease to denature other allergens in mixtures.

Microbial allergens are addressed in detail in the IOM reports *Clearing the Air* (IOM, 2000) and *Indoor Allergens* (IOM, 1993), which should be consulted for additional information.

Structural Components of Fungi and Bacteria

Some components of microbial cells have been investigated for their possible role in human health effects. Three have attracted particular attention from researchers.

Fungal cell walls are composed of acetylglucosamine polymer fibrils embedded in a matrix of glucose polymers formally referred to as $\beta(1{\rightarrow}3)$-glucans. Potent T-cell adjuvants, the $\beta(1{\rightarrow}3)$-glucans have been investigated as antitumor agents (Kiho et al., 1991; Kitamura et al., 1994; Kraus and Franz, 1991). They increase resistance to gram-negative bacterial infection by stimulating macrophages and effecting the release of tumor-necrosis factor α mediated by endotoxin (Adachi et al., 1994a,b; Brattgjerd et al., 1994; Saito et al., 1992; Sakurai et al., 1994; Zhang and Petty, 1994). Soluble glucans have an effect in the lung similar to that of endotoxin (Fogelmark et al., 1994).

Endotoxins—biologically active lipopolysaccharides—are components of some bacterial cell walls that are released when the bacteria die or the cell walls are damaged. They are responsible for some characteristic toxic effects of gram-negative bacteria. Endotoxin exposure has been associated with occupational lung disease among workers exposed at high levels (Douwes and Heederik, 1997; Milton, 1999). Rylander's literature review (2002) notes that studies report both adverse and beneficial effects from low-level exposure to endotoxins, and suggests further research to clarify the role of other agents found in connection with them—$\beta(1{\rightarrow}3)$-glucans in particular—in health outcomes attributed to endotoxin exposure.

Peptidoglycans are the chemical substances that make up the rigid cell walls of eubacteria (bacteria with rigid cell walls, also called true bacteria). They are a major component of the cell walls of gram-positive bacteria and, like endotoxins, may be released into the environment when the cells die or are damaged. One study of classrooms in two elementary schools noted that high concentrations of a biomarker of the presence of peptidoglycans were associated with a teacher's perception of the severity of indoor-air quality problems (Liu et al., 2000). Their possible role in adverse health outcomes related to damp indoor environments is otherwise unexplored.

Microbial Volatile Organic Compounds

MVOCs are small-molecule, volatile substances that are typically released by growing fungi and bacteria as end products of their metabolism. They are often odorous, causing the typical smell of "mold," "cellar," or organic soil. Chemically, they are usually alcohols, aldehydes, ketones, esters, lactones, hydrocarbons, terpenes, and sulfur and nitrogen compounds (Korpi, 2001). However, most of them have sources in addition to microbial growth, so their occurrence is not specific for damp indoor environments with microbial growth. Among the several substances generally considered MVOCs are 3-methylfuran, 3-methyl-1-butanol, 1-octen-3-ol, 2-methylisoborneol, and geosmin (Smedje et al., 1996).

Although the odor of mold has often been associated with respiratory symptoms in damp buildings, the specific role of individual MVOCs or their mixtures in adverse health outcomes has not been studied.

Toxic Products of Microbial Secondary Metabolism [9]

Many fungi and bacteria are able to produce compounds called secondary metabolites. The compounds are not produced in all growth conditions but are often produced in cases of nutrient starvation, in the presence of other environmental stressors, or in the presence of competing organisms. Many secondary metabolites are toxic or otherwise biologically active. Commonly known microbial secondary metabolites are mycotoxins, bacterial toxins, antibiotics, and antimicrobial agents (Demain, 1999). Microbial toxins are not volatile, but they may be carried by spores (Sorenson et al., 1987).

Numerous studies have examined the fungi and bacteria that may produce toxins while growing on building materials (Andersson et al., 1997; Nielsen et al., 1998; Nikulin et al., 1994; Pitt et al., 2000; Tuomi et al., 2000). The same bacterial strain has been shown to express different degrees of toxicity and inflammatory potential while growing on different building materials (Roponen et al., 2001); this supports the view that the substrate is important in the regulation of secondary metabolism.

Production of secondary metabolites, including microbial toxins, may

---

[9]Chapter 4 addresses the toxic potential of fungi and bacteria found in damp indoor environments in greater detail.

vary within a single toxigenic strain (Jarvis and Hinckley, 1999; Larsen et al., 2001; Vesper and Vesper, 2002; Vesper et al., 2001). Variable production of toxins while microorganisms are growing on building materials has been shown experimentally (Murtoniemi et al., 2002, 2003a,b; Ren et al., 1999b). The identity of a mold species thus is insufficient information on which to predict its toxic potential.

Mycotoxin production depends on a number of factors, including the availability of nutrients and water activity of the substrate on which the mold grows, temperature (Gqaleni et al., 1997), the sporulation cycle of the organisms (Larsen and Frisvad, 1994), and the presence of other organisms that are in competition for the moisture, nutrients, and other aspects of the growth environment (Wicklow and Shotwell, 1983). The presence of competing organisms appears to be important, and toxins seem to be produced to inhibit the growth of or kill competitors (Wicklow and Shotwell, 1983). Smith and Moss (1986) found that some molds stop making toxins after a few generations when grown in isolation; if generally true, this suggests that testing to determine whether a microorganism might have produced mycotoxins is best conducted in the early stages of growth after isolation from their environment.

The time in the organism's life cycle also appears to influence toxin production. *Aspergillus* and *Penicillium* species are known to produce potent toxins with sporulation (Larson and Frisvad, 1994). The large energy demands of sporulation require an available supply of nutrients and precursors for structural molecules, such as proteins, nucleic acids, and lipids. Germination of spores likewise requires a large amount of energy. Reducing competition for nutrients, water, oxygen, or other resources by inhibiting the growth of other occupiers of the mold's ecologic niche gives a toxigenic mold a competitive edge toward survival of its offspring (Wicklow and Shotwell, 1983).

One potentially toxigenic fungus found in water-damaged buildings is *Stachybotrys chartarum,* formerly referred to as *Stachybotrys atra* or *Stachybotrys alternans.* It is a cellulose-degrading fungus that grows well on wetted paper, gypsum board, and the paper liner and gypsum core of plasterboard (Hyvärinen et al., 2002; Murtoniemi et al., 2002; Nielsen et al., 1998). *Stachybotrys* may also occur on other types of materials, although less frequently (Hyvärinen et al., 2002). The cellulolytic properties of the fungus explain its occurrence on the wetted paper liner of plasterboard, but it is not fully understood why the gypsum core alone also supports the growth of *Stachybotrys* and its toxin production, as assessed by the in vitro cytotoxicity of the spores (Murtoniemi et al., 2002). A study did find a decrease in *S. chartarum* growth and sporulation (compared with a reference board) when desulfurization gypsum was used in the core and when the liner was treated with biocide or starch was removed from the plasterboard (Murtoniemi et al., 2003b). The same study found that treating plasterboard liner with biocide did not decrease growth and sporulation but did increase the cytotoxicity of the spores produced.[10]

Other possibly toxigenic fungi found in buildings or building materials include *Aspergillus versicolor*, *A. fumigatus*, *A. flavus*, and some species of *Penicillium*, *Trichoderma*, *Fusarium*, and *Chaetomium* (Gravesen et al., 1994). Their toxins have been isolated in mold-infested building materials (Nielsen et al., 1999; Tuomi et al., 2000) and in house dust or carpet dust of damp houses (Engelhart et al., 2002; Richard et al., 1999). However, toxin-producing fungal species produce toxins of varied potency (Abbas et al., 2002; Jarvis, 2002; Nielsen et al., 2002).

Some bacteria found in damp indoor environments are also capable of producing toxins. Among the bacterial types that are potentially toxic while growing on building materials are species of *Streptomyces*, *Bacillus*, and *Nocardiopsis* (Andersson et al., 1998; Jussila et al., 2001; Peltola et al., 2001a,b).

Although mycotoxins or bacterial toxins have often been shown to occur in mold-infested materials, as well as house dust in damp buildings, they have seldom been isolated directly from the air. Spores and fragments of toxigenic fungi may carry these toxins (Sorenson et al., 1987), and this speaks for possible airborne exposure, but little information is available on the degree to which the occupants might be exposed to them. That is partly because of methodologic problems of exposure assessment in general (see Chapter 3). Chapter 4 addresses toxins produced by microbial agents in greater detail.

Gaps in Building Microbiology Science

The great variations in environmental mycoflora in indoor spaces and the large number of variables that affect their occurrence and measurement are among the factors that make it difficult to set quantitative or qualitative guidelines or standards for the microbial quality of indoor air. However, there is evidence of clear differences in harmful potential between different microbes (Huttunen et al., 2003) and more such research would elucidate connections between agents and effects. Chapters 4 and 5 address those concepts in greater detail.

---

[10]When Murtoniemi et al. (2003c) examined growth of the bacteria *Streptomyces californicus* on various plasterboards, they found that removal of starch from the liner and core inhibited growth and sporulation almost completely; spore cytotoxicity was not affected by the presence of a biocide.

Building Materials and Microbial Growth

High moisture content is commonly observed in building materials (Haverinen et al., 2001b). That is not necessarily abnormal, nor does it necessarily mean that there will be microbial exposure. Care must be exercised in the interpretation of indications of high moisture content. Some signs of moisture may indicate old damage, already dried out, that may or may not still be a possible source of exposure. The signs may also indicate problems below the surface or periodic problems. Visible mold, although not a precise measure of exposure, is probably the clearest risk indicator for potential exposure.

Building materials differ in the degree to which their constituents support microbial growth. Below are brief descriptions of the characteristics of some common materials that influence microbial growth.

Wood has a cellular structure, the cell walls being made up of two natural polymers, cellulose and lignin. Water in wood is present as free water in cell cavities and in combination with cellulose in the cell walls. Wood is hygroscopic and always tends to achieve a moisture content in balance with its environment (Oliver, 1997). Wood is also used in various composite products, which traditionally are poor at resisting moisture unless they are bound together with waterproof products, such as glues. The variability in the properties of those products is high; factors that affect their susceptibility to moisture include environmental conditions, component properties, manufacturing processes, preservative treatments, and chemical modification of raw materials (Wang, 1992).

Many fungi use cellulose as a source of nutrients. However, they vary in their ability to degrade cell walls of wood. Fungal growth on wooden material depends on species, surface characteristics of the material, air humidity, and temperature (Viitanen, 1994, 2002). Pasanen et al. (2000b) studied microbial growth in wood-based materials collected from buildings with moisture problems and found high median concentrations of viable fungi in all wood-based materials regardless of whether the damage was considered current or complete. Tuomi et al. (2000) analyzed the occurrence of mycotoxins in moisture-damaged material samples and found them in most of the material categories tested, but 82% of the mycotoxin-positive samples contained cellulose matter, such as paper, board, wood, or paper-covered gypsum board.

Chang et al. (1995) evaluated growth of fungi on cellulose ceiling tiles and found that although dust deposited on old used tiles provided valuable nutrients, even new ceiling tiles could support growth when ERH was above 85%; fungal growth could be limited only if the wetted tiles were dried quickly and thoroughly. Doll (2002) did not observe growth on ceiling tile kept for 8 weeks in an environmental chamber at 85% RH and 72°F (22°C), but did see growth after 3–6 weeks at 95% RH. The samples were not inoculated with fungi in the laboratory—contamination was from natural sources.

Insulation materials include a wide array of wood-based, mineral, and organic materials. The wood-based materials are more hygroscopic than the mineral or organic materials; that is, their moisture content is much higher at a given ERH. Therefore, the moisture behavior may vary substantially among the materials (Nevander and Elmarsson, 1994). Pasanen et al. (2000a) studied the occurrence of microbial growth in insulation-material samples, including glass wool, polystyrene foam, and granulated cork. They observed a correlation between total spores and fungal concentration and the RH of the materials but usually not the %-MC of the materials. Ezeonu et al. (1994) observed no fungal colonization in fiberglass insulation below 50% RH and delayed colonization below 90% RH.

Compared with wood-based materials, masonry and cementitious materials are low in nutrients and biologically inert. That does not necessarily mean that they are immune to problems. Clay brick, for example, is a fast-wetting material because of its powerful capillary suction, and cementitious materials are hygroscopic and slow in drying. Therefore, if wetted, those materials may support microbiologic and chemical deterioration through their interaction with other materials (Oliver, 1997). Pasanen et al. (2000b) showed that the culturable fungal concentrations correlated with %-MC but not with the RH of the material in concrete, cement, mortar, and plaster-based finishing coating. Stone or mineral-based materials are commonly used for interior finishing in facilities with high moisture loads. Those materials are resistant to microbial growth and are not biodegradable. However, nutrients from water and air can accumulate on them and support microbial growth.

Polyvinyl chloride (PVC) materials are among the most frequently used wall and floor finishing materials because they provide inexpensive, easy-to-clean surfaces. They typically resist microbial growth, but (as discussed below) they may degrade in the presence of moisture.

Paint, varnish, and similar materials are often used to protect other materials from water absorption, as well as for aesthetic reasons. Different types of paints differ in their water permeability and capacity to tolerate moisture (Oxley and Gobert, 1994). Peeling or blistering of a painted surface is often a sign of excess moisture in the structure underneath.

DAMPNESS-RELATED PROBLEMS NOT ASSOCIATED WITH BIOLOGIC SOURCES

Apart from mold, bacteria, and mite-related contaminants, moisture sometimes contributes to the release of nonmicrobial chemicals into the indoor air. It has been known for many years that the rate of release of formaldehyde from composite building materials that contain urea-formaldehyde resins, such as particle board, increases with the humidity of the surrounding air (van Netten et al., 1989). The emission of formaldehyde occurs, in part, as a consequence of hydrolysis of the resin. In chamber studies, Andersen et al. (1975) found that increasing the RH from 30% to 70% doubled the rate of formaldehyde emission from particle board.

There have been numerous anecdotal reports of indoor odor and irritation complaints associated with moist building materials, particularly plastic materials on moist alkaline substrates, such as concrete. The phenomenon has also been investigated in a few scientific studies. Offermann et al. (2000) reported increased emission rates of potentially odorous and irritating alcohols from the PVC backing of carpet tiles placed on a concrete slab that had a high water content. Lorenz et al. (2000) reviewed four case studies of health symptoms thought to be caused by chemicals that were emitted when high moisture content was combined with building materials that contained plasticizers. When materials were moistened and heated, they measured high emission rates of alcohols, phthalic anhydride, and other compounds thought to be irritating. In a study of four geriatric hospitals, dampness-related and moisture-related degradation of a plasticizer in PVC flooring was strongly associated with asthma symptoms (OR, 8.6; CI, 1.3–57) (Norbäck et al., 2000). In the same study, the dampness and degradation of the plasticizer were less strongly but still statistically significantly associated with increases in ocular symptoms, nasal symptoms, and lysozyme in nasal lavage (an indicator of inflammation) and with a decrease in tear-film stability (Wieslander et al., 1999). The degradation of the plasticizer was indicated by increased indoor airborne concentrations of 2-ethyl-1-hexanol. Wålinder et al. (2001) also reported a higher concentration of 2-ethyl-1-hexanol in the air of a water-damaged office building with PVC flooring, relative to a control office building in the same complex.

More recently, Sjoberg and Nilsson (2002) offered a theoretical analysis of how heating systems embedded in concrete slabs can exacerbate emissions from alkaline hydrolysis of floor coverings. Two problematic scenarios were identified. In the first, the heating system drives construction moisture out of the concrete slab to moisten the floor covering. In the second, moisture from the soil is driven through the concrete slab because of the temperature gradient that occurs when the heating system is turned off, for example, in the summer after an extended period of heating that has warmed the soil beneath the slab.

It can be concluded only that dampness-related emissions of chemicals have been confirmed and linked in a few studies with health symptoms and odor complaints. The available data are too sparse to support conclusions about health implications of dampness-related emissions of chemicals from materials.

SUMMARY

Moisture and microbial growth are present in all buildings, and there is no widely accepted definition of the conditions that constitute a "dampness problem." Moisture-damage observations may include visual observations of dampness or microbial growth, readings of moisture measurements, and other signs that can be associated with excess moisture in building construction. The reported prevalence of signs of dampness in buildings varies widely. In most datasets, at least 20% of buildings have signs of a dampness problem. The available dampness data are primarily from studies of homes; however, some data suggest that dampness in workplaces, schools, and HVAC systems should not be ignored. The extent, location, and duration of building dampness are important for an understanding of its role in health problems, but there has not been much research to evaluate their influence.

Water problems in buildings originate in rainwater, groundwater, plumbing, construction, water use by occupants, and condensation of water vapor. Moisture problems begin when materials stay wet long enough for microbial growth, physical deterioration, or chemical reactions to occur. The important variables are the rate of wetting and the rate of drying. A complex set of moisture-transport and air-transport processes related to building design, construction, operation, and maintenance and to climate determine whether a building will have a moisture problem. Below-grade spaces are particularly prone to moisture problems (Lstiburek, 2002a).

Dampness has been associated with an array of building characteristics, including age of the building, lack of central heating, humidifiers, presence of pets, low temperatures, and crowding. During the life span of a building, weather changes and other events cause at least temporary wetting of some of its parts; signs of microbial growth can thus be detected on many parts of a structure. Microbial growth is regulated by the available resources, conditions, and competing organisms. Indoor microbial concentrations are also influenced by indoor temperature and humidity, building materials, type of foundation, and HVAC system characteristics.

Environmental microorganisms are diverse. They differ enormously in their needs for particular environmental conditions, so there will almost always be some fungi or bacteria that do well in any microenvironmental conditions. In buildings, the general microbial needs of temperature, nutrients, oxygen, and light are usually met; therefore, the availability of moisture is the primary limiting factor in microbial growth. The materials used in the building determine both the amount of moisture needed to support growth and the type of microorganisms whose growth will be favored. Moisture may also trigger degradation of building materials and so contribute to the release of nonmicrobial chemicals into the indoor air.

Research on the biologic and human health effects of microbial and other agents associated with damp indoor environments is discussed in the following chapters. It should be noted that there is very little documentation of interactions of these various agents. While it is evident that various pollutants occur simultaneously in these environments, an overall risk assessment of the combined exposures is not possible with the present knowledge.

FINDINGS, RECOMMENDATIONS, AND RESEARCH NEEDS

On the basis of the review of the papers, reports, and other information presented in this chapter, the committee has reached the following findings and recommendations and identified the following research needs regarding damp buildings. The committee's discussion of the public health response to damp indoor spaces (Chapter 7) provides additional observations on how some of the recommendations for actions might be accomplished.

Findings

• The term *dampness* has been used to define a variety of moisture problems in buildings, including high RH, condensation, and signs of excess moisture or microbial growth. However, there is no generally accepted definition of *dampness* or of what constitutes a "dampness problem" and no generally accepted metric for characterizing dampness.

• Dampness—as defined and documented in studies using a wide variety of metrics—is prevalent in residential housing in a wide array of climates. The prevalence and significance of dampness are less well understood in nonresidential buildings like office buildings and schools than in residential buildings. Relatively little information is available on the prevalence and importance of dampness and microbial growth in HVAC systems.

• Environmental microorganisms require moisture and nutrients to grow. The range of temperatures in buildings permits the growth of many microorganisms even if it is not optimal for a particular genus or species.

• Dampness increases the risk of microbial contamination and can cause or exacerbate the release of chemical emissions from building materials and furnishings.

• Dampness problems in buildings result from failures in design, construction, operation, maintenance, and use. The prevalence and nature of dampness problems suggest that what is known about their causes and prevention is not consistently applied in building design, construction, maintenance, and use.

• The prevalence of dampness problems appears to increase as buildings age and deteriorate, but some modern construction techniques and materials and the presence of air-conditioning probably increase the risk of dampness problems. Scientific studies have not, in general, provided data to confirm or refute this idea.

• Changes in building design, operation, maintenance, and use are the key to preventing the manifestation of dampness-related building damage and microbial growth.

Recommendations

• Precise, agreed-on definitions of *dampness* should be developed to allow important information to be gathered about mechanisms by which dampness and dampness-related effects and exposures affect occupant health. More than one definition may be required to meet the specific needs of health researchers (epidemiologists, physicians, and public health practitioners) in contrast with those involved in preventing or remediating dampness (architects, engineers, and builders). However, definitions should be standardized to the extent possible. Any efforts to establish common definitions should be international in scope.

• Increased attention should be paid to HVAC systems as a potential site for the growth and dispersal of microbial contaminants that may result in adverse health effects in building occupants.

• Building professionals (architects, home builders, facility managers and maintenance staff, code officials, and insurers) should receive better training in how and why dampness problems occur and their prevention.

• Current building codes should be reviewed and modified as necessary to reduce dampness problems.

Research Needs

As noted above, standardized dampness metrics and associated dampness-assessment protocols should be developed to characterize the nature, severity, and spatial extent of dampness. Using the standardized metrics, the determinants of dampness problems in buildings should be studied to ascertain where to focus intervention efforts and health-effects research.

In addition, the committee identified the following research needs:

• Economic research is needed to determine the societal cost of dampness problems and to quantify the economic impact of design, construction, and maintenance practices that prevent or limit dampness problems.

• New and continuing research is needed to better characterize
— The presence and health effects of bacteria that grow on damp materials indoors.

— Dampness-related emissions of spores, bacteria, and smaller particles of biologic origin, dampness-related chemical emissions from building materials and furnishings and any role these emissions may have in adverse health outcomes.

— The nature and significance of dampness-related microbial contamination in HVAC systems.

— The microbial ecology of buildings, that is, the link between dampness, different building materials, microbial growth, and microbial interactions.

— The impact of the duration of moisture damage on materials and its possible influence on occupant health.

— The effectiveness of various changes in building designs, construction methods, operation, and maintenance in reducing dampness problems.

• Research should be performed to develop design, construction, and maintenance practices for buildings and HVAC systems that reduce moisture problems.

Chapter 7 operationalizes some of these research needs by suggesting specific actions and actors to implement them.

REFERENCES

Abbas HK, Johnson BB, Shier WT, Tak H, Jarvis BB, Boyette CD. 2002. Phytotoxicity and mammalian cytotoxicity of macrocyclic trichothecene mycotoxins from *Myrothecium verrucaria*. Phytochemistry 59(3):309–313.

ACGIH (American Conference of Governmental Industrial Hygienists). 1999. Bioaerosols: Assessment and Control. Cincinnati, OH: ACGIH.

Adachi Y, Ohno N, Yadomae T. 1994a. Preparation and antigen specificity of an anti-(1→3)-β-D-glucan antibody. Biological and Pharmaceutical Bulletin 17(11):1508–1512.

Adachi Y, Okazaki M, Ohno N, Yadomae T. 1994b. Enhancement of cytokine production by macrophages stimulated with (1→3)-β-D-glucan, grifolan (GRN) isolated from *Grifola frondosa*. Biological and Pharmaceutical Bulletin 17(12):1554–1560.

Advanced Energy. 2001. Technology Assessment Report: A Field Study Comparison of the Energy and Moisture Performance Characteristics of Ventilated Versus Sealed Crawl Spaces in the South. Instrument #DE-FC26-00NT40995. http://www.advancedenergy.org/root/buildings/crawlspaces/reports/tech_assess.pdf.

Andersen B, Nissen AT. 2000. Evaluation of media for detection of *Stachybotrys* and *Chaetomium* species associated with water-damaged buildings. International Biodeterioration & Biodegradation 46(2):111–116.

Andersen I, Lindgvist GR, Molhave L. 1975. Indoor air pollution due to chipboard use as a construction material. Atmospheric Environment 9:1121–1127.

Andersson MA, Nikulin M, Köljalg U, Andersson MC, Rainey F, Reijula K, Hintikka EL, Salkinoja-Salonen M. 1997. Bacteria, molds and toxins in water-damaged building materials. Applied and Environmental Microbiology 63(2):387–393.

Andersson MA, Mikkola R, Kroppenstedt RM, Rainey FA, Peltola J, Helin J, Sivonen K, Salkinoja-Salonen MS. 1998. The mitochondrial toxin produced by Streptomyces griseus strains isolated from an indoor environment is valinomycin. Applied and Environmental Microbiology 64(12):4767–4773.

Angell W. 1988. Home Moisture Sources. Minnesota Cooperative Extension, Cold Climate Housing Information Center, University of Minnesota, St. Paul.

Arruda LK, Platts-Mills TAE, Fox JW, Chapman MD. 1990. *Aspergillus fumigatus*. Allergen I, a major Ige-binding protein, is a member of the mitogillin family of cytotoxins. Journal of Experimental Medicine 172(5):1529–1532.

Aukrust L, Borch SM. 1979. Partial purification and characterization of two *Cladosporium herbarum* allergens. International Archives of Allergy and Applied Immunology 60(1): 68–79.

Batterman SA, Burge H. 1995. HVAC systems as emission sources affecting indoor air quality—a critical review. International Journal of HVAC and Research 1(1):61–80.

Beguin H, Nolard N. 1994. Mould biodiversity in homes I. Air and surface analysis of 130 dwellings. Aerobiology 10:157–166.

Bencko V, Maelichercik J, Melichercikova V, Wirth Z. 1993. Microbial growth in spray humidifiers of health facilities. Indoor Air 3(1):20–25.

Björkholtz D. 1987. Lämpö ja kosteus-rakennusfysiikka (Temperature- and moisture-building physics, in Finnish). Rakentajain Kustannus Oy, Vammalan Kirjapaino Oy, Vammala.

Bornehag C-G, Blomquist G, Gyntelberg F, Järvholm B, Malmberg P, Nordvall L, Nielsen A, Pershagen G, Sundell J. 2001. Dampness in buildings and health. Nordic interdisciplinary review of the scientific evidence on associations to "dampness" in buildings and health effects (NORDDAMP) Indoor Air 11:72–86.

Brattgjerd S, Evensen O, Lauve A. 1994. Effect of injected yeast glucan on the activity of macrophages in Atlantic salmon, *Salmo salar L.*, as evaluated by in vitro hydrogen peroxide production and phagocytic capacity. Immunology 83(2):288–294.

Brennan T, Cummings J, Lstiburek J. 2002. Unplanned airflows and moisture problems. ASHRAE Journal 44(11):44–52.

Brunekreef B, Dockery DW, Speizer FE, Ware JH, Spengler JD, Ferris BG. 1989. Home dampness and respiratory morbidity in children. American Review of Respiratory Diseases 140:1363–1367.

Burge HA, Hoyer ME, Solomon WR, Simmons EG, Gallup J. 1989. Quality control factors for *Alternaria* allergens. Mycotaxon 34(1):55–63.

Burge HA, Macher JM, Milton DK, Ammann HM. 1999. Data Evaluation. In: Bioaerosols: Assessment and Control, JM Macher, HA Ammann, HA Burge, DK Milton, PR Morey, eds. Cincinnati, OH: ACGIH. pp. 14-1–14-11.

Buttner MP, Stetzenbach LD. 1993. Monitoring airborne fungal spores in an experimental indoor environment to evaluate sampling methods and the effects of human activity on air sampling. Applied and Environmental Microbiology 59(1):219–226.

Chang JCS, Foarde KK, Vanosdell DW. 1995. Growth evaluation of fungi (*Penicillium* and *Aspergillus* spp.) on ceiling tiles. Atmospheric Environment 29(17):2331–2337.

Christian JE. 1994. Moisture sources. In: Moisture Control in Buildings. HR Trechsel, ed. Philadelphia, PA: ASTM.

CMHC (Canadian Mortgage and Housing Corporation). 1996. Molds in Finished Basements. Ottawa, Ontario, Canada: Canadian Mortgage and Housing Corporation.

Crameri R. 1998. Recombinant *Aspergillus fumigatus* allergens: from nucleotide sequences to clinical applications. International Archives of Allergy and Immunology 115(2):99–114.

Cruz A, Saenz de Santamaria M, Martinez J, Martinez A, Guisantes J, Palacios R. 1997. Fungal allergens from important allergenic fungi imperfecti [Review]. Allergologia et Immunopathologia 25(3):153–158.

Daisy JM, Angell WJ. 1998. A Survey and Critical Review of the Literature on Indoor Air Quality, Ventilation, and Health Symptoms in Schools. Lawrence Berkeley National Laboratories. March.

Dales RE, Miller D, McMullen E. 1997. Indoor air quality and health: validity and determinants of reported home dampness and moulds. International Journal of Epidemiology 26(1):120–125.

Dales RE, Miller D, White J. 1999. Testing the association between residential fungus and health using ergosterol measures and cough recordings. Mycopathologia 147(1):21–27.

DeKoster JA, Thorne PS. 1995. Bioaerosol concentrations in noncomplaint, complaint and intervention homes in the Midwest. American Industrial Hygiene Association Journal 56:573–580.

Demain AL. 1999. Pharmaceutically active secondary metabolites of microorganisms. Applied Microbiology and Biotechnology 52:455–463.

Deuell B, Arruda LK, Hayden ML, Chapman MD, Platts-Mills TAE. 1991. *Trichophyton tonsurans* allergen I. Characterization of a protein that causes immediate but not delayed hypersensitivity. Journal of Immunology 147(1):96–101.

Dharmage S, Bailey M, Raven J, Mitakakis T, Guest D, Cheng A, Rolland J, Thien F, Abramson M, Walters EH. 1999a. A reliable and valid home visit report for studies of asthma in young adults. Indoor Air 9:188–192.

Dharmage S, Bailey M, Raven J, Mitakakis T, Thien F, Forbes A, Guest D, Abramson M, Walters H. 1999b. Prevalence and residential determinants of fungi within homes in Melbourne, Australia. Clinical and Epidemiological Allergy 29:1481–1489.

Dill I, Niggemann B. 1996. Domestic fungal viable propagules and sensitization in children with IgE mediated allergic diseases. Pediatric Allergy and Immunology 7(3):151–155.

Doll SC. 2002. Determination of Limiting Conditions for Fungal Growth in the Built Environment. Thesis (Doctor of Science). Harvard School of Public Health.

Douwes J, Heederik D. 1997. Epidemiologic investigations of endotoxins. International Journal of Occupational Environmental Health 3(1):S26–S31.

Douwes J, van der Sluis B, Doekes G, van Leusden F, Wijnands L, van Strien R, Verhoeff A, Brunekreef B. 1999. Fungal extracellular polysaccharides in house dust as a marker for exposure to fungi: relations with culturable fungi, reported home dampness, and respiratory symptoms. Journal of Allergy and Clinical Immunology 103(3/1):494–500.

Engelhart S, Loock A, Skutlarek D, Sagunski H, Lommel A, Farber H, Exner M. 2002. Occurrence of toxigenic *Aspergillus* versicolor isolates and sterigmatocytins in carpet dust from damp indoor environments. Applied and Environmental Microbiology 68(8): 3886–3890.

Engvall K, Norrby C, Norbäck D. 2001. Asthma symptoms in relation to building dampness and odour in older multifamily houses in Stockholm. International Journal of Tuberculosis and Lung Disease 5(5):468–477.

Evans J, Hyndman S, Steward-Brown S, Smith S, Petersen S. 2000. An epidemiological study of the relative importance of damp housing in relation to adult health. Journal of Epidemiology and Community Health 54:677–686.

Ezeonu IM, Noble JA, Simmons RB, Price DL, Crow SA, Ahearn DG. 1994. Effect of relative humidity on fungal colonization of fiberglass insulation. Applied and Environmental Microbiology 60(6):2149–2151.

Fadel R, David B, Paris S, Guesdon JL. 1992. *Alternaria* spore and mycelium sensitivity in allergic patients: in vivo and in vitro studies. Annals of Allergy 69(4):329–335.

Flannigan B, Miller JD. 2001. Microbial growth in indoor environments. In: Microorganisms in Home and Indoor Work Environments. Flannigan B, Samson RA, Miller JD, eds. New York : Taylor & Francis. pp. 35–67.

Fogelmark B, Sjostrand M, Rylander R. 1994. Pulmonary inflammation induced by repeated inhalations of beta(1,3)-D-glucan and endotoxin. International Journal of Experimental Pathology 75(2):85–90.

Gallup J, Kozak P, Cummins L, Gillman S. 1987. Indoor mold spore exposure: characteristics of 127 homes in Southern California with endogenous mold problems. Advances in Aerobiology 51:139–142.

Garrett MH, Rayment PR, Hooper MA, Abramson MJ, Hooper BM. 1998. Indoor airborne fungal spores, house dampness and associations with environmental factors and respiratory health in children. Clinical and Experimental Allergy 28:459–467.

Górny RL, Reponen T, Willeke K, Schmechel D, Robine E, Boissier M, Grinshpun SA. 2002. Fungal fragments as indoor air biocontaminants. Applied and Environmental Microbiology 68(7):3522–3531.

Górny RL, Mainelis G, Grinshpun SA, Willeke K, Dutkiewicz J, Reponen T. 2003. Release of *Streptomyces albus* propagules from contaminated surfaces. Environmental Research 91(1):45–53.

Gqaleni N, Smith JE, Lacey J, Gettinby G. 1997. Effects of temperature, water activity, and incubation time on production of aflatoxins and cyclopiazonic acid by an isolate of *Aspergillus flavus* in surface agar culture. Applied and Environmental Microbiology 63(3):1048–1053.

Grant C, Hunter CA, Flannigan B, Bravery AF. 1989. The moisture requirements of moulds isolated from domestic dwellings. International Biodeterioration & Biodegradation 25: 259–284.

Gravesen S, Frinsvad JC, Samson RA. 1994. Microfungi. Munksgaard, Copenhagen, Denmark. 168 p.

Gravesen S, Nielsen PA, Iversen R, Nielsen KF. 1999. Microfungal contamination of damp buildings—examples of risk constructions and risk materials. Environmental Health Perspectives 107(3):505–508.

Harriman LG, Brundrett GW, Kittler R. 2001. "The new ASHRAE design guide for humidity control in commercial buildings", Indoor Air Quality 2001—Moisture, Microbes, and Heath Effects: Indoor Air Quality and Moisture in Buildings Conference Papers.

Haverinen U. 2002. Modeling moisture damage observations and their association with health symptoms. Doctoral dissertation. National Public Health Institute, Department of Environmental Health, Kuipio, Finland. http://www.ktl.fi/publications/2002/a10.pdf.

Haverinen U, Husman T, Vahteristo M, Koskinen O, Moschandreas D, Nevalainen A, Pekkanen J. 2001a. Comparison of two-level and three-level classifications of moisture-damaged dwellings in relation to health effects. Indoor Air 11(3):192–199.

Haverinen U, Vahteristo M, Husman T, Pekkanen J, Moschandreas D, Nevalainen A. 2001b. Characteristics of moisture damage in houses and their association with self-reported symptoms of the occupants. Indoor and Built Environment 10(2):83–94.

Haverinen U, Vahteristo M, Moschandreas D, Husman T, Nevalainen A, Pekkanen J. 2003. Knowledge-based and statistically modelled relationships between residential moisture damage and occupant reported health symptoms. Atmospheric Environment. 37(4):577–585.

Horner WE, Reese G, Lehrer SB. 1995. Identification of the allergen *Psi c* 2 from the basidiomycete *Psilocybe cubensis* as a fungal cyclophilin. International Archives of Allergy and Immunology 107(1–3):298–300.

Hung LL, Terra JA. 1996. A case of fungal proliferation in a computer facility under construction: Part 1—The contamination. Proceedings of IAQ'96 Paths to Better Building Environments, K Teichman, ed. Atlanta, GA: ASHRAE.

Hunt DRG, Gidman AMIA. 1982. A national field survey of house temperatures. Building and Environment 17(2):102–124.

Hunter CA, Grant C, Flannigan B, Bravery AF. 1988. Mould in buildings: the air spora of domestic dwellings. International Biodeterioration 24:81–101.

Husman H. 1996. Health effects of indoor-air microorganisms. Scandinavian Journal of Work, Environment and Health 22:5–13.

Huttunen K, Hyvärinen A, Nevalainen A, Komulainen H, Hirvonen MR. 2003. Production of proinflammatory mediators by indoor air bacteria and fungal spores in mouse and human cell lines. Environmental Health Perspectives 111(1):85–92.

Hyvärinen A. 2002. Characterizing moisture damaged buildings—environmental and biological monitoring. Academic Dissertation. Department of Environmental Sciences, University of Kuopio, Kuopio, Finland, and the National Public Health Institute.

Hyvärinen A, Meklin T, Vepsäläinen A, Nevalainen A. 2002. Fungi and actinobacteria in moisture-damaged building materials—concentrations and diversity. International Biodeterioration & Biodegradation 49:27–37.

ICBO (International Conference of Building Officials). 2000. International Residence Code 2000. May.

IOM (Institute of Medicine). 1993. Indoor Allergens: Assessing and Controlling Adverse Health Effects. Washington, DC: National Academy Press.

IOM. 2000. Clearing the Air: Asthma and Indoor Air Exposures. Washington, DC: National Academy Press.

Jaakkola MS, Nordman H, Piipari R, Uitti J, Laitinen J, Karjalainen A, Hahtola P, Jaakkola JJ. 2002. Indoor dampness and molds and development of adult-onset asthma: a population-based incident case-control study. Environmental Health Perspectives 110(5): 543–547.

Jarvis BB. 2002. Chemistry and toxicology of molds isolated from water-damaged buildings. Advances in Experimental Medicine and Biology. 504:43–52.

Jarvis BB, Hinckley SF. 1999. Analysis for *Stachybotrys* toxins. In: Bioaerosols, Fungi and Mycotoxins: Health Effects, Assessment, Prevention and Control. Eckardt Johanning, ed. Albany, NY: Eastern New York Environmental Health Center.

Jarvis JQ, Morey PR. 2001. Allergic respiratory disease and fungal remediation in a building in a subtropical climate. Applied Occupational and Environmental Hygiene 16(3):380–388.

Johanning E, Landsbergis P, Gareis M, Yang CS, Olmsted E. 1999. Clinical experience and results of a sentinel health investigation related to indoor fungal exposure. Environmental Health Perspectives 107(3):489–494.

Jussila J, Komulainen H, Huttunen K, Roponen M, Halinen A, Hyvärinen A, Kosma VM, Pelkonen J, Hirvonen MR. 2001. Inflammatory responses in mice after intratracheal instillation of spores of Streptomyces californicus isolated from indoor air of a moldy building. Toxicology and Applied Pharmacology 171(1):61–69.

Karagiozis AC. 2001. Advanced hygrothermal models and design models. eSim 2001—The Canadian conference on building energy simulation, June 13th–14th, Ottawa, Canada. http://www.esim.ca/2001/documents/proceedings/Session3-4.pdf.

Kiho T, Sakushima M, Wang SR, Nagai K, Ukai S. 1991. Polysaccharides in fungi. XXVI. Two branched $(1{\rightarrow}3)$- β-D-glucans from hot water extract of Yu er. Chemical and Pharmaceutical Bulletin 39(3):798–800.

Kilpeläinen M, Terho EO, Helenius H, Koskenvuo M. 2001. Home dampness, current allergic diseases, and respiratory infections among young adults. Thorax 56(6):462–467.

Kitamura S, Hori T, Kurita K, Takeo K, Hara C, Itoh W, Tabata K, Elgasaeter A, Stokke BT. 1994. An antitumor, branched $(1{\rightarrow}3)$-beta-D-glucan from a water extract of fruiting bodies of *Cryptoporus volvatus*. Carbohydrate Research 263(1):111–121.

Klánová K. 2000. The concentrations of mixed populations of fungi in indoor air: rooms with and without mould problems; rooms with and without health complaints. Central European Journal of Public Health 8(1):59–61.

Korpi A. 2001. Fungal volatile metabolites and biological responses to fungal exposure. Kuopio University Publications C. Natural and Environmental Sciences 129. Kuopio, Finland. (PhD Thesis).

Kraus J, Franz G. 1991. β(1→3)Glucans: anti-tumor activity and immunostimulation. In: Fungal Cell Wall and Immune Response. Latge JP, Boucias D, Eds. Berlin, Heidelberg: SpringerVerlag. NATO ASI series, H53:431–444.

Kurnitski J. 2000. Humidity control in outdoor-air-ventilated crawl spaces in cold climate by means of ventilation, ground covers and dehumidification. Doctoral Dissertation. Department of Mechanical Engineering, Laboratory of Heating, Ventilating and Air Conditioning, Helsinki University of Technology. Report A3.

Larsen TO, Frisvad JC. 1994. Production of volatiles and presence of mycotoxins in conidia of common indoor Penicillia and Aspergilli. In: Health Implications of Fungi in Indoor Environments, pp. 251–279. Air Quality Monograph, Vol. 2. RA Samson, B Flannigan, ME Flannigan et al., eds. New York: Elsevier.

Larsen TO, Svendsen A, Smedsgaard J. 2001. Biochemical characterization of Ochratoxin A-producing strains of the genus Penicillium. Applied and Environmental Microbiology 67(8):3630–3635.

Lawton MD, Dales RE, White J. 1998. The influence of house characteristics in a Canadian community on microbiological contamination. Indoor Air 8:2–11.

Li D-W, Kendrick B. 1995. A year-round study on functional relationships of airborne fungi with meteorological factors. International Journal of Biometeorology 39:74–80.

Liu LJ, Krahmer M, Fox A, Feigley CE, Featherstone A, Saraf A, Larsson L. 2000. Investigation of the concentration of bacteria and their cell envelope components in indoor air in two elementary schools. Journal of the Air and Waste Management Association 50(11): 1957–1967.

Lorenz W, Sigrist G, Otto H-H. 2000. Moisture indicating emissions of phthalates and their effects. Proceedings of Healthy Buildings 2000, vol. 4, Indoor Air Information, Oy, Finland. pp. 405–410.

Lstiburek J. 2000. Builder's Guide. Building Science Corporation.

Lstiburek J. 2001. Moisture, building enclosures and mold. Part 1 of 2. HPAC Engineering. Dec.:22–26.

Lstiburek J. 2002a. Moisture control for buildings. ASHRAE Journal Feb.:36–41.

Lstiburek J. 2002b. Moisture, building enclosures and mold. Part 2 of 2. HPAC Engineering. Jan.:77–81.

Lstiburek J, Carmody J. 1996. Moisture Control Handbook: Principles and Practices for Residential and Small Commercial Buildings. New York: John Wiley & Sons, Inc.

Lstiburek J, Pettit B. 2000. EEBA Builder's Guide: Hot-Dry/Mixed-Dry Climate (Revised). Bloomington, MN: Energy and Environmental Building Association.

Lstiburek J, Pettit B. 2001. EEBA Builder's Guide: Mixed-Humid Climate (Revised). Bloomington, MN: Energy and Environmental Building Association.

Lstiburek J, Pettit B. 2002a. EEBA Builder's Guide: Cold Climate (Revised). Bloomington, MN: Energy and Environmental Building Association.

Lstiburek J, Pettit B. 2002b. EEBA Builder's Guide: Hot-Humid Climates (Revised). Bloomington, MN: Energy and Environmental Building Association.

Martin CJ, Platt SD, Hunt SM. 1987. Housing conditions and ill health. British Medical Journal 294:1125–1127.

Martiny H, Moritz M, Ruden H. 1994. Occurrence of Microorganisms in Different Filter Media of Heating, Ventilation and Air Conditioning (HVAC) Systems. In: Proceedings of IAQ 94-Engineering Indoor Environments. Atlanta, GA: ASHRAE. pp. 131–137.

Meklin T, Husman T, Vepsäläinen A, Vahteristo M, Koivisto J, Halla-Aho J, Hyvärinen A, Moschandreas D, Nevalainen A. 2002. Indoor air microbes and respiratory symptoms of children in moisture damaged and reference schools. Indoor Air 12(3):175–183.

Meklin T, Hyvärinen A, Toivola M, Reponen T, Koponen V, Husman T, Taskinen T, Korppi M, Nevalainen A. 2003. Effect of building frame and moisture damage on microbiological indoor air quality in school buildings. AIHA Journal 64(1):108–116.

Mendell MJ, Cozen M. 2002. Building-related symptoms among U.S. office workers and risks factors for moisture and contamination: preliminary analyses of U.S. EPA BASE data. Lawrence Berkeley National Laboratory Report, LBNL-51567, Berkeley, CA.

Mendell MJ, Naco GM, Wilcox TG, Sieber WK. 2003. Environmental risk factors and work-related lower respiratory symptoms in 80 office buildings: an exploratory analysis of NIOSH data. American Journal of Industrial Medicine 43(6):630–641.

Menzies D, Popa J, Hanley JA, Rand T, Milton DK. 2003. Effect of ultraviolet germicidal lights installed in office ventilation systems on workers' health and wellbeing: double-blind multiple crossover trial. Lancet 362(9398):1785–1791.

Miller JD, Laflamme AM, Sobol Y, Lafontaine P, Greenhalgh R. 1988. Fungi and fungal products in some Canadian houses. International Biodeterioration 24:103–120.

Miller JD, Haisley PD, Reinhardt JH. 2000. Air sampling results in relation to extent of fungal colonization of building materials in some water-damaged buildings. Indoor Air 10(3):146–151.

Milton DK. 1999. Endotoxin and other bacterial cell-wall components. In: Bioaerosols: Assessment and Control. Macher J, Milton DK, Burge HA, Morey P, eds. Cincinnati, OH: American Conference of Governmental Industrial Hygienists.

Morey PR. 1988. Microorganisms in buildings and HVAC systems: a summary of 21 environmental studies. In: Proceedings of IAQ'88—Engineering Solutions to Indoor Air Problems. Atlanta, GA: ASHRAE. pp. 10–24.

Morey PR. 1994. Suggested guidance on prevention of microbial contamination for the next revision of ASHRAE Standard 62. In: Proceedings of IAQ '94—Engineering Indoor Environments. Atlanta, GA: ASHRAE. pp. 139–148.

Morey PR, Williams CM. 1991. Is Porous Insulation Inside an HVAC System Compatible with a Healthy Building? In: Proceedings of IAQ '91–Healthy Buildings, Atlanta, GA: ASHRAE. pp. 128-135.

Mullins J. 2001. Microorganisms in outdoor air. In: Microorganisms in Home and Indoor Work Environments. B. Flannigan, RA Samson, JD Miller, eds. New York: Taylor & Francis. pp. 3–16.

Murtoniemi T, Nevalainen A, Suutari M, Toivola M, Komulainen H, Hirvonen M-R. 2001. Induction of cytotoxicity and production of inflammatory mediators in RAW264.7 macrophages by spores grown in six different plasterboards. Inhalation Toxicology 13:233–247.

Murtoniemi T, Nevalainen A, Suutari M, Hirvonen MR. 2002. Effect of liner and core materials of plasterboard on microbial growth, spore-induced inflammatory responses and cytotoxicity in macrophages. Inhalation Toxicology 14(11):1087–1101.

Murtoniemi T, Hirvonen MR, Nevalainen A, Suutari M. 2003a. The relation between growth of four microbes on six different plasterboards and biological activity of spores. Indoor Air 13(1):65–73.

Murtoniemi T, Nevalainen A, Hirvonen MR. 2003b. Effect of plasterboard composition on *Stachybotrys chartarum* growth and biological activity of spores. Applied and Environmental Microbiology 69(7):3751–3757.

Murtoniemi T, Keinänen MM, Nevalainen A, Hirvonen MR. 2003c. Starch in plasterboard sustains *Streptomyces californicus* growth and bioactivity of spores. Journal of Applied Microbiology 94(6):1059–1065.

Nevalainen A, Pasanen A-L, Niininen M, Reponen T, Kalliokoski P. 1991. The indoor air quality in Finnish homes with mold problems. Environment International 17:299–302.

Nevalainen A, Partanen P, Jääskeläinen E, Hyvärinen A, Koskinen O, Meklin T, Vahteristo M, Koivisto J, Husman T. 1998. Prevalence of moisture problems in Finnish houses. Indoor Air (Supplement 4):45–49.

Nevander LE, Elmarsson B. 1994. Fukt handbook. Practik och teori (Moisture handbook, In Swedish). AB Svensk Byggtjänst och Författarna Andra, reviderade utgåvan, Svenskt Tryck AB, Stockholm.

Nielsen KF, Hansen M, Larsen T, Thrane U. 1998. Production of trichothecene mycotoxins in water damaged gypsum boards in Danish buildings. International Biodeterioration & Biodegradation 42:1–7.

Nielsen KF, Gravesen S, Nielsen PA, Andersen B, Thrane U, Frisvad JC. 1999. Production of mycotoxins on artificially and naturally infested building materials. Mycopathologia 145(1):43–56.

Nielsen KF, Huttunen K, Hyvärinen A, Andersen B, Jarvis BB, Hirvonen MR. 2002. Metabolite profiles of *Stachybotrys* isolates from water-damaged buildings and their induction of inflammatory mediators and cytotoxicity in macrophages. Mycopathologia 154(4): 201–206.

Nikulin M, Pasanen A-L, Berg S, Hintikka E-L. 1994. *Stachybotrys atra* growth and toxin production in some building materials and fodder under different relative humidities. Applied and Environmental Microbiology 60(9):3421–3424.

Norbäck D, Björnsson E, Janson C, Palmgren U, Boman G. 1999. Current asthma and biochemical signs of inflammation in relation to building dampness in dwellings. The International Journal of Tuberculosis and Lung Disease 3(5):368–376.

Norbäck D, Wieslander G, Nordström K, Walinder R. 2000. Asthma symptoms in relation to measured building dampness in upper concrete floor construction, and 2-ethyl-1-hexanol in indoor air. The International Journal of Tuberculosis and Lung Disease 4(11):1016–1025.

Offermann FJ, Hodgson AT, Robertson JP. 2000. Contaminant emission rates from PVC backed carpet tiles on a damp concrete. Proceedings of Healthy Buildings 2000, vol. 4, Indoor Air Information, Oy, Finland. pp. 379–384.

Oliver A. 1997. Dampness in Buildings. Second Edition revised by James Douglas and J. Stewart Stirling. Blackwell Science Ltd.

ORNL/IBP (Oak Ridge National Laboratory/ Fraunhofer Institute for Building Physics). 2003. WUFI ORNL/IBP moisture design tools for architects and engineers. http://www.ornl.gov/sci/btc/apps/moisture/.

Ottney TC. 1993. Particle Management for HVAC Systems. ASHRAE Journal 35(7):14–23.

Oxley TA, Gobert EG. 1994. The professionals and home owners guide to dampness in buildings, second edition. Guildford and King's Lynn, England: Biddles Ltd.

Pasanen A-L. 1992. Airborne mesophilic fungal spores in various residential environments. Atmospheric Environments 26A(16):2861–2868.

Pasanen A-L, Heinonen-Tanski H, Kalliokoski P, Jantunen MJ. 1992a. Fungal micro-colonies on indoor surfaces—an explanation for the base level fungal spore counts in indoor air. Atmospheric Environment 26B(1):121–124.

Pasanen A-L, Niininen M, Kalliokoski P, Nevalainen A, Jantunen MJ. 1992b. Airborne *Cladosporium* and other fungi in damp versus reference residences. Atmospheric Environment 26B(1):117–120.

Pasanen A-L, Kasanen JP, Rautiala S, Ikäheimo M, Rantamäki J, Kääriäinen H, Kalliokoski P. 2000a. Fungal growth and survival in building materials under fluctuating moisture and temperature conditions. International Biodeterioration and Biodegradation 46(2):117–127.

Pasanen A-L, Rautiala S, Kasanen J-P, Raunio P, Rantamäki J, Kalliokosli P. 2000b. The relationship between measured moisture conditions and fungal concentrations in water-damaged building materials. Indoor Air 10(2):111–120.

Pearce M, Huelman PH, Janni KA, Olsen W, Seavey RT, Velsey D. 1995. Long-term monitoring of mold contamination in flooded homes. Journal of Environmental Health 58(3):6–11.

Peltola JSP, Andersson MA, Haahtela T, Mussalo-Rauhamaa H, Rainey FA, Kroppenstedt RM, Samson RA, Salkinoja-Salonen M. 2001a. Toxic metabolite producing bacteria and fungus in an indoor environment. Applied and Environmental Microbiology 67(7):3269–3274.

Peltola JSP, Andersson MA, Kämpfer P, Auling G, Kroppenstedt RM, Busse HJ, Salkinoja-Salonen M, Rainey FA. 2001b. Isolation of toxigenic Nocardiopsis strains from indoor environments and description of two new Nocardipsis species N. exhalans sp. nov. and N. umidischolae sp. nov. Applied and Environmental Microbiology 67(9):4293–4304.

Pessi A-M, Suonketo J, Pentti M, Kurkilahti M, Peltola K, Rantio-Lehtimäki A. 2002. Microbial growth inside insulated external walls as an indoor air biocontamination source. Applied and Environmental Microbiology 68(2):963–967.

Pitt JI, Basilico JC, Abarca ML, Lopez C. 2000. Mycotoxins and toxigenic fungi. Medical Mycology 38(Supplement 1):41–46.

Rautiala S, Reponen T, Hyvärinen A, Nevalainen A, Husman T, Vehviläinen A, Kalliokoski P. 1996. Exposure to airborne microbes during the repair of moldy buildings. American Industrial Hygiene Association Journal 57:279–284.

Rautiala S, Reponen T, Nevalainen A, Husman T, Kalliokoski P. 1998. Control of exposure to airborne viable microorganisms during remediation of moldy buildings; report of three case studies. American Industrial Hygiene Association Journal 59:455–460.

Rautiala SH, Nevalainen AI, Kalliokoski PJ. 2002. Firefighting efforts may lead to massive fungal growth and exposure within one week. A case report. International Journal of Occupational Medicine and Environmental Health 15(3):303–308.

Ren P, Jankun TM, Leaderer BP. 1999a. Comparisons of seasonal fungal prevalence in indoor and outdoor air and in house dusts of dwellings in one Northeast American county. Journal of Exposure Analysis & Environmental Epidemiology 9(6):560–568.

Ren P, Ahearn DG, Crow SA. 1999b. Comparative study of Aspergillus mycotoxin production on enriched media and construction material. Journal of Industrial Microbiology & Biotechnology 23(3):209–213.

Reponen T, Nevalainen A, Raunemaa T. 1989. Bioaerosol and particle mass levels and ventilation in Finnish homes. Environment International 15:203–208.

Reponen T, Willeke K, Grinshpun S, Nevalainen A. 2001. Biological particle sampling. In: Aerosol measurement, Principles, techniques and applications. Second edition. Baron PA, Willeke K, eds. New York; John Wiley and Sons. pp. 751–778.

Reynolds SJ, Streifel AJ, McJilton CE. 1990. Elevated airborne concentrations of in residential and office environments. American Industrial Hygiene Association Journal 51(11):601–604.

Richard JL, Plattner RD, May J, Liska SL. 1999. The occurrence of ochratoxin A in dust collected from a problem household. Mycopathologia 146(2):99–103.

Roponen M, Toivola M, Meklin T, Ruotsalainen M, Komulainen H, Nevalainen A, Hirvonen M-R. 2001. Differences in inflammatory responses and cytotoxicity in RAW264.7 macrophages induced by Streptomyces anulatus grown on different building materials. Indoor Air 11(3):179–184.

Rose W. 1997. Control of moisture in the modern building envelope: the history of the vapor barrier in the United States 1923–1952. APT Bulletin 18(4):13–19.

Ross MA, Curtis L, Scheff PA, Hryhorczuk DO, Ramakrishnan V, Wadden RA, Persky VW. 2000. Association of asthma symptoms and severity with indoor bioaerosols. Allergy 55(8):705–711.

Rylander R. 2002. Endotoxin in the environment—exposure and effects. Journal of Endotoxin Research 8(4):241–252.

Saito K, Nishijima M, Ohno N, Nagi N, Yadomae T, Miyazaki T. 1992. Activation of complement and limulus coagulation system by an alkali-soluble glucan isolated from *Omphalia lapidescens* and its less branched derivatives. Chemical and Pharmaceutical Bulletin 40:1227–1230.

Sakurai T, Ohno N, Yadomae T. 1994. Changes in immune mediators in mouse lung produced by administration of soluble $(1\rightarrow3)$-β-D-glucan. Biological and Pharmaceutical Bulletin 17(5):617–622.

Salo P. 1999. Identifying and preventing fungal contamination problems in new home construction. In: Bioaerosols, Fungi and Mycotoxins Health Effects, Assessment, Prevention, and Control. E Johanning, ed. Albany, NY: New York Occupational and Environmental Health Center.

Samson RA, Flannigan B, Flannigan ME, Verhoeff A, Adan O, Hoekstra E, eds. 1994. Health implications of fungi in indoor environments. New York: Elsevier.

Sanchez H, Bush RK. 1994. Complete sequence of a cDNA encoding an *Alternaria* allergen. Journal of Allergy and Clinical Immunology 93:208.

Schmidt M, Zargari A, Holt P, Lindbom L, Hellman U, Whitley P, van der Ploey I, Harfast B, Scheynius A. 1997. The complete cDNA sequence and expression of the first major allergenic protein of *Malasseziu furfur*, Mal f I. European Journal of Biochemistry 246(1): 181–185.

Seppänen O, Fisk WJ. 2002. Association of ventilation system type with SBS symptoms in office workers. Indoor Air 12(2):98–112.

Shaughnessy RJ, Levetin E, Rogers C. 1998. The Effects of UV-C on Biological Contamination of AHUs in a Commercial Office Building: Preliminary Results. In: Proceedings of IAQ and Energy 98–Using ASHRAE Standards 62 and 90.1. Atlanta, GA: ASHRAE. pp. 229–236.

Shelton BG, Kirkland KH, Flanders WD, Morris GK. 2002. Profiles of airborne fungi in buildings and outdoor environments in the United States. Applied and Environmental Microbiology 68(4):1743–1753.

Shen HD, Liaw SF, Lin WL, Ro LH, Yang HL, Han SH. 1995. Molecular cloning of cDNA encoding for the 68 kDa allergen of *Penicillium notatum* using MoAbs. Clinical and Experimental Allergy 25(4):350–356.

Shen HD, Au LC, Lin WL, Liaw SF, Tsai JJ, Han SH. 1997. Molecular cloning and expression of *Penicillium citrinum* allergen with sequence homology and antigenic cross-reactivity to a hsp 70 human heat shock protein. Clinical and Experimental Allergy 27(6):682–690.

Shen HD, Lin WL, Tam MF, Wang SR, Tsai JJ, Chou H, Han SH. 1998. Alkaline serine proteinase: a major allergen of *Aspergillus oryzae* and its cross-reactivity with *Penicillium citrinum*. International Archives of Allergy and Immunology 116(1): 29–35.

Sjoberg A, Nilsson L-O. 2002. Floor heating may cause IAQ problems. Proceedings of Indoor Air 1:968–973. Indoor Air 2002, Inc., Santa Cruz, CA.

Smart Growth in America. 1999. Rehabilitation Codes. http://wwwsmartgrowthamerica/rehabcodes.html.

Smedje G, Norbäck D, Wessén B, Edling C. 1996. Asthma among school employees in relation to the school environment. In: Proceedings of the Indoor Air '96 Conference, Nagoya, Japan, vol.1, Seec Ishibashi Inc., Tokyo. pp. 611–616.

Smith JE, Moss MO. 1986. Mycotoxins Formation, Analysis, and Significance. New York: John Wiley and Sons.

Sorenson WG, Frazer DG, Jarvis BB, Simpson J, Robinson VA. 1987. Trichothecene mycotoxins in aerosolized conidia of *Stachybotrys atra*. Applied and Environmental Microbiology 53(6):1370–1375.

Spengler J, Neas L, Nakai S, Dockery D, Speizer F, Ware J, Raizenne M. 1994. Respiratory symptoms and housing characteristics. Indoor Air 4:72–82.

Strachan DP, Flannigan B, McCabe EM, McGarry F. 1990. Quantification of airborne moulds in the homes of children with and without wheeze. Thorax 45:382–387.

Straube J. 2001. Wetting, Storage and Drying Processes. Westford Symposium V. Westford, MA: Building Science Corporation.

Straube J. 2002. Moisture, materials and buildings. Heating Piping and Air Conditioning Magazine. April.

Straube J, Burnett E. 1997. Rain control and screened wall systems. Proceedings of the Seventh Building Science and Technology Conference. Toronto, Ontario. March.

Su HJ, Rotnitzky A, Burge HA, Spengler JD. 1992. Examination of fungi in domestic interiors by using factor analysis: correlations and associations with home factors. Applied and Environmental Microbiology 58(1):181–186.

Tariq SM, Matthews SM, Stevens M, Hakim EA. 1996. Sensitization to *Alternaria* and *Cladosporium* by the age of 4 years. Clinical and Experimental Allergy 26:794–798.

TenWolde A, Rose WB. 1994. Criteria for humidity in the building and building envelope. In: Bugs, Mold & Rot II. Proceedings of a Workshop on Residential Moisture Problems, Health Effects, Building Damage, and Moisture Control. Washington, DC: National Institute of Building Sciences.

Thatcher TL, Layton DW. 1995. Deposition, resuspension, and penetration of particles within a residence. Atmospheric Environment 29:1487–1497.

Tuomi T, Reijula K, Johnsson T, Hemminki K, Hintikka E-L, Lindroos O, Kalso S, Koukila-Kähkölä P, Mussalo-Rauhamaa H, Haahtela T. 2000. Mycotoxins in crude building materials from water-damaged buildings. Applied and Environmental Microbiology 66(5):1899–1904.

U.S. GAO (United States General Accounting Office). 1995. School facilities: Condition of America's Schools. Washington, DC: U.S. General Accounting Office. GAO/HEHS-95-61.

van Netten C, Shirtliffe C, Svec J. 1989. Temperature and humidity dependence of formaldehyde release from selected building materials. Bulletin of Environmental Contamination and Toxicology 42(4):558–565.

Verhoeff AP, van Wijinen JH, Brunekreef B, Fischer P, van Reenen-Hoekstra ES, Samson RA. 1992. Presence of viable mould propagules in indoor air in relation to house damp and outdoor air. Allergy 47:83–91.

Verhoeff AP, van Wijinen JH, van Reenen-Hoekstra ES, Samson RA, van Strien RT, Brunekreef B. 1994. Fungal propagules in house dust. II. Relation with residential characteristics and respiratory symptoms. Allergy 49:540–547.

Vesper S, Vesper MJ. 2002. Stachylysin may be a cause of hemorrhaging in humans exposed to *Stachybotrys chartarum*. Infection and Immunity 70(4):2065–2069.

Vesper S, Magnuson ML, Dearborn DG, Yike I, Haugland RA. 2001. Initial characterization of the hemolysis from *Stachybotrys chartarum*. Infection and Immunity 69(2):912–916.

Viitanen H. 1994. Factors affecting the development of biodeterioration in wooden constructions. Materials and Structures 27:483–493.

Viitanen H. 1997. Modelling the time factor in the development of mould fungi—the effect of critical humidity and temperature conditions on pine and spruce sapwood. Holzforschung 51:6–14.

Viitanen H. 2002. Mould growth on painted wood. Seminar on Bio-deterioration of Coated Wood—Coating and Substrate, COST E18 Joint Working Group Meeting. April, Lisbon, Portugal.

Waegemaekers M, van Wageningen N, Brunekreef B, Boleij JSM. 1989. Respiratory symptoms in damp homes. Allergy 44:192–198.

Wålinder R, Norbäck D, Wessen B, Venge P. 2001. Nasal lavage biomarkers: effect of water damage and microbial growth in an office building. Archives of Environmental Health 56(1):30–36.

Wang Q. 1992. Wood-based boards—Response to attack by mould and stain fungi. Doctoral dissertation. Department of Forest Products, The Swedish University of Agricultural Sciences. Sveriges lantbruksuniversitet, Istitutionen för virkeslärä, Uppsala.

Wicklow DT, Shotwell OL. 1983. Intrafungal distribution of aflatoxin among conidia and sclerotia of *Aspergillus flavus* and *Aspergillus parasiticus*. Canadian Journal of Microbiology 29(1):1–5.

Wickman M, Gravesen S, Nordvall SL, Pershagen G, Sundell J. 1992. Indoor viable dustbound microfungi in relation to residential characteristics, living habits, and symptoms in atopic and control children. Journal of Allergy and Clinical Immunology 89(3): 752–759.

Wieslander G, Norbäck D, Nordström K, Wålinder R, Venge P. 1999. Nasal and ocular symptoms, tear film stability and biomarkers in nasal lavage, in relation to building-dampness and building design in hospitals. International Archives of Occupational and Environmental Health 72:451–461.

Williamson IJ, Martin CJ, McGill G, Monie RDH, Fennerty AG. 1997. Damp housing and asthma: a case-control study. Thorax 52:229–234.

Yunginger JW, Jones RT, Nesheim ME, Geller M. 1980. Studies on *Alternaria* allergens. III. Isolation of a major allergenic fraction (ALT-I). Journal of Allergy and Clinical Immunology 66(2):138–147.

Zhang K, Petty HR. 1994. Influence of polysaccharides on neutrophil function: Specific antagonists suggest a model for cooperative saccharide-associated inhibition of immune complex-triggered superoxide production. Journal of Cellular Biochemistry 56(2):225–235.

Zock JP, Jarvis D, Luczynska C, Sunyer J, Burney P. European Community Respiratory Health Survey. 2002. Housing characteristics, reported mold exposure, and asthma in the European Community Respiratory Health Survey. Journal of Allergy and Clinical Immunology 110(2):285–292.

3

Exposure Assessment

INTRODUCTION

Assessments of exposure to environmental agents in indoor air play a central role in epidemiologic studies that seek to characterize population risks, in screening studies aimed at identifying individuals at risk, and in interventions designed to reduce risk. Because of the central importance of exposure assessment, there is a need to understand the strengths and limitations of the approaches that are available to assess exposures in those contexts. Indoor dampness may be associated with some respiratory health effects (Chapter 5), and a causal role for microorganisms has been suggested. However, the specific roles of infectious and noninfectious microorganisms and their components in diseases related to indoor environments are poorly understood. The lack of knowledge regarding the role of microorganisms in the development and exacerbation of those diseases is due largely to the lack of valid quantitative exposure-assessment methods and knowledge of which specific microbial agents may primarily account for the presumed health effects. In most studies, exposure is assessed by means of questionnaires, and relatively few studies have attempted to measure exposure to microorganisms.

Indoor environments contain a complex mixture of live (viable) and dead (nonviable) microorganisms, fragments thereof, toxins, allergens, microbial volatile organic compounds (MVOCs), and other chemicals. Sensitive and specific methods are available for the quantification of some biologic agents, such as endotoxins, but not for others. Many of the newly developed methods—for example, measurement of microbial agents, such as $\beta(1\rightarrow3)$-glucans or fungal extracellular polysaccharides (EPSs)—have not been well validated and are not commercially available. Even for some well-established methods, such as the *Limulus amebocyte* lysate (LAL) assay for measuring bacterial endotoxins, substantial variations in exposure assessment between laboratories have been demonstrated (Chun et al., 2000; Reynolds et al., 2002; Thorne et al., 1997). It is known that the conditions of storage and transport of bioaerosol samples and extraction of dust samples may affect the activity of some biologic agents, such as endotoxins, and thus their measured concentrations, but those conditions are not often addressed (Douwes et al., 1995; Duchaine et al., 2001; Thorne et al., 1994). Finally, there may be biologic agents whose health effects have not been identified. Microbial exposure assessment in the indoor environment is therefore associated with large uncertainties, which potentially result in large measurement errors and biased exposure–response relationships.

This chapter focuses on exposure assessment of microorganisms and microbial agents that occur in damp indoor environments. It discusses issues related to dampness in general only briefly.

DEFINITIONS

Exposure[1]

Two classes of exposure measures can be distinguished: the theoretically ideal (and typically unknown) *risk-relevant exposure* metric (E_{RR}) that represents the individual breathing-zone concentration of an agent of interest over a period that is relevant to the risk of developing the health outcome of interest and the practical and available *exposure surrogate* that correlates to some extent with the E_{RR}. When used without qualification in this report, *exposure* refers to surrogate exposure measures.

The E_{RR} is the theoretical measure of exposure that best represents the risk of adverse health consequences. Researchers often do not know enough about the specific pathogenesis of indoor-related diseases to identify the appropriate E_{RR} confidently. One possible E_{RR} for the exacerbation of asthma, for example might be a short-term average that captures peak agent exposures in the breathing zone immediately before the exacerbation. Relevant averaging times might range from about 20 min to 48 hours.

Direct exposure surrogates include personal monitoring involving the measurement of agent concentrations with monitors carried by individual subjects. These offer more proximal measures of individual exposure than

---

[1]This section is derived from *Clearing the Air* (IOM, 2000), pages 51–54.

do the indirect approaches, but usually at the expense of sample size or ability to characterize long-term exposures. Indirect measures include environmental area monitoring (airborne or dust sampling), recall questionnaires, real-time diaries, and biologic response markers (IgG against fungal antigens, for example). These approaches tend to be more practical in large-scale studies and often are better suited to long-term exposure characterization than are direct measures.

Exposure Mechanisms

Inhalation is usually presumed to be the most important mechanism of exposure to fungi and other dampness-related microbial agents in indoor environments. It is also generally believed that the most harmful agents are within particles, such as fungal spores; however, although this has been the general assumption, recent studies have identified hyphal fragments (Górny et al., 2002) and dust (Englehart et al., 2002) as potential carriers of harmful agents. This section briefly discusses the process of exposure; it focuses on exposures to fungal spores, but the same exposure mechanisms and associated questions apply to other microbial particles of similar size.

Fungal growth occurs on indoor surfaces—including surfaces in heating, ventilating, and air-conditioning systems—and an inhalation exposure to a fungal spore requires that the spore be initially aerosolized at the site of growth and transported to the inhaled parcel of air. Some fungi actively (forcibly) discharge spores into the air (Burge, 2000). In other cases, the initial aerosolization is likely to be caused by indoor air movements or physical disturbances caused by people. After initial aerosolization, a spore may be transported by air motion to the inhaled air parcel.

Most fungal spores have aerodynamic diameters of 2–10 μm (American Thoracic Society, 1997) and deposit quickly on indoor surfaces because of gravitational settling. For example, a 10-μm particle with unit density will fall 1 meter in 5.5 minutes in still air, and a 5-μm particle will fall 1 meter in 21 minutes (Hinds, 1982). Because the deposition rates of these large particles caused by gravitational settling exceed typical ventilation and filtration rates in houses,[2] most spores deposit on indoor surfaces after aerosolization. The deposition of spores is confirmed by their detection in dust samples taken from a broad array of indoor surfaces, including surfaces that are too dry to support fungal growth.

---

[2]Deposition on surfaces will cause 5-μm-aerodynamic-diameter particles to be removed from indoor air at a rate equivalent to 1.5–5 air changes per hour of ventilation (Thatcher et al., 2001). For a 10-μm particle, removal by deposition may be as high as the equivalent of 10 air changes per hour of ventilation. Thus, in most buildings, deposition on surfaces is the largest removal process for particles of 5–10 μm.

Once deposited, spores can be resuspended by disturbances, such as walking and cleaning. Thus, the inhalation-exposure process for fungal spores (and other microbial particles of similar size) may be largely a consequence of resuspension. Thatcher and Layton (1995) have shown that resuspension occurs predominantly for particles larger than 1 μm and that the amount of resuspension increases with particle size. In experiments, such activities as walking, sitting, and house-cleaning increased air concentrations of 5- to 10-μm particles by a factor of 1.5–11. The surface properties of spores may affect their adherence to surfaces and the probability of their resuspension. There is evidence that human activities, including particle resuspension, cause a "personal cloud" of particles, whereby people's exposures to particles exceed those indicated by measurements at a fixed location (Özkaynak et al., 1996). The same personal cloud would be expected for fungal spores. The spores that deposit on surfaces can also be transported to other locations by tracking, for example, sticking to shoes and then detaching at another location.

Many of the above comments also apply to the process of inhalation exposure to fungal spores that are transported to the indoors from outdoors. Those spores can be brought into a building with outdoor air by natural ventilation through open windows and by air infiltration through unintentional cracks and holes in the building envelope and can be tracked in by people and pets. Once they are inside, the processes of spore settling, resuspension, and tracking would be expected to influence inhalation exposures as they do exposure to fungal spores from indoor sources.

Because spores and other components of molds are present on indoor surfaces and people have contact with these surfaces, exposures to fungal agents may occur through dermal contact and transport of lipid-soluble chemicals through the skin. In addition, incidental ingestion of fungal constituents on surfaces and in household dust may occur as a consequence of hand-to-mouth activity. Exposures via dermal contact or ingestion are known to be important for some chemicals and for lead. Infants are generally affected more than adults because of their contact with floors and their high level of hand-to-mouth activity. However, the significance of those routes of exposure to indoor fungi and other dampness-related microbial pollutants is not known.

In summary, the entire process of fungal-spore aerosolization, transport, deposition, resuspension, and tracking, all of which determine inhalation exposure, is poorly understood. A better understanding of the process would enable a better assessment of exposures and might elucidate better means of reducing them. The significance of exposures to fungi in normal indoor environments through dermal contact and ingestion is also not well understood.

Dose[3]

Dose is the amount of an agent that is absorbed or deposited in the body of an exposed organism at a given time (NRC, 1991). *Internal dose* is the amount of an agent that is absorbed into the body, whereas *biologically effective dose* is the amount of an agent or its metabolites that interacts with a target site.

The primary determinants of where an inhaled gas, such as an MVOC, makes contact with the respiratory system are its solubility and reactivity. Reactive gases tend to reach the upper respiratory system. The primary determinant of deposition of airborne particles is the aerodynamic particle diameter (d_{ae}). Aerodynamic particle diameter, as distinct from physical diameter, determines the motion of particles in air. The d_{ae} of a particle is defined as the diameter of the unit density sphere that has the same terminal settling velocity as the particle of interest (ICRP, 1994). Particles with d_{ae} larger than 15 μm are captured preferentially (but not exclusively) in the upper respiratory tract (nose and throat). Particles with d_{ae} of 2.5–15 μm enter the lungs but tend to deposit in the upper conducting airways, where their mass and high velocities favor inertial impaction. Because they lack inertia, smaller particles move with the inhaled air stream into the alveolar region, where they may or may not deposit. The fraction of particles that deposit in the deep lung increases with decreasing d_{ae} below 0.5 μm because of the high diffusion constants of very small particles.

The role of particle density in determining d_{ae} is critical. A spherical particle with a physical diameter of 16 μm but a density of 0.1 will behave aerodynamically like a 5-μm water droplet. That property helps to explain the ability of large-diameter, low-density pollen grains to penetrate and deposit in the lung. Once deposited in the lungs, airborne agents may react with biomolecules, be absorbed into the blood, or be cleared from the lungs. From the viewpoint of indoor-related symptoms and diseases, the relevant sites and nature of interactions between inhaled agents and the human body remain uncertain, and this uncertainty limits our ability to define biologically effective dose in this context. It is important to note that all measures of dose, like those of exposure, can be viewed as surrogates of the theoretical risk-relevant dose measure.

SAMPLING STRATEGIES

Several strategies are available for exposure assessment conducted for risk-assessment purposes.[4] In epidemiology, questionnaires are the most

---

[3]This section is derived from *Clearing the Air* (IOM, 2000), pages 55–56.

[4]Sampling strategies or diagnostic tools to assess whether a building has dampness or mold problems or to assess potential sources of exposure are discussed separately in Chapters 2 and 6.

commonly-used instrument for gathering exposure information (for example, by asking about the presence of dampness or visible mold in the home). For individual patients with suspected indoor-related health problems, a home visit by an occupational hygienist with experience in this field may be the method of choice. Alternatively, personal or environmental monitoring can be used to measure agents of interest in the home. The latter approach has the potential to result in a more valid and accurate exposure assessment; however, this depends heavily on the chosen sampling strategy, which in turn depends on many factors, including

- Specific disease or symptoms.
- Acute vs chronic health outcomes (for example, disease exacerbation vs disease development).
- Population vs patient-based approach.
- Suspected exposure variation in time and space and between controls and cases.
- Available methods to measure individual agents.
- Costs of sampling and analyses.

For indoor-associated health problems, many exposures have to be considered because it is often not clear which specific microorganisms or agents cause symptoms or diseases. In fact, studies are often conducted with the specific aim of assessing which exposures may contribute to the development of symptoms. However, in practice, the funding and availability of methods of measuring specific agents (many methods are not commercially available and are applied only in research settings) severely limit the potential to measure all agents of interest.

Settled Dust vs Airborne Measurements

Indoor exposure assessment may use air or surface sampling or both. Swab samples can be taken, but they have limited value in quantitative exposure assessment and are usually used only as a diagnostic tool to characterize whether buildings have dampness- or mold-related problems (see Chapter 6).

In most studies, dust samples from dust reservoirs, such as living-room and bedroom floors and mattresses, are collected for analysis of microbial content (with or without prior sieving or extraction). A theoretical advantage of settled-dust sampling is the presumed time integration that occurs in the deposition of bioaerosols on surfaces. Surface sampling may thus be the method of choice for assessing the association between exposure and the development of chronic conditions, such as asthma. The method is fast, easy, and inexpensive, using only a vacuum cleaner and filters or nylon sampling bags to collect dust, so it is particularly useful in large epidemiologic studies (focusing on chronic diseases), in which airborne measurements often are not feasible. One example in which this method is widely applied is the routine measurement of settled dust allergens. Allergen concentrations are usually expressed in units of allergen per gram of dust.

One limitation of the common practice of reporting concentrations of allergen or specific microbial agents per gram of dust collected should be noted: by dividing by total amount of dust collected, this expression of exposure does a poor job of characterizing the total burden of a specific agent in a building. For example, homes A and B could have the same amount of an agent (fungal allergen, endotoxin, viable microorganisms, or the like) per gram of dust by the conventional measure, whereas home A might have 10 times more dust than home B, so the average exposure of occupants of home A could be 10 times that of occupants of home B. For exposure-assessment purposes, it may therefore be more accurate to express exposure as floor-dust concentration per square meter sampled than as concentration per gram of sampled dust.

It is critical that surface sampling procedures be standardized so that sample results can be compared between sampling sessions. This requires standardization with regard to the selection of the sampling location, the technique of vacuuming, vacuum suction and the duration of sampling. Provided that sampling procedures are standardized, sampling of settled dust is reproducible as has been demonstrated for samples taken repeatedly over time (Heinrich et al., 2003).

Although surface sampling has advantages in many situations (particularly when a proxy of long-term average exposure is required), airborne measurements may be more desirable in others. Airborne measurements allow fluctuations in exposure to be assessed over the course of a week, a day or even hours; this can be essential in studying acute adverse effects such as daily lung-function changes with such metrics as FEV_1 (forced expiratory volume in 1 sec) or PEF (peak expiratory flow). Airborne sampling is also likely to capture the more appropriate dust fraction; that is, inhalable particles. Chew et al. (2003) propose that reservoir dust and air sampling represent different types of potential exposure to residents, suggesting that collection of both air and dust samples may be essential. However, airborne concentrations of specific agents are generally low in the residential indoor environment, and for many laboratory-based methods analytic sensitivity is not sufficient, so short-term airborne sampling is impossible for most agents. "Aggressive air sampling" has been suggested to overcome the problem of low indoor-air concentrations under "routine" conditions (IOM, 1993; Rylander, 1999; Rylander et al., 1992). Aggressive sampling involves activities intended to encourage the generation of biologic aerosols during sampling by agitating floor dust with devices that mimic people walking on carpets (Buttner et al., 2002) or by rapping on ventilation ducts (Dillon et al., 1999). Its usefulness in exposure assessment, however, is not clear. Viable microorganisms in the air can be identified with great sensitivity, provided that one is able to capture them alive and select a medium that can support their growth so that they can be measured under normal circumstances with methods for airborne sampling. However, sampling of viable microorganisms in the air with culture techniques will provide at best a "snapshot" of current exposure, given the high variability of microbial concentrations, the episodic nature of emissions from some microbial agents, and the relatively short sampling time allowed for this method. Thus, assessing the "true exposure" (E_{RR}) requires many samples and is not possible in most population studies.

In summary, airborne measurements may be a good indicator of exposure from a theoretical point of view, particularly for assessing acute short-term exposures, but detection problems limit their use for most biological agents in practice. Surface sampling is often the only alternative. When long-term exposures are being assessed, surface sampling may have an additional advantage over airborne measurements in that airborne measurements require a much larger number of samples to be taken because of the expected large variation in airborne concentrations. Nonetheless, it should be stressed that surface sampling is crude and is expected to yield a poor surrogate of airborne concentrations and the theoretical risk-relevant dose measure. Results of surface sampling as a measure of exposure should be interpreted with caution (Chew et al., 1996).

Personal vs Area Sampling

Assuming that airborne sampling is the desirable choice in a particular situation, personal measurements best represent the current airborne E_{RR}. Therefore, personal sampling is preferred to area sampling. Modern sampling equipment is now sufficiently light and small to use for personal sampling, and several studies of chemical air pollution have demonstrated its feasibility in both the indoor and outdoor environments (Janssen et al., 1999, 2000). However, practical constraints may make personal sampling impossible: it might be too cumbersome for the study subjects, or there might be no portable equipment to make the desired measurements (such as measurements of viable microorganisms).

If personal sampling is not possible, area sampling can be applied to reconstruct personal exposure with the "microenvironmental model" approach.[5] The microenvironmental model of human exposure is widely ac-

---

[5]Addressed in greater detail in *Clearing the Air* (IOM, 2000), page 54, from which this discussion is derived.

cepted for environmental exposure assessment (Sexton and Ryan, 1988). In that model, exposure of a person to an airborne agent is defined as the time-weighted average of agent concentrations encountered as the person passes through a series of microenvironments. However, exposures to microbial agents—such as particulate allergens, endotoxins, and fungal spores—often occur episodically because of inadvertent disturbance and resuspension of reservoirs of biologic agents by human activities (vacuum cleaning, handling of bedding, and the like) or because of mold blooms. Those episodic exposure patterns are not likely to be accurately captured by environmental area samplers. In addition, it is practically impossible to measure all the relevant microenvironments. Given those uncertainties, personal sampling is, despite some practical problems, a preferred method.

When, Where, and How Often to Sample

To the extent to which it is possible, samples should be taken to represent E_{RR} at the appropriate time. In the case of acute effects, exposure measurements taken shortly (up to 8 or 12 hours) before the effects take place would clearly be the most useful. However, it is not always possible to collect such information. Personal sampling is preferable, but if it cannot be performed, ambient sampling can be conducted where the person in question spends the most time. If air sampling is impossible for the reasons mentioned above, settled-dust samples can be taken in the same areas.

The case of chronic effects is more complicated because ideally exposure should be assessed before the occurrence of the effects and preferably at the time that is biologically most relevant, that is, when the exposure is thought to be the most problematic (such as when fungi are releasing spores) or when subjects are most susceptible to exposure. That is possible only in longitudinal cohort studies, and even then it often is not clear when people are most susceptible to the exposure of interest, although it is generally assumed that early childhood is the most relevant period for allergens. Cohort studies, however, are time-consuming and expensive. Most often, case-control studies are conducted; in these studies, exposure can be assessed only retrospectively. Settled-dust sampling (which is reviewed in Macher, 2001a,b) may be the best option because microbial agents in house dust appear to be relatively stable over long periods, and current concentrations may be a reasonable proxy for past exposures, assuming that the subjects have not moved homes or substantially changed the home conditions. It is not clear which sampling site best represents exposure; therefore, often a combination of bedroom and living-room floor dust samples and mattress dust samples is taken, sometimes including samples from the kitchen floor.

For risk-assessment purposes, measures of exposure need to be both accurate and precise so that the effect of exposure on disease can be estimated with minimal bias and maximal efficiency. Therefore, exposure must be assessed with a minimal measurement error. Precision can be gained (that is, measurement error can be reduced) by increasing the number of samples taken in each home. In population studies, repeated sampling within the home as a proxy for within-subject variation in exposure is particularly effective for exposures that are known to vary widely in the home relative to the variation observed between homes. If the within-home variation is smaller than the between-home variability, however, repeated sampling will not significantly reduce the measurement error, and one or a few samples will be sufficient. If within- and between-home variations are known (from previous surveys or pilot studies, for example), the number of samples required to obtain a given reduction in risk-estimate bias can be computed in the manner described by Cochran (1968). A within-home to between-home variance ratio of 3:1 to 4:1—which is not uncommon in airborne sampling of viable microorganisms—implies that 27–36 samples per home are required to estimate the average exposure reliably for an epidemiologic study with no more than a 10% bias in the relationship between some health end point and the exposure (Heederik and Attfield, 2000; Heederik et al., 2003).

Studies that include repeated measurements are scarce, so within-home and between-home variation cannot be accurately assessed. However, data are available on some agents. For example, it is well known that the concentration of total airborne viable fungi varies widely within a building even over very short periods (Hunter et al., 1988; Verhoeff et al., 1994). Viable mold counts in house-dust samples taken from the same location within a 6-week interval also showed very poor reproducibility (Verhoeff et al., 1994). In the same study, the variation in isolated genera and species between duplicate samples was even more substantial, with a very high within-home to between-home variance ratio of 3:1 to 4:1. More recently, that was confirmed in another study focusing on dustborne concentrations (Chew et al., 2001). It was demonstrated further that measurements of markers of fungal exposure in house dust, such as fungal EPSs were more reproducible, with an estimated within-home to between-home variance ratio of only 0.5:1. The estimated within-home variation of $\beta(1\rightarrow3)$-glucans and *total* culturable fungi was similar to the between–home variations, with ratios close to 1:1. Endotoxin concentrations in house dust in 20 homes in the United States measured repeatedly during a period of 12 months were significantly correlated ($r = 0.76$ for bed dust and 0.40 for bedroom-floor dust); this suggests average to good reproducibility for this measure (Park et al., 2000). In addition, a much larger study in Germany involving repeated dust sampling in 745 homes with a median interval of 7 months between first and second sampling periods showed that allergen (mite and cat) and endotoxin concentrations were well correlated over time, with crude correlation coefficients of 0.65–0.75 for the allergens and 0.59 for endotoxins (Heinrich et al., 2003). Viable-spore counts were, however, very poorly correlated—a correlation coefficient of only 0.06. On the basis of that limited experience, within-home variability of indoor-air concentrations of biologic agents are expected to be generally high and within-home variability of concentrations of these components in settled house dust generally low (compared with between-home variation). An exception is viable microorganisms, the concentration of which is highly variable in both indoor air and settled dust.

Little is known about spatial variation—that is, variation in concentrations between sampling locations at the same site, such as, in the case of surface sampling, on the same floor or bed. For example, studies have shown that house dust mite and cat allergen distribution is highly variable in settled dust (Hirsch et al., 1998; Loan et al., 2003). Expression as allergen mass did not reduce this variability (Hirsch et al., 1998). Isolated sampling of settled dust thus does not necessarily characterize the total burden of a specific agent in a building. However, in the case of floor dust, samples taken from the center of the room (as is commonly done in studies) have been shown to yield concentrations very similar to the mean concentration level for the whole floor, indicating that a single sample taken in this manner may be representative (Loan et al., 2003). Similar studies for other microbial agents have not yet been conducted.

Thus, because only sparse data are available on variation in exposure to biologic agents in the home environment, it is not possible to recommend how many samples should be taken to produce an accurate assessment of the E_{RR}. However, there is a strong suggestion that airborne concentrations are characterized by high variability over time, an indication that one sample per home is unlikely to be sufficient even when acute health effects are being considered, because variations in exposure occur over very short periods. Measurements of specific microbial agents in house dust generally appear to vary less and seem stable even over relatively long periods (up to 12 months and perhaps even longer), so one or a few samples may be sufficient. If only one floor sample is to be collected, research suggests that it be taken from the center of the floor (in front of a couch or a chair); for mattresses, the whole mattress should be sampled. Although measurements of dust can be more precise, it is not clear how well they represent *airborne* exposure. Measurements of viable microorganisms vary greatly over time regardless of whether they are sampled in air or in floor or bed dust, and many samples might be required.

In most circumstances, the only reason to go to the expense to measure specific taxa or the presence of glucan, ergosterol and the like is for the purposes of research into the health effects of exposure to those agents.

However, persons experiencing health outcomes with suspected or established links to a specific agent (aspergillosis, for example) may gain useful information from a more detailed characterization. Testing in those circumstances could be used to identify problematic environments and inform remediation decisions.

ASSESSING MICROORGANISMS

Measurement of microorganisms relies on collection of a sample into or onto solid, liquid, or agar media and then microscopic, microbiologic, biochemical, immunochemical, or molecular biologic analysis (Eduard and Heederik, 1998). Two distinctly different approaches are used for evaluation of microbial exposure: culture-based and nonculture methods.

Culture-Based Methods

Exposure to microorganisms in the indoor environment can be studied by counting culturable propagules in settled-dust samples. Alternatively, airborne exposure can be studied with various devices for microbial bioaerosol sampling (these are reviewed at length by Eduard and Heederik, 1998). There are three standard methods of active sampling of airborne culturable bioaerosols (Heederik et al., 2003):

- *Impactor methods.* With impactor sampling, bioaerosols moving in the air stream pass through a round jet or a slit to a culture medium. Multistage devices allow some size discrimination by sequentially increasing the velocity through the jet and decreasing the jet-to-plate spacing.
- *Liquid impinger methods.* Liquid impingers collect microorganisms by directing the air stream into a liquid collection fluid. Bacteria, viruses, and fungal spores are retained in the collection fluid and can subsequently be plated onto appropriate culture media or evaluated with other analytical techniques, although some re-entrainment and losses occur (Grinshpun et al., 1997; Willeke et al., 1988).
- *Air filtration methods.* Several sampling methods in common use rely on filtration to collect bioaerosols from a sampled air volume. After sampling, filters are agitated or sonicated in a solution. The solution is then serially diluted and plated on culture media or examined with other analytical techniques (biotechnology-based, immunological, or chemical assay, for example).

After sample collection (either airborne or from surfaces), bacterial growth media is incubated at a defined temperature for 3 days while fungal growth media may require incubation of 10 days. Colonies are counted and identified manually or with the aid of image-analysis techniques (reviewed in Eduard and Heederik, 1998). Concentrations are expressed as colony-forming units (CFUs) per sampled cubic meter of air or, in the case of surface sampling, per gram of sampled dust or per square meter.

Counting of culturable microorganisms has some serious drawbacks, including poor reproducibility; selection toward certain species because of chosen culture media, temperature, and the like; and the fact that dead microorganisms, cell debris, and microbial components are not detected although they may have toxic or allergenic properties. In addition, good methods for personal air sampling of culturable microorganisms are not available, and, although air concentrations usually vary widely in time, air sampling during a period of more than 15 minutes is often not possible and repeated sampling may be difficult logistically. But counting of culturable microorganisms is potentially a very sensitive technique that can identify many species.

Traditionally used culture methods to assess concentrations of culturable microorganisms in indoor air or settled dust have proved to be of little use for quantitative exposure assessment. They usually provide qualitative, rather than quantitative, information that can be important in risk assessment in that not all fungal and bacterial species pose the same hazard. A more extensive review of culture-based methods is available in the American Conference of Industrial Hygienists (ACGIH) publication "Bioaerosols, Assessment and Control" (1999).

Nonculture Methods

Nonculture methods enumerate organisms without regard to viability. Microorganisms in dust samples can be stained with a fluorochrome, such as acridine orange, and counted with an epifluorescence microscope (Thorne et al., 1994). Taxonomic classification of microorganisms is limited because little detail can be observed. Scanning electron microscopy (SEM) can also be used and allows better determination (Eduard et al., 1988; Karlsson and Malmberg, 1989) but it is expensive. Simple light microscopy may be used to count microorganisms, but counting is based only on morphologic recognition, which may result in large measurement errors. Bacteria collected with impingers or filters can be counted using flow cytometry after staining with 4',6-diamino-2-phenylindole (DAPI) or with fluorescent in situ hybridization (FISH) (Lange et al., 1997).

The main advantages of microscopy or flow cytometry are that both dead and living microorganisms are counted, selection effects are reduced, personal air sampling is possible, and sampling time can be varied over a wide range. Disadvantages include laborious and complicated procedures, high costs per sample, unknown validity, no detection of possibly relevant toxic or allergenic components or cell debris, and few possibilities of determination of microorganisms. Eduard and Heederik (1998) have published a more extensive review on microscopy and flow-cytometry methods for counting nonculturable microorganisms; the ACGIH (1999) review is another valuable reference for these methods.

Little or no experience is available with the more recently developed and more advanced nonculture based methods (scanning electron and epifluorescence microscopy, and flow cytometry, for example) in the nonindustrial indoor environment so the usefulness of these methods in indoor risk assessment is unknown.

ASSESSING MICROBIAL CONSTITUENTS

Instead of counting culturable or nonculturable microbial propagules, constituents or metabolites of microorganisms can be measured as a surrogate of microbial exposure. Toxic components (for example, from mycotoxins) or proinflammatory components (from endotoxins) can be measured, but nontoxic molecules may also serve as markers of large groups of microorganisms or of specific microbial genera or species. The use of advanced methods, such as polymerase chain reaction (PCR) technologies and immunoassays, have opened new avenues for detection and speciation regardless of whether the organisms are culturable (Buttner et al., 2001; Cruz-Perez et al., 2001a,b).

Markers for assessment of fungal biomass include ergosterol measured by gas chromatography-mass spectrometry (GCMS) (Miller and Young, 1997) and fungal EPSs measured with specific enzyme immunoassays (Douwes et al., 1999), which allow partial identification of the mold genera present. Volatile organic compounds produced by fungi may be suitable markers of fungal growth (Dillon et al., 1996). Other agents, such as $\beta(1\rightarrow3)$-glucans (Aketagawa et al., 1993; Douwes et al., 1996) and bacterial endotoxins are being measured on the basis of their toxic potency. Endotoxins are measured with an LAL assay prepared from blood cells of the horseshoe crab, *Limulus polyphemus* (Bang, 1956). Analytical chemistry methods for measurement of lipopolysaccharides (LPSs) have also been developed by using GCMS (Sonesson et al., 1988, 1990); however, these methods require special LPS extraction procedures and have not been widely used. Three methods for measuring $\beta(1\rightarrow3)$-glucans have been described, of which one is based on the LAL assay (Aketagawa et al., 1993) and two are enzyme immunoassays (Douwes et al., 1996; Milton et al., 2001). Finally, PCR techniques have been developed for the identification of specific species of bacteria and fungi (Alvarez et al., 1994; Haugland et al., 1999; Khan and Cerniglia, 1994). Application of quantitative PCR for analysis of environmental samples that contain microorganisms is still under development (Buttner et al., 2001; Cruz-Perez et al., 2001a,b).

A number of methods—including GC, GCMS, high-performance liquid chromatography (HPLC), capillary electrophoresis, thin-layer chromatography, enzyme-linked immunosorbent assays (ELISA), and cell-culture cytotoxicity testing—have been described to measure a large number of mycotoxins. Recent advances in technology have given laboratories the ability to test for specific mycotoxins without using cost-prohibitive GC or HPLC techniques. One source indicates that surface, bulk, food and feeds, and air samples can be analyzed relatively inexpensively for the following mycotoxins: aflatoxin; ochratoxin; trichothecenes, including T-2 toxin; fumonisins; deoxynivalenol or DON (vomitoxin); satratoxins; verrucarins; zearalenone; citrinin; alternariol; gliotoxin; patulin; and sterigmatocystin (Adler, 2002).

Most of the methods for measuring microbial constituents (an exception is the method for measuring bacterial endotoxins and some mycotoxins) are in an experimental phase and have not yet been routinely applied in epidemiologic studies or are not commercially available. Important advantages of those methods include the stability of most of the measured components, which allows longer sampling times for airborne measurements and frozen storage of samples before analysis; the use of standards in most of the methods; and the enhanced possibility of testing for reproducibility.

ASSESSING BIOALLERGENS

Antibody-based immunoassays, particularly ELISA, are widely used for the measurement of aeroallergens and allergens in settled dust in buildings. To date, the house dust mite allergens Der p 1, Der f 1, and Der p/f 2 have been most widely investigated, and the methods have been well described (Platts-Mills and Chapman, 1987; Platts-Mills and de Weck, 1989; Price et al., 1990). Methods for measuring fungal allergens are not widely available, primarily because fungal allergen production in nature is highly variable and depends on many factors, including substrate and temperature. The variability makes it difficult to develop specific antibody-based immunoassays that detect the relevant fungal allergens in a specific environment.

INDIRECT EXPOSURE-ASSESSMENT METHODS

Signs and Measurements of Dampness, Moisture, or Mold

Humans are poor humidity sensors, but some signs of inappropriate moisture can be directly perceived. Such perceptions are the basis of most epidemiologic studies, in which data on moisture conditions are collected by questionnaire. Questions are typically formulated to seek information on whether leaks, floods, wet basements, window condensation, visible fungal growth, or moldy odors are present. However, there is considerable variation in how the questions are framed. In some studies, the dampness indicator is limited to recent experience, such as "presence of damp stains or mold growth on indoor surfaces in last 2 years" (Brunekreef, 1992); others record experience with building dampness over the subjects' lifetimes. Assessment of dampness may also be classified according to specific building environments circumstances, for example, "Have you previously or do you currently notice moisture stains in the structures of your home?" (Pirhonen et al., 1996). Some have collected information about specific areas, such as "mold in a child's bedroom" or the living room. In many cases, the questions are collapsed into broader categories of dampness (Kilpeläinen et al., 2001; Yang et al., 1998). Table 3-1 provides examples of the studies. It should be noted that reporting bias may be a source of error in such research. Dales and colleagues (1997) report that under some conditions allergy patients may be more likely than nonallergic people to report visible fungal growth. However, other studies have demonstrated that such bias is unlikely (Verhoeff et al., 1995; Zock et al., 2002).

Moisture conditions in buildings may be best discovered through direct observation and inspection. Home inspectors are known to rely on smell to supplement visual inspection. Among the items typically included in an inspection report are presence of mold, water stains, evidence of leaks or flooding, current leaks, crawl space conditions, attic sheathing condition, and overall stoutness or dilapidation of the building. Characterization of rainwater discharge and management is also necessary, given the importance and prevalence of foundation leakage to the overall moisture balance of a building. In one of the earlier studies of building dampness, Platt et al. (1989) trained surveyors to assess dampness by severity and type, mold by severity and location, and details of building structure. Air samples were also taken from rooms, and spore counts were estimated and fungi identified where possible. Koskinen et al. (1999) reported a study in which civil engineers recorded signs of leakage, presence of moist spots, detachment of paint or other surface material, and deformation of wood or discoloration and then categorized the findings into "moisture absent" or "moisture present." Mohamed and colleagues (1995) described a study in Nairobi, Kenya, in which interviewers assessed the home with a standardized checklist; many of the homes lacked solid floors. Results of those and other studies will probably encourage others to provide more information about dampness in buildings in developing countries.

Attempts to quantify building characteristics with engineering protocols have been limited to a few epidemiologic studies. One of the most comprehensive attempts to describe dampness in relation to both indoor

TABLE 3-1 Dampness Definitions and Associated Environmental
Assessments in Selected Cross-Sectional and Analytic Epidemiologic
Studies in Which "Dampness" Was a Key Risk Factor in Health
Outcomes

Reference	Questions Used to Define Dampness
Åberg et al., 1996	By self-report: 1) Do you usually have moisture (dampness/ice) far down inside windowpanes in winter (does not apply to bath/shower room and not between double panes) —never? —sometimes? —often (every week)? 2) Have you had damp or mold damage in your home after/during the child's first year?
Andriessen et al., 1998	By self-report: Reports of moisture stains and mold presence during last 2 years
Brunekreef, 1992	By self-report: Damp stains or mold growth on indoor surfaces in last 2 years
Brunekreef et al., 1992	By self-report: 1) Presence of mold or mildew in home (ever) 2) Water damage to home (ever) 3) Occurrence of water on basement floor (ever)
Cuijpers et al., 1995	By self-report: 1) Damp stains in last 2 years 2) Mold growth in last 2 years
Dales et al., 1991	Mold sites—number of sites with visible mold or mildew in last year Moisture—appearance of wet or damp spots excluding basement in last year Flooding—appearance of flooding, water damage, or leaks in basement in last year Dampness/mold—any one of above variables positive
Dales and Miller, 1999	By self-report: 1) Ever have mold or mildew on any surface inside present home? 2) Did this occur during last 12 months?
Engvall et al., 2001	By self-report: 1) Episodes of water leakage in last 5 years 2) Condensation on windows 3) Slow drying of wet towels in bathroom 4) Perception of pungent, moldy musty and stuffy odors in dwelling
Evans et al., 2000	By self-report: Is damp or condensation a serious problem in your home? (Do not include normal condensation on windows.) A serious problem? A minor problem? No problem?

TABLE 3-1 continued

Reference	Questions Used to Define Dampness
Hu et al., 1997	By self-report: 1) Visible mold growth? 2) Basement water damage? 3) Leaking, wet or damp spots on indoor surfaces?
Jaakkola et al., 2002	By self-report: Presence of visible mold and/or mold odor in workplace?
Jedrychowski and Flak, 1998	By self-report: Presence of dampness or molds on walls: 0 = no moisture stains or mold growth 1 = only small moisture stains (up to 1 m²) 2 = larger moisture stains 3 = mold visible on smaller surfaces (up to 1 m²) 4 = mold visible on larger surfaces (more than 1 m²) (Codes 1–4 were collapsed into one group as "present.")
Kilpeläinen et al., 2001	By self-report: 1) Have you had mold growth on the surfaces of any of your dwellings during the last year? 2) Have you had damp stains on the walls or on the ceilings of any of your dwellings in the last year? 3) Has there been a leak or water damage in any of your dwellings in the last year? Categorized into: 1) Presence of visible mold (yes to question 1) 2) Visible mold or damp stains or water damage during the last year (yes to questions 1, 2, or 3)
Koskinen et al., 1999	By self-report: 1) Presence of visible signs of moisture 2) Water damage 3) Observations of mold
Mohamed et al., 1995	By self-report: Damage caused by dampness in child's sleeping area
Nafstad et al., 1998	By interview: Presence of water damage, damp stain, or visible mold/mildew growth during last 2 years
Pirhonen et al., 1996	By self report: 1) Have you had or are you able to see visible mold growth on the walls of your home? 2) Have you or are you aware of an odor of mold or cellar-like air in your home? 3) Have you or do you notice moisture stains in the structures of your home? 4) Have you or are you suffering from water/moisture damage in your home?

(continued on next page)

TABLE 3-1   continued

Reference	Questions Used to Define Dampness
Platt et al., 1989	Surveyor assessment of dampness and mold
Strachan and Carey, 1995	Self-reports of visible mold growth in child's bedroom
Strachan et al., 1990	Self-reports of visible mold in home
Tariq et al., 1996	Self-reported dampness Lack of central heating
Waegemaekers et al., 1989	Home dampness assessed: 1) According to five dampness characteristics— visible mold growth, damp spots, silverfish or sow bugs (wood lice), stale odor, or wet crawl space. If a home had two or more, it was' "damp." 2) On basis of responder's own perception: "Do you generally consider your home to be dry or damp?" Damp—if responded damp or rather damp Dry—if answered dry or rather dry 3) On basis of results of viable fungal-spore measurements in living rooms of 36 homes selected according to questionnaire to ensure that a relatively large number of "damp" homes were measured
Wever-Hess et al., 2000	Self-reported damp housing at time of enrollment
Williamson et al., 1997	By self-report (structured interview with trained researcher): 1) Presence of current dampness or condensation in the home 2) Exposure to dampness and mould in previous dwellings
Yang et al., 1998	By self-report: Home dampness defined as presence of any of following during last year: 1) Visible mold or mildew growth on surfaces inside house 2) Standing water inside the home 3) Water damage 4) Leakage of water into building

and outdoor characteristics has been reported by Jedrychowski and Flak (1998). Williamson et al. (1997) also applied an extensive assessment protocol to assess building dampness. Surveyors measured spot temperatures and relative humidity outdoors and in each room of the dwelling. An electronic resistance moisture meter was also used to measure dampness just above skirting-board height in the rooms. Dampness was coded as 0 (<10%), 1 (11–25%), 2 (26–50%), 3 (51–75%), or 4 (>75%) and the scores were summed for a total dampness score. The presence and severity of visible mold growth on each wall were graded subjectively on a four-point scale: 0 = absent, 1 = trace, 2 = obvious but localized, and 3 = obvious and widespread. Dwellings with a total mold score of 3 or more were classified as having significant mold contamination.

Biomarkers

For biologic agents, very few biomarkers of exposure or dose have been identified, and their validity for exposure assessment in the indoor environment is often not known. Adducts formed by aflatoxins and ochratoxins as they damage DNA, RNA, and proteins can be measured from the body fluids of exposed persons (Bechtel, 1989; Sabbioni and Wild, 1991). They reflect repair activity more than they reflect degree of exposure, however, and do not accurately quantify exposure. But DNA adducts do indicate past presence of and damage to nucleic acids and constitute a limited biomarker of exposure, effect, and susceptibility (Miraglia et al., 1996). A 2003 study in rats suggested that measurement of stachylysin (a proteinaceous hemolysin) in serum with an ELISA technique could be used as a biomarker of exposure to *Stachybotrys chartarum* (Van Emon et al., 2003). However, although the method is sensitive and specific, it is not clear whether it is useful for quantitatively assessing indoor exposures to that mold. The same is true of the measurement of trichothecene mycotoxins in serum (Croft et al., 2002).

To the committee's knowledge, no other *direct* methods for measuring biologic agents or metabolites thereof in blood or urine have been described. IgG antibodies in serum have been suggested as an *indirect* marker of recent occupational exposure to fungi (Burrell and Rylander, 1981; Eduard et al., 1992), but little is known about the quantitative relation between serum IgG and airborne exposure. A few studies involving children and school indoor environments suggested that the correlation between IgG and mold exposure is poor (Immonen et al., 2002; Taskinen et al., 2002).

IgE and inflammatory markers in blood, sputum, nasal-lavage fluid, and exhaled breath condensate have been suggested as biomarkers of exposure (Hirvonen et al., 1999; Roponen et al., 2001, 2003), but these are more appropriately addressed as markers (or intermediates) of effect since they indicate susceptible persons and play a major role in the pathophysiological events leading to symptoms and disease. Therefore, they should not be considered markers of exposure.

Predictive Exposure Models

If the factors that explain the variation in indoor microbial exposure were known, mathematical models could be developed to predict exposure in homes where no exposure measurements were taken, provided that valid information on determinants of exposure were available. Various studies demonstrate that home and occupant characteristics assessed by questionnaire (such as age of building, presence of pets in the home, type of carpet, type of heating and ventilation, and damp or mold spots) are associated with indoor microbial exposures (Bischof et al., 2002; Douwes et al., 1998, 1999; Gehring et al., 2001). However, not all studies find housing characteristics to be predictive (Wood et al., 1988, for example), and the explained variance is too small to predict exposure reliably on the basis of these factors. Therefore, no such predictive models can now be used to assess exposure to biologic agents in the indoor environment accurately.

CONCENTRATIONS IN THE ENVIRONMENT

This section briefly discusses reported indoor concentrations of some biologic agents. However, the concentrations should be interpreted with caution because the studies used different sampling and analytic procedures that potentially could result in large differences in exposure assessment.

Fungi

The concentrations of viable fungi in indoor environments are usually a few to several thousand CFUs per cubic meter of air. Those concentrations are highly variable and depend on such factors as climate and season; type, construction, age, and use of the building; and ventilation rate. The observed concentrations also depend on the sampling and analytic methods used. Table 3-2 summarizes fungal concentrations that have been observed in buildings in different countries; these data should be interpreted with caution because of the methodologic limitations described above, and they cannot be used as reference values.

Bacteria

Relatively few studies have reported concentrations of bacteria in indoor air. The problems of accurate exposure assessment discussed for fungi also apply to measurements of airborne bacteria. New methods focusing on quantification of bacterial biomass with chemical markers (Szponar and Larsson, 2000, 2001) or, more specifically, DNA-based methods (Macneil et al., 1995) will probably help to solve some of these limitations. For some airborne pathogens, specific methods based on gene-amplification reactions have been developed (Pena et al., 1999), but applications of DNA-based methods for larger-scale analyses of airborne bacteria still need more validation.

The reported concentrations of viable bacteria in indoor air are summarized in Table 3-2. Usually, the total concentrations measured on standard media are reported, and few studies have characterized the bacterial flora of

TABLE 3-2   Reported Concentrations of Airborne Bacteria in Indoor Air in Selected Studies

Country	Indoor Space	Concentrations Observed,[a] CFU/m^3	Reference
Poland	Healthy homes (n = 27)	GM 1,021	Pastuszka et al., 2000
	Moldy homes (n = 43)	GM 980	
USA	Homes (n = 44)	GM 1,258 (220–4,006)	Ross et al., 2000
USA	Home	Median 98	Macher et al., 1991
USA	Homes 1 year after flood (n = 46)	Mean 880	Curtis et al., 2000
Finland	Homes and day-care centers with moisture problems		Hyvärinen et al., 2001
	Fall	Range 36–4,600	
	Winter	Range 14–35,000	
	Reference buildings		
	Fall	Range 220–10,000	
	Winter	Range 79–18,000	
Finland	Moldy homes (n = 32)	GM 1,011	Nevalainen et al., 1991
Hong Kong	Homes	<1,000	Lee et al., 1999
	Reference homes (n = 18)	GM 678	
UAE	High-quality housing	Mean 5,471	Jaffal et al., 1997
	Medium-quality housing	Mean 9,871	
	Low-quality housing	Mean 15,179	
France	Office	Mean 447	Parat et al., 1997
Taiwan	Day-care centers	GM 735	Li et al., 1997

[a]GM: geometric mean.

indoor air. Table 3-3 shows a rough summary of those bacteria found and indicates that gram-positive cocci usually dominate the airborne bacterial flora. Again, the data in these tables cannot be used as reference values.

Endotoxins

Endotoxin concentrations measured within particular nonindustrial indoor spaces vary widely, from a few to several thousand endotoxin units[6] (EU) per milligram of house dust (Table 3-4). Concentrations measured as EU per square meter vary even more widely. However, ranges of endotoxin

---

[6]Endotoxin unit is a standardized unit of biologic activity, measured with the LAL test and calibrated to the U.S. Pharmacopoeia reference endotoxin; it is equal to the international unit of activity used by the World Health Organization.

TABLE 3-3   Bacterial Types Found in Different Indoor Environments

Group of Bacteria	Indoor Air	Outdoor Air	Hospital Air	Subway Air
Micrococci	+++	+++	+++	+++
Staphylococci	++	++		
Aerococci	+	++		
Streptococci	+	+		++
Gram-positive rods			+	
Corynebacteria	++			
Spore-forming rods	++	++		+
Actinomycetes			+	
Gram-negative rods	+	++	+	++

KEY:
+++ dominating group, >40% of isolates.
  ++ found frequently, 10–40% of isolates.
   + found regularly, <10% of isolates.
SOURCE: Nevalainen, 1989.

TABLE 3-4   Overview of Epidemiologic Studies Indicating Adverse or Protective Effects on Respiratory Health Related to Indoor Endotoxin Exposure

Reference	Population	N
Michel et al., 1991	Adult asthma patients	28
Michel et al., 1996	Adult asthma (40) and rhinitis (29) patients	69
Rizzo et al., 1997	Children (50% with asthma)	20
Douwes et al., 2000	Children (50% with airway symptoms)	148
Park et al., 2001[b]	Infants	499
Gehring et al., 2001[b]	Infants (6–12 months)	1,884
Gereda et al., 2000	Infants with wheeze (9–24 months)	61
Gehring et al., 2001[b]	Infants (6–12 months)	1,884
Gehring et al., 2002	Children (50% atopy or asthma)	454
Braun-Fahrländer et al., 2002	Children (in farming and nonfarming families)	812

[a]Mean exposure (or range of mean exposures if no overall mean given); 1 endotoxin unit is about 0.1 ng (exact conversion factor depends on source of endotoxin for calibration).
[b]Longitudinal study (all other studies were cross-sectional studies).
SOURCE: Excerpted and adapted from Douwes et al., 2002.

concentrations reported by researchers appear to vary only moderately between studies regardless of the geographic area. That is remarkable, considering that analytic methods applied in those studies were not standardized. Only a few studies have focused on airborne concentrations in the indoor environment. Park et al. (2000) reported a mean airborne endotoxin concentration of 0.64 EU/m^3 measured in 15 homes in Boston, Massachusetts (and mean dust endotoxin concentrations of 44–105 EU/mg), indicating that time-weighted mean exposures are very low compared with the work environment, where airborne endotoxin concentrations can range from several tens to thousands of endotoxin units per cubic meter (Douwes et al., 2002).

β(1→3)-Glucans

Methods used to analyze β(1→3)-glucans in environmental (settled or airborne) dust samples have not been standardized and are therefore not comparable across studies. In Sweden and Switzerland, typical exposures

Exposure[a]	Health effect
	Adverse effects
2.59 ng/mg	Decline in FEV_1 and FEV_1/FVC; increase in asthma medication and symptoms
1.78 ng/mg	Decline in FEV_1, and FEV_1/FVC; increase in asthma medication and symptoms
1–100 EU/mg	Increase in asthma medication and symptoms in asthmatic children
24.9 EU/mg	Increased PEF variability in atopic children with asthma symptoms
100 EU/mg	Increased prevalence of wheeze during first year of life
2.9 EU/mg	Increased prevalence of wheeze and respiratory infections
	Protective effects
20–2,000 EU/mg	Decreased prevalence of atopic sensitization; increased type 1 helper T cell (Th1) and decreased type 2 helper T cell (Th2) expression
2.9 EU/mg	Decreased risk of eczema
24,221 EU/m^2	Decreased prevalence of atopic sensitization
23–38 EU/mg	Decreased prevalence of hay fever, atopy sensitization, and atopic asthma

in buildings with mold problems ranged from about 10 to more than 100 ng/m^3 according to an LAL assay of $\beta(1\rightarrow3)$-glucans in airborne dust samples that were generated by rigorous agitation of settled dust in those buildings (Rylander, 1999). Exposures in buildings that had no obvious mold problems were close to 1 ng/m^3. In the Netherlands and Germany, mean $\beta(1\rightarrow3)$-glucans concentrations in house dust determined with a specific enzyme immunoassay were highly comparable at around 1,000–2,000 µg/g of dust and 500–1,000 µg/m^2 (Chew et al., 2001; Douwes et al., 1996, 1998, 2000; Gehring et al., 2001). Samples were also taken in homes that were not selected specifically on the basis of mold problems and were analyzed in the same laboratory with identical procedures. No airborne samples were taken.

EVALUATION OF EXPOSURE DATA

No health-based recommended exposure limits for indoor biologic agents exist, and this makes the interpretation of exposure difficult, particularly in case studies. Strategies to evaluate exposure data (either quantitatively or qualitatively) should include comparison of exposure data with background concentrations or, better, a comparison of exposures between symptomatic and nonsymptomatic subjects. A quantitative evaluation involves comparing exposures, whereas a qualitative evaluation could consist of comparing species or genera of microorganisms in different environments. Because of differences in climatic and meteorological conditions and differences in measurement protocols used in various studies (viable versus non-viable microorganism sampling, sampler type, analysis, and so on), reference material from the literature can seldom be used. Thus, to draw valid conclusions, it is important in each study to include measurements in indoor environments of subjects without symptoms. Furthermore, interpretations of airborne sampling should be based on multiple samples because space–time variability in the environment is high. Finally, the proper interpretation of exposure results requires detailed information about sampling and analytic procedures (including quality control) and knowledge of the potential problems associated with those procedures.

It is not possible to reach a general conclusion on whether total fungal counts represent a meaningful measure of exposure for indoor-related health effects. In cases where health outcomes have established links to a specific agent or microorganism, it is appropriate to focus on measurement of that agent or microorganism. If, on the other hand, agents such as $\beta(1\rightarrow3)$-glucans are involved, then a total fungi count may be a relevant measure as almost all fungi contain $\beta(1\rightarrow3)$-glucans. Given the present state of knowledge, it may be appropriate to make both specific and total fungi measures when this is possible.

Further, it is currently not clear whether fungal counting methods do a better job of characterizing a person's or population's true exposure than the traditionally-applied culture methods: this is largely dependent on the aim of the study, the specific health outcome(s) of interest, and the nature and source of the exposure. For some health outcomes—those involving allergic sensitization, for example—the identity of the microbial agent may be as important as the amount of agent present. These gaps in the knowledge base create a potential for misinterpretation and misuse of results that must be kept in mind whenever sampling is conducted. More research is needed to further our understanding of which exposure assessment methods are most relevant in assessing health risks from indoor exposures. General recommendations with regard to exposure assessment methods for the purpose of risk assessment can therefore not be given, particularly since indoor-related symptoms or diseases may be caused by multiple exposures.

FINDINGS, RESEARCH NEEDS, AND RECOMMENDATIONS

Based on the review of the papers, reports and other information presented in this chapter, the committee has reached the following findings and recommendations, and has identified the following research needs regarding exposure assessment for damp indoor environments.

- The evaluation of exposure characterization results should, whenever possible, be based on:
  — Comparison of exposure data with background concentrations or, better, a comparison of exposures between symptomatic and nonsymptomatic subjects.
  — Multiple samples, because space–time variability in the environment is high.
  — Detailed information about sampling and analytic procedures (including quality control) and knowledge of the potential problems associated with those procedures.
- The lack of knowledge regarding indoor microbial exposures and related health problems is due primarily to a lack of valid quantitative methods for assessing exposure.
- There are several methods for measuring and characterizing fungal populations, but methods for assessing human exposure to fungal agents are poorly developed. Part of the difficulty is related to the large number of fungal species that are measurable indoors and the fact that fungal allergen content and toxic potential varies among species and among morphologic forms within species. In addition, the most common methods for fungal assessment—counting cultured colonies and identifying and counting

spores—have variable and uncertain relationships to allergen, toxin, and irritant content of exposures.

• Existing exposure assessment methods for fungal and other microbial agents need rigorous validation and further refinement to make them more suitable for large-scale epidemiologic studies. This includes standardization of protocols for sample collection, transport of samples, extraction procedures, and analytical procedures and reagents. Such work should result in concise, internationally accepted protocols that will allow measurement results to be compared both within and across studies.

• Research is needed to develop improved exposure assessment methods, particularly methods based on nonculture techniques and techniques for measuring constituents of microorganisms—allergens, endotoxins, $\beta(1\rightarrow3)$-glucans, fungal extracellular polysaccharides (EPSs), fungal spores, other particles and emissions of microbial origin. These needs include:

— Further improvement of light and portable personal airborne exposure measurement technology.

— More rapid development of measurement methods for specific microorganisms that use DNA-based and other technology.

— Rapid and direct-reading assays for bioaerosols for the immediate evaluation of potential health risks.

• Application of the new or improved methods will allow more valid exposure assessment of microorganisms and their components, which should facilitate more-informed risk assessments.

REFERENCES

Åberg N, Sundell B, Eriksson B, Hesselmar B, Åberg B. 1996. Prevalence of allergic diseases in school children in relation to family history, upper respiratory infections, and residential characteristics. Allergy 51(4):232–237.

ACGIH (American Conference of Governmental Industrial Hygienists). 1999. Bioaerosols: Assessment and Control. Macher JM, ed. Cincinnati, OH: American Conference of Governmental Industrial Hygienists.

Adler CM. 2002. Mycotoxins: characteristics, sampling methods, and limitations. ENVIRO-CHECK, Inc. Winter 2002–2003. http://www.envirocheckonline.com/docs/mycotoxins.pdf. accessed June 16, 2003.

Aketagawa J, Tanaka S, Tamura H, Shibata Y, Sait H. 1993. Activation of limulus coagulation factor G by several $(1\rightarrow3)$-β-D-glucans: comparison of the potency of glucans with identical degree of polymerization but different conformations. Journal of Biochemistry 113:683–686.

Alvarez AJ, Buttner MP, Toranzos GA, Dvorsky EA, Toro A, Heikes TB, Mertikas-Pifer LE, Stetzenbach LD. 1994. Use of solid-phase PCR for enhanced detection of airborne micro-organisms. Applied Environmental Microbiology 60:374–376.

American Thoracic Society. 1997. American Thoracic Society Workshop, Achieving Healthy Indoor Air. American Journal of Respiratory and Critical Care Medicine 156(Supplement 3):534–564.

Andriessen J, Brunekreef B, Roemer W. 1998. Home dampness and respiratory health status in European children. Clinical and Experimental Allergy 28:1191–1200.

Bang FB. 1956. A bacterial disease of Limulus polyphemus. Bulletin of the John Hopkins Hospital 98:325–350.

Bechtel DH. 1989. Molecular dosimetry of hepatic aflatoxin B_1-DNA adduct: linear correlation with hepatic cancer risk. Regulatory Toxicology and Pharmacology 10(1):74–81.

Bischof W, Koch A, Gehring U, Fahlbusch B, Wichmann HE, Heinrich J. 2002. Predictors of high endotoxin concentrations in the settled dust of German homes. Indoor Air 12:2–9.

Braun-Fahrländer C, Riedler J, Herz U, Eder W, Waser M, Grize L, Maisch S, Carr D, Gerlach F, Bufe A, Lauener RP, Schierl R, Renz H, Nowak D, von Mutius E, Allergy and Endotoxin Study Team. 2002. Environmental exposure to endotoxin and its relation to asthma in school-age children. New England Journal of Medicine 347(12):869–877.

Brunekreef B. 1992. Damp housing and adult respiratory symptoms. Allergy 47(5):498–502.

Brunekreef B, Groot B, Rijcken B, Hoek G, Steenbekkers A, de Boer A. 1992. Reproducibility of childhood respiratory symptom questions. The European Respiratory Journal 5(8): 930–935.

Burge HA. 2000. The fungi. In: Indoor Air Quality Handbook. JD Spengler, JM Samet, JF McCarthy, eds. New York: McGraw-Hill.

Burge HA, Pierson DL, Groves TO, Strawn KF, Mishra SK. 2000. Dynamics of airborne fungal populations in a large office building. Current Microbiology 40:10–16.

Burrell R, Rylander R. 1981. A critical review of the role of precipitins in hypersensitivity pneumonitis. European Journal of Respiratory Diseases 62(5):332–343.

Buttner MP, Cruz-Perez P, Stetzenbach LD. 2001. Enhanced detection of surface-associated bacteria in indoor environments by quantitative PCR. Applied Environmental Microbiology 67(6):2564–2570.

Buttner MP, Cruz-Perez P, Stetzenbach LD, Garrett PJ, Lutke AE. 2002. Measurement of airborne fungal spore dispersal from three types of flooring materials. Aerobiologia 18(1):1–11.

Chew GL, Muilenberg ML, Gold D, Burge HA. 1996. Is dust sampling a good surrogate for exposure to airborne fungi? Journal of Allergy and Clinical Immunology 97:419.

Chew GL, Douwes J, Doekes G, Higgins KM, Strien R, Spithoven J, Brunekreef B. 2001. Fungal extracellular polysaccharides, $\beta(1\rightarrow3)$-glucans, and culturable fungi in repeated sampling of house dust. Indoor Air 11:171–178.

Chew GL, Rogers C, Burge HA, Muilenberg ML, Gold DR. 2003. Dustborne and airborne fungal propagules represent a different spectrum of fungi with differing relations to home characteristics. Allergy 58(1):13–20.

Chun DT, Chew V, Bartlett K, Gordon T, Jacobs RR, Larsson BM, Larsson L, Lewis DM, Liesivuori J, Michel O, Milton DK, Rylander R, Thorne PS, White EM, Brown ME. 2000. Preliminary report on the results of the second phase of a round-robin endotoxin assay study using cotton dust. Applied Occupational and Environmental Hygiene 15: 152–157.

Cochran WG. 1968. Errors of measurement in statistics. Technometrics 10:637–666.

Croft WA, Jastromski BM, Croft AL, Peters HA. 2002. Clinical confirmation of trichothecene mycotoxicosis in patient urine. Journal of Environmental Biology 23(3):301–320.

Cruz-Perez P, Buttner MP, Stetzenbach LD. 2001a. Specific detection of Aspergillus fumigatus in pure culture using quantitative polymerase chain reaction. Molecular and Cellular Probes 15:81–88.

Cruz-Perez P, Buttner MP, Stetzenbach LD. 2001b. Specific detection of Stachybotrys chartarum in pure culture using quantitative polymerase chain reaction. Molecular and Cellular Probes 15:129–138.

Cuijpers CE, Swaen GM, Wesseling G, Sturmans F, Wouters EF. 1995. Adverse effects of the indoor environment on respiratory health in primary school children. Environmental Research 68(1):11–23.

Curtis L, Ross M, Persky V, Scheff P, Wadden R, Ramakrishnan V, Hryhorczuk D. 2000. Bioaerosol concentrations in the quad cities 1 year after the Mississippi river floods. Indoor and Built Environment 9(1):35–43.

Dales RE, Miller D. 1999. Residential fungal contamination and health: microbial cohabitants as covariates. Environmental Health Perspectives 107(Supplement 3):481–483.

Dales RE, Burnett R, Zwanenburg H. 1991. Adverse health effects among adults exposed to home dampness and molds. American Review of Respiratory Disease 143(3):505–509.

Dales RE, Miller D, McMullen E. 1997. Indoor air quality and health: validity and determinants of reported home dampness and moulds. International Journal of Epidemiology 26:120–125.

Dillon HK, Heinsohn PA, Miller JD, eds. 1996. Field Guide for the Determination of Biological Contaminants in Environmental Samples. Fairfax, VA: American Industrial Hygiene Association.

Dillon HK, Miller JD, Sorenson WG, Douwes J, Jacobs RR. 1999. A review of methods applicable to the assessment of mold exposure to children. Environmental Health Perspectives 107(Supplement 3):473–480.

Douwes J, Versloot P, Hollander A, Heederik D, Doekes G. 1995. Influence of various dust sampling and extraction methods on the measurement of airborne endotoxin. Applied Environmental Microbiology 61:1763–1769.

Douwes J, Doekes G, Montijn R, Heederik D, Brunekreef B. 1996. Measurement of $\beta(1\rightarrow3)$-glucans in the occupational and home environment with an inhibition enzyme immunoassay. Applied Environmental Microbiology 62:3176–3182.

Douwes J, Doekes G, Heinrich J, Koch A, Bischof W, Brunekreef B. 1998. Endotoxin and $\beta(1\rightarrow3)$-glucan in house dust and the relation with home characteristics: a pilot study in 25 German houses. Indoor Air 8:255–263.

Douwes J, van der Sluis B, Doekes G, van Leusden F, Wijnands L, van Strien R, Verhoeff A, Brunekreef B. 1999. Fungal extracellular polysaccharides in house dust as a marker for exposure to fungi: relations with culturable fungi, reported home dampness and respiratory symptoms. Journal of Allergy and Clinical Immunology 103:494–500.

Douwes J, Zuidhof A, Doekes G, van der Zee S, Wouters I, Boezen HM, Brunekreef B. 2000. $(1\rightarrow3)$-β-D-glucan and endotoxin in house dust and peak flow variability in children. American Journal of Respiratory and Critical Care Medicine 162:1348–1354.

Douwes J, Pearce N, Heederik D. 2002. Does bacterial endotoxin prevent asthma? Thorax 57:86–90.

Duchaine C, Thorne PS, Mériaux A, Grimard Y, Whitten P, Cormier Y. 2001. Comparison of endotoxin exposure assessment by bioaerosol impinger and filter sampling methods. Applied Environmental Microbiology 67(6):2775–2780.

Eduard W, Heederik D. 1998. Methods for quantitative assessment of airborne levels of noninfectious micro-organisms in highly contaminated work environments. American Industrial Hygiene Association Journal 59:113–127.

Eduard W, Sandven P, Johansen BV, Bruun R. 1988. Identification and quantification of mould spores by scanning electron microscopy (SEM): analysis of filter samples collected in Norwegian saw mills. Annals of Occupational Hygiene 31(Supplement 1):447–455.

Eduard W, Sandven P, Levy F. 1992. Relationships between exposure to spores from *Rhizopus microsporus* and *Paecilomyces variotti* and serum IgG antibodies in wood trimmers. International Archives of Allergy and Immunology 97:274–282.

Englehart S, Loock A, Skutlarek D, Sagunski H, Lommel A, Färber H, Exner M. 2002. Occurrence of toxigenic *Aspergillus versicolor* isolates and sterigmatocystin in carpet dust from damp indoor environments. Applied and Environmental Microbiology 68(8): 3886–3890.

Engvall K, Norrby C, Norbäck D. 2001. Asthma symptoms in relation to building dampness and odour in older multifamily houses in Stockholm. International Journal of Tuberculosis and Lung Disease 5(5):468–477.

Evans J, Hyndman S, Stewart-Brown S, Smith D, Petersen S. 2000. An epidemiological study of the relative importance of damp housing in relation to adult health. Journal of Epidemiology and Community Health 54:677–686.

Gehring U, Douwes J, Doekes G, Koch A, Bischof W, Wichmann HE, Heinrich J. 2001. β(1→3)-glucan in house dust of German homes related to culturable mold spore counts, housing and occupant characteristics. Environmental Health Perspectives 109:139–144.

Gehring U, Bischof W, Fahlbusch B, Wichmann HE, Heinrich J. 2002. House dust endotoxin and allergic sensitization in children. American Journal of Respiratory and Critical Care Medicine 166(7):939–944. [Erratum: American Journal of Respiratory and Critical Care Medicine (2003) 167(1):91.]

Gereda JE, Leung DYM, Thatayatikom A, Streib JE, Price MR, Klinnert MD, Liu AH. 2000. Relation between house-dust endotoxin exposure, type 1 T-cell development, and allergen sensitization in infants at high risk of asthma. Lancet 355(9216):1680–1683.

Górny RL, Reponen T, Willeke K, Schmechel D, Robine E, Boissier M, Grinshpun SA. 2002. Fungal fragments as indoor air biocontaminants. Applied and Environmental Microbiology 68(7):3522–3531.

Grinshpun SA, Willeke K, Ulevicius V, Juozaitiis A, Terzieva S, Donnelly J, Stelma GA, Brenner K. 1997. Effect of impaction, bounce and reaerosolization on collection efficiency of impingers. Aerosol Science and Technology 26(4):326–342.

Haugland RA, Heckman JL, Wymer LJ. 1999. Evaluation of different methods for the extraction of DNA from fungal conidia by quantitative competitive PCR analysis. Journal of Microbiological Methods 37(2):165–176.

Heederik D, Attfield M. 2000. Characterization of dust exposure for the study of chronic occupational lung disease—a comparison of different exposure assessment strategies. American Journal of Epidemiology 151(10):982–990.

Heederik D, Douwes J, Thorne PS. 2003. Biological Agents—evaluation. In: Modern Industrial Hygiene. J Perkins, ed. Cincinnati, OH: ACGIH.

Heinrich J, Hölscher B, Douwes J, Richter K, Koch A, Bischof W, Fahlbusch B, Kinne R, Wichmann HE. 2003. Reproducibility of allergen, endotoxin and fungi measurements in the indoor environment. Journal of Exposure Analysis and Environmental Epidemiology 13:152–160.

Hinds WC. 1982. Aerosol Technology. New York: John Wiley and Sons.

Hirsch T, Kuhlisch E, Soldan W, Leupold W. 1998. Variability of house dust mite allergen exposure in dwellings. Environmental Health Perspectives 106(10):659–664.

Hirvonen MR, Ruotsalainen M, Roponen M, Hyvärinen A, Husman T, Kosma V-M, Komulainen H, Savolainen K, Nevalainen A. 1999. Nitric oxide and proinflammatory cytokines in nasal lavage fluid associated with symptoms and exposure to moldy building microbes. American Journal of Respiratory and Critical Care Medicine 160:1943–1946.

Hu F, Persky V, Flay B, Phil D, Richardson PH. 1997. An epidemiological study of asthma prevalence and related factors amoung young adults. Journal of Asthma 34(1):67–76.

Hunter CA, Grant C, Flannigan B, Bravery AF. 1988. Mould in buildings: the air spora of domestic dwellings. International Biodeterioration & Biodegradation 24:81–101.

Hyvärinen A, Reponen T, Husman T, Nevalainen A. 2001. Comparison of indoor air quality in mold problem and reference buildings in subarctic climate. Central European Journal of Public Health 9(3):133–139.

ICRP (International Commission on Radiological Protection). 1994. ICRP 66: Human Respiratory Tract Model for Radiological Protection. Annals of the ICRP 24(1–3).

Immonen J, Laitinen S, Taskinen T, Pekkanen J, Nevalainen A, Korppi M. 2002. Mould-specific immunoglobulin G antibodies in students from moisture and mould-damaged schools: a 3-year follow-up study. Pediatric Allergy and Immunology 13:125–128.

IOM (Institute of Medicine). 1993. Indoor Allergens: Assessing and Controlling Adverse Health Effects, Washington, DC: National Academy Press.

IOM. 2000. Clearing the Air: Asthma and Indoor Air Exposures. Washington, DC: National Academy Press.

Jaakkola MS, Nordman H, Piipari R, Uitti J, Laitinen J, Karjalainen A, Hahtola P, Jaakkola JJ. 2002. Indoor dampness and molds and development of adult-onset asthma: a population-based incident case-control study. Environmental Health Perspectives 110(5): 543–547.

Jaffal AA, Banat IM, El Mogleth AA, Nsanze H, Bener A, Ameen AS. 1997. Residential indoor airborne microbial populations in the United Arab Emirates. Environment International 23(4):529–533.

Janssen NAH, Hoek G, Harssema G, Brunekreef B. 1999. Personal exposure to fine particles in children correlates closely with ambient fine particles. Archives of Environmental Health 54:95–101.

Janssen NAH, de Hartog JJ, Hoek G, Brunekreef B, Lanki T, Timonen KL, Pekkanen J. 2000. Personal exposure to fine particulate matter in elderly subjects: relation between personal, indoor, and outdoor concentrations. Journal of the Air & Waste Management Association 50:1133–1143.

Jedrychowski W, Flak E. 1998. Separate and combined effects of the outdoor and indoor air quality on chronic respiratory symptoms adjusted for allergy among preadolescent children. International Journal of Occupational Medicine and Environmental Health 11(1):19–35.

Karlsson K, Malmberg P. 1989. Characterization of exposure to molds and actinomycetes in agricultural dusts by scanning electron microscopy, fluorescence microscopy and the culture method. Scandinavian Journal of Work, Environment, and Health 15:353–359.

Khan AA, Cerniglia CE. 1994. Detection of *Pseudomonas aeruginosa* from clinical and environmental samples by amplification of the exotoxin A gene using PCR. Applied Environmental Microbiology 60:3739–3745.

Kilpeläinen M, Terho EO, Helenius H, Koskenvuo M. 2001. Home dampness, current allergic diseases, and respiratory infections among young adults. Thorax 56(6):462–467.

Koskinen OM, Husman TM, Meklin TM, Nevalainen AI. 1999. The relationship between moisture or mould observations in houses and the state of health of their occupants. European Respiratory Journal 14(6):1363–1367.

Lange JL, Thorne PS, Lynch NL. 1997. Application of flow cytometry and fluorescent in situ hybridization for assessment of exposures to airborne bacteria. Applied Environmental Microbiology 63:1557–1563.

Lee SC, Chang M, Chan KY. 1999. Indoor and outdoor air quality investigation at six residential buildings in Hong Kong. Environment International 25(4):489–496.

Li CS, Hsu CW, Tai ML. 1997. Indoor pollution and sick building syndrome symptoms among workers in day-care centers. Archives of Environmental Health 52(3):200–207.

Loan R, Siebers R, Fitzharris P, Crane J. 2003. House dust-mite allergen and cat allergen variability within carpeted living room floors in domestic dwellings. Indoor Air 13(3): 232–236.

Macher JM. 2001a. Review of methods to collect settled dust and isolate culturable microorganisms. Indoor Air 11(2):99–110.

Macher JM. 2001b. Evaluation of a procedure to isolate culturable microorganisms from carpet dust. Indoor Air 11(2):134–140.

Macher JM, Huang FY, Flores M. 1991. A two-year study of microbiological indoor air quality in a new apartment. Archives of Environmental Health 46(1):25–29.

Macneil L, Kauri T, Robertson W. 1995. Molecular techniques and their potential application in monitoring the microbiological quality of indoor air. Canadian Journal of Microbiology 41(8):657–665.

Michel O, Ginanni R, Duchateau J, Vertongen F, Le Bon B, Sergysels R. 1991. Domestic endotoxin exposure and clinical severity of asthma. Clinical and Experimental Allergy 21:441–448.

Michel O, Kips J, Duchateau J, Vertongen F, Robert L, Collet H, Pauwels R, Sergysels R. 1996. Severity of asthma is related to endotoxin in house dust. American Journal of Respiratory and Critical Care Medicine 154:1641–1646.

Miller JD, Young JC. 1997. The use of ergosterol to measure exposure to fungal propagules in indoor air. American Industrial Hygiene Association Journal 58:39–43.

Milton DK, Alwis KU, Fisette L, Muilenberg M. 2001. Enzyme linked immunosorbent assay specific for (1→6) branched, (1→3)-β-D-glucan detection in environmental samples. Applied and Environmental Microbiology 67(12):5420–5424.

Miraglia M, Brera C, Colatosti M. 1996. Application of biomarkers to assessment of risk to human health from exposure to mycotoxins. Microchemical Journal 54(4):472–477.

Mohamed N, Ng'ang'a L, Odhiambo J, Nyamwaya J, Menzies R. 1995. Home environment and asthma in Kenyan schoolchildren: a case-control study. Thorax 50(1):74–78.

Nafstad P, Øie L, Mehl R, Gaarder P, Lodrup-Carlsen K, Botten G, Magnus P, Jaakkola J. 1998. Residential dampness problems and symptoms and signs of bronchial obstruction in young Norwegian children. American Journal of Respiratory and Critical Care Medicine 157:410–414.

Nevalainen A. 1989. Bacterial aerosols in indoor air. (doctoral dissertation). Publications of the National Public Health Institute A3/1989, Kuopio, Finland.

Nevalainen A, Pasanen A-L, Niininen M, Reponen T, Kalliokoski P. 1991. The indoor air quality in Finnish homes with mold problems. Environment International 17:299–302.

NRC (National Research Council). 1991. Human Exposure Assessment for Airborne Pollutants. Washington, DC: National Academy Press.

Özkaynak H, Xue J, Spengler J, Wallace L, Pellizzari E, Jenkins P. 1996. Personal exposure to airborne particles and metals: results from the Particle TEAM Study in Riverside, California. Journal of Exposure Analysis and Environmental Epidemiology 6(1):57–78.

Parat S, Perdrix A, Fricker-Hidalgo H, Saude I. 1997. Multivariate analysis comparing microbial air content of an air-conditioned building and a naturally ventilated building over one year. Atmospheric Environment 31:441–449.

Park JH, Spiegelman DL, Burge HA, Gold DR, Chew GL, Milton DK. 2000. Longitudinal study of dust and airborne endotoxin in the home. Environmental Health Perspectives 108:1023–1028.

Park JH, Gold DR, Spiegelman DL, Burge HA, Milton DK. 2001. House dust endotoxin and wheeze in the first year of life. American Journal of Respiratory and Critical Care Medicine 163(2):322–328.

Pastuszka JS, Paw UKT, Lis DO, Wlazło A, Ulfig K. 2000. Bacterial and fungal aerosol in indoor environment in Upper Silesia, Poland. Atmospheric Environment 34(22):3833–3842.

Pena J, Ricke SC, Shermer CL, Gibbs T, Pillai SD. 1999. A gene amplification-hybridization sensor based methodology to rapidly screen aerosol samples for specific bacterial gene sequences. Journal of Environmental Science and Health Part A—Toxic/Hazardous Substances & Environmental Engineering 34(3):529–556.

Pirhonen I, Nevalainen A, Husman T, Pekkanen J. 1996. Home dampness, moulds and their influence on respiratory infections and symptoms in adults in Finland. European Respiratory Journal 9(12):2618–2622.

Platt SD, Martin CJ, Hunt SM, Lewis CW. 1989. Damp housing, mould growth, and symptomatic health state. British Medical Journal 298(6689):1673–1678.

Platts-Mills TAE, Chapman MD. 1987. Dust mites: immunology, allergic disease, and environmental control. Journal of Allergy and Clinical Immunology 80:755–775.

Platts-Mills TAE, de Weck AL. 1989. Dust mite allergens and asthma—a worldwide problem. Journal of Allergy and Clinical Immunology 83:416–427.

Price JA, Pollock I, Little SA, Longbottom JL, Warner JO. 1990. Measurement of airborne mite antigen in homes of asthmatic children. Lancet 336:895–897.

Reynolds S, Thorne P, Donham K, Croteau EA, Kelly KM, Lewis D, Whitmer M, Heederik D, Douwes J, Connaughton I, Koch S, Malmberg P, Larsson BM, Milton DK. 2002. Interlaboratory comparison of endotoxin assays using agricultural dusts. American Industrial Hygiene Association Journal 63:430–438.

Rizzo MC, Naspitz CK, Fernandez-Caldas E, Lockey RF, Mimica I, Sole D. 1997. Endotoxin exposure and symptoms in asthmatic children. Pediatric Allergy and Immunology 8(3):121–126.

Roponen M, Kiviranta H, Seuri M, Tukiainen H, Myllykangas-Luosujärvi R, Hirvonen MR. 2001. Inflammatory mediators in nasal lavage, induced sputum and serum of employees with rheumatic and respiratory disorders. European Respiratory Journal 18:542–548.

Roponen M, Toivola M, Alm S, Nevalainen A, Jussila J, Hirvonen MR. 2003. Inflammatory and cytotoxic potential in the airborne particle material assessed by nasal lavage and cell exposure methods. Inhalation Toxicology 15(1):23–38.

Ross MA, Curtis L, Scheff PA, Hryhorczuk DO, Ramakrishnan V, Wadden RA, Persky VW. 2000. Association of asthma symptoms and severity with indoor bioaerosols. Allergy 55:705–711.

Rylander R. 1999. Indoor air-related effects and airborne $(1{\rightarrow}3)$-β-D-glucan. Environmental Health Perspectives 107(Supplement 3):501–503.

Rylander R, Persson K, Goto H, Yuasa K, Tanaka S. 1992. Airborne β,1-3-glucan may be related to symptoms in sick buildings. Indoor Environment 1:263–267.

Sabbioni G, Wild CP. 1991. Identification of an aflatoxin G_1-serum albumin adduct and its relevance to the measurement of human exposure to aflatoxins. Carcinogenesis 12(1):97–103.

Sexton K, Ryan PB. 1988. Assessment of Human Exposure to Air Pollution: Methods, Measurements, and Models. In: Air Pollution, the Automobile, and Public Health. AY Watson, RR Bates, D Kennedy, Eds. Sponsored by the Health Effects Institute, Cambridge, MA. Washington, DC: National Academy Press.

Sonesson A, Larsson L, Fox A, Westerdahl G, Odham G. 1988. Determination of environmental levels of peptidoglycan and lipopolysaccharide using gas chromatography-mass spectrometry utilizing bacterial amino acids and hydroxy fatty acids as biomarkers. Journal of Chromatography 431(1):1–15.

Sonesson A, Larsson L, Schütz A, Hagmar L, Hallberg T. 1990. Comparison of the Limulus Amebocyte Lysate Test and gas chromatography-mass spectrometry for measuring lipopolysaccharides (endotoxins) in airborne dust from poultry-processing industries. Applied Environmental Microbiology 56:1271–1278.

Strachan DP, Carey I. 1995. Home environment and severe asthma in adolescence: a population based case-control study. British Medical Journal 311:1053–1060.

Strachan DP, Flannigan B, McCabe EM, McGarry F. 1990. Quantification of airborne moulds in the homes of children with and without wheeze. Thorax 45(5):382–387.

Szponar B, Larsson L. 2000. Determination of microbial colonisation in water-damaged buildings using chemical marker analysis by gas chromatography-mass spectrometry. Indoor Air 10:13–18.

Szponar B, Larsson L. 2001. Use of mass spectrometry for characterising microbial communities in bioaerosols. Annals of Agricultural and Environmental Medicine 8(2):111–117.

Tariq S, Matthews SM, Stevens M, Hakim EA. 1996. Sensitization to *Alternaria* and *Cladosporium* by the age of 4 years. Clinical and Experimental Allergy 26:794–798.

Taskinen TM, Laitinen S, Nevalainen A, Vepsäläinen A, Meklin T, Reiman M, Korppi M, Husman T. 2002. Immunoglobulin G antibodies to moulds in school-children from moisture problem schools. Allergy 57:9–16.

Thatcher TL, Layton DW. 1995. Deposition, resuspension, and penetration of particles within a residence. Atmospheric Environment 29(13):1487–1497.

Thatcher, TL, McKone TE, Fisk WJ, Sohn MD, Delp WW, Riley WJ, Sextro RG. 2001. Factors affecting the concentration of outdoor particles indoors (COPI): Identification of data needs and existing data. LBNL-49321. Berkeley, CA: Lawrence Berkeley National Laboratory Report.

Thorne PS, Lange JL, Bloebaum PD, Kullman GJ. 1994. Bioaerosol sampling in field studies: can samples be express mailed? American Industrial Hygiene Association Journal 55:1072–1079.

Thorne PS, Reynolds SJ, Milton DK, Bloebaum PD, Zhang X, Whitten P, Burmeister LF. 1997. Field evaluation of endotoxin air sampling assay methods. American Industrial Hygiene Association Journal 58:792–799.

Van Emon JM, Reed AW, Yike I, Vesper SJ. 2003. Measurement of Stachylysin™ in serum to quantify human exposures to the indoor mold *Stachybotrys chartarum*. Journal of Occupational and Environmental Medicine 45(6):582–591.

Verhoeff AP, van Reenen-Hoekstra ES, Samson RA, van Strien RT, Brunekreef B, van Wijnen JH. 1994. Fungal propagules in house dust. I. Comparison of analytic methods and their value as estimators of potential exposure. Allergy 49:533–539.

Verhoeff AP, van Strien RT, van Wijnen JH, Brunekreef B. 1995. Damp housing and childhood respiratory symptoms: the role of sensitization to dust mites and molds. American Journal of Epidemiology 141:103–110.

Waegemaekers M, van Wageningen N, Brunekreef B, Boleij JS. 1989. Respiratory symptoms in damp homes. A pilot study. Allergy 44(3):192–198.

Wever-Hess J, Kouwenberg JM, Duiverman EJ, Hermans J, Wever AM. 2000. Risk factors for exacerbations and hospital admissions in asthma of early childhood. Pediatric Pulmonology 29(4):250–256.

Willeke K, Lin X, Grinshpun SA. 1998. Improved aerosol collection by combined impaction and centrifugal motion. Aerosol Science and Technology 29(5):439–456.

Williamson IJ, Martin C, McGill G, Monie RDH, Fennery AG. 1997. Damp housing and asthma: a case-control study. Thorax 52(3):229–234.

Wood RA, Eggleston PA, Lind P, Ingemann L, Schwartz B, Graveson S, Terry D, Wheeler B, Adkinson NF Jr. 1988. Antigenic analysis of household dust samples. The American Review of Respiratory Disease 137(2):358–363.

Yang CY, Tien YC, Hsieh HJ, Kao WY, Lin MC. 1998. Indoor environmental risk factors and childhood asthma: a case-control study in a subtropical area. Pediatric Pulmonology 26(2):120–124.

Zock JP, Jarvis D, Luczynska C, Sunyer J, Burney P. 2002. Housing characteristics, reported mold exposure, and asthma in the European Community Respiratory Health Survey. Journal of Allergy and Clinical Immunology 110(2):285–292.

4

Toxic Effects of Fungi and Bacteria

Although a great deal of attention has focused on the effects of bacteria and fungi mediated by allergic responses, these microorganisms also cause nonallergic responses. Studies of health effects associated with exposure to bacteria and fungi show that respiratory and other effects that resemble allergic responses occur in nonatopic persons. In addition, outcomes not generally associated with an allergic response—including nervous-system effects, suppression of the immune response, hemorrhage in the mucous membranes of the intestinal and respiratory tracts, rheumatoid disease, and loss of appetite—have been reported in people who work or live in buildings that have microbial growth. This chapter discusses the available experimental data on those nonallergic biologic effects. It first discusses the bioavailability of the toxic components of fungi and bacteria and the routes of exposure to them and then summarizes the results of research on various toxic effects—respiratory, immunotoxic, neurotoxic, sensory, dermal, and carcinogenic—seen in studies of microbial contaminants found indoors. It does not address possible toxic effects of nonmicrobial chemicals released under damp conditions by building components, furniture, and other items in buildings; chemical releases from such materials are discussed in Chapter 2. Except for a few studies on cancer, toxicologic studies of mycotoxins are acute or short-term studies that use high exposure concentrations to reveal immediate effects in small populations of animals. Chronic studies that use lower exposure concentrations and approximate human exposure more closely have not been done except for a small number of cancer studies.

Chapter 5 discusses human health effects and includes some case reports relevant to toxic end points.

CONSIDERATIONS IN EVALUATING THE EVIDENCE

Most of the information reviewed in this chapter is derived from studies in vitro (that is, studies in an artificial environment, such as a test tube or a culture medium) or animal studies. In vitro studies, as explained below, are not suitable for human risk assessment. Risk can be extrapolated from animal studies to human health effects only if chronic animal exposures have produced sufficient information to establish no-observed-adverse-effect levels (NOAELs) and lowest-observed-adverse-effect levels (LOAELs). Extrapolation of risk exposure from animal experiments must always take into account species differences between animals and humans, sensitivities of vulnerable human populations, and gaps in animal data. Risk assessment requires not only hazard identification but also dose-response evaluation and exposure assessment in humans whose risk is being evaluated. Estimates of exposures of humans to spores, bacteria, microbial fragments, and dust that contains mycotoxins are inherently imprecise and imperfect; biomarkers of exposure to toxins are few, and exposures to single or multiple mycotoxins carried by such agents have not been measured indoors. Thus results of animal studies cannot be used by themselves to draw conclusions about human health effects. However, animal studies are important in identifying hazardous substances, defining their target organs or systems and their routes of exposure, and elucidating their toxicokinetics and toxicodynamics, the mechanisms that account for biologic effects, and the metabolism and excretion of toxic substances. Animal studies are also useful for generating hypotheses that can be tested through studies of human health outcomes in controlled exposures, clinical studies, or epidemiologic investigations, and they are useful for risk assessment that informs regulatory and policy decisions.

BIOAVAILABILITY AND ROUTE OF EXPOSURE

Issues That Affect Bioavailability

Some molds found in damp indoor spaces can produce mycotoxins. Table 4-1 lists a number of mycotoxins and the organisms that produce them. Bacteria can also produce toxins. Although there has not been a large amount of research conducted on the effects of those toxins in the context of their growth in damp buildings, mycotoxins and bacterial toxins have been studied for several decades because of their role in outbreaks of illness associated with the ingestion of moldy food (Etzel, 2002). More recently,

TABLE 4-1 Some Mycotoxins and the Microorganisms That Produce Them

Chemical Compound	Microorganisms That Produce Mycotoxins	References
Ergot alkaloids	*Claviceps purpurea*, species of *Aspergillus, Rhizopus, Penicillium*	Sorensen, 1993; Larsen et al., 2001
Substituted coumarins, for example, aflatoxins from *Aspergillus flavus, Aspergillus parasiticus*; ochratoxins	Several species of *Aspergillus, Penicillium*	van Walbeek et al., 1969; Sorensen, 1993
Quinones, for example, citrinins	Several species of *Aspergillus, Penicillium*	Sorenson, 1993; Malmstrom et al., 2000
Anthoquinones, for example, rugulosin	*Penicillium islandicum*	
Trichothecenes (sesquiterpenes with trichothecene skeleton, olefinic group at C-9, 10, epoxy at C-12, 13), for example, T-2 toxin; DON (deoxynivalenol or vomitoxin)	*Fusarium* and *Stachybotrys* species	Etzel, 2002
Macrocyclic trichothecenes (having carbon chain between C-4 and 15 in ester or ether linkage, for example, satratoxins G, H; verrucarins B, J; trichoverrins A, B)	*Stachybotrys, Myrothecium*, others	Sorensen, 1993; Jarvis, 1991
Substituted furans, for example, citreoviridin	*Penicillium citreoviride*	Nishie et al., 1988
Epipolythiodioxopiperazines, for example, gliotoxin	At least six species of *Aspergillus, Penicillium*	Waring and Beaver, 1996
Lactones, lactams, for example, patulin, stachybotrylactones, stachybotrylactams	*Penicillium, Stachybotrys*	Jarvis et al., 1995, 1998
Estrogenic compounds, for example, zearalenone	Many species of *Fusarium*	Betina, 1989; Kuiper-Goodman et al., 1987

concerns that toxins from microorganisms that grow in damp indoor environments may play a role in illnesses reportedly associated with living or working in damp buildings have focused attention on the adverse health effects of inhaling mycotoxins.

The degree to which a toxin can harm tissues varies with a number of factors, including the chemical nature of the toxin, the route of entry into the body, the amount to which the target organism and organ are exposed, and the susceptibility of the target species (Coulombe, 1993; Eaton and Klaassen, 2001; Filtenborg et al., 1983; Vesper and Vesper, 2002). Interspecies differences in susceptibility can result from differences in absorption, distribution, metabolism, excretion, and the effectiveness of a toxin at its receptor (site of action) (Eaton and Klaassen, 2001; Fink-Gremmels, 1999; Russell, 1996).

Once produced, mycotoxins must be airborne to be inhaled. Mycotoxins are found in and on the spores of molds that produce them, on hyphal fragments, and in dust from substrates on which mold grows and carpet dust (Englehart et al., 2002; Górny et al., 2002; Larsen and Frisvad, 1994; Sorenson, 1993, 1995; Sorenson et al., 1987). They are exuded into the substrate on which a microbial agent is growing, for instance, growth medium in the laboratory and gypsum board, wood, paper, and other building materials in damp or wet buildings (Andersen et al., 2002; Andersson et al., 1997; Buttner et al., 2001; Gravesen and Nielsen, 1999; Nieminen et al., 2002). Mycotoxins have also been isolated from dust sampled in moldy buildings that did not contain mold spores (Englehart et al., 2002; Gravesen et al., 1999; Nielsen et al., 1998). Mycotoxins are found in and on materials that can be aerosolized as particles, so such aerosols can become a source of mycotoxin exposure. But particles are not the only vehicle of exposure to mycotoxins. Mycotoxins are not generally thought to be volatile (Jarvis et al., 1995), but some, such as sesquiterpenes, are semivolatile, and others are at least partially water-soluble and thus able to enter the air in droplet aerosols (Harrach et al., 1982; Peltola et al., 1999, 2002).

The bioavailability of aerosols (including mold spores, contaminated dust, bacteria, and microbial fragments) in the respiratory tract after inhalation depends in part on the size of the particles formed, because their size determines where they are deposited in the respiratory tract and this determines bioavailability. Figure 4-1 shows the relationship between spore diameter and respiratory deposition of a number of mold genera. Figure 4-2 shows the percentage of inhaled spores that are deposited in the respirable (alveolar) area of the lung. The size of mold spores depends on the species that produce them. Spores of genera that use the air pathway for dispersion, including *Aspergillus* and *Penicillium,* are in the range of 1–2 μm and are respirable. Some molds (such as *Stachybotrys chartarum* and *Memnoniella echinata*) that do not spread their spores through aerosol dispersion are wet

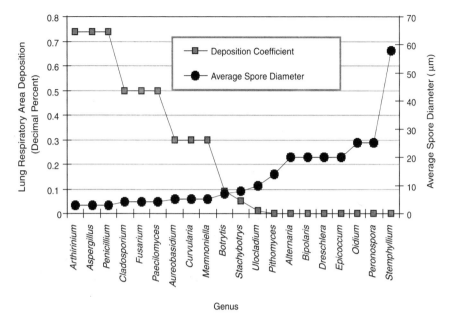

FIGURE 4-1   Spore-deposition coefficients of mold genera in indoor environments.
SOURCE: Miller et al., 2001.

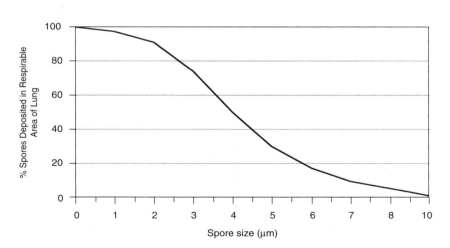

FIGURE 4-2   Percentage of inhaled spores that are deposited in respirable (alveolar) area of lung.
SOURCE: Miller et al., 2001.

and slimy during sporulation; once dry, the spores can be dispersed into air through disturbance of contaminated surfaces and are of inhalable size (5–7 μm) (Sorenson et al., 1987). Such bacteria as *Streptomyces californicus* isolated from damp indoor spaces are about 1 μm in diameter and can reach the lower airways and alveoli when inhaled (Jussila et al., 2001). Furthermore, Wainman and colleagues (2000) have shown that semivolatile chemicals, such as terpenes and limonene (which can be produced by molds that also produce trichothecene mycotoxins and are also often used indoors as cleaning solvents) react with ozone in indoor air and form particles of a respirable size, 0.2–0.3 μm diameter.

Apart from particle size, determining the bioavailability of mycotoxins found on or in particles is complicated because even toxins from spores that lodge in the nasal mucous membranes can damage cells locally or be absorbed into the systemic circulation (Morgan et al., 1993). Lipid-soluble toxins pass readily through membranes, and the degree of their absorption depends on the blood supply to the tissue (Rozman and Klaassen, 1996).

The bioavailability of mycotoxins and bacterial toxins also depends on residence time and clearance mechanisms. Many mycotoxins affect residence time and clearance by inhibiting phagocytic activity of macrophages or reducing ciliary beat rate (Amitani et al., 1995; Coulombe et al., 1991; Jakab et al., 1994; Sorenson and Simpson, 1986; Sorenson et al., 1986; Wilson et al., 1990). The toxic effect of spores and other particles on alveolar macrophages can impair the ability of these cells to protect against not only mycotoxins but also other bacteria and infectious particles. Slowed ciliary clearance allows longer residence time in the airway and increases the time for absorption of toxins from mold spores, fragments, or dust (Coulombe et al., 1991).

Because the respiratory system is the primary route of entry for gases and particles suspended in air, determination of exposure to air contaminants is complicated because air contains a mixture of substances and the concentration of individual toxicants changes with time and location in the exposure mixture. That is particularly true for toxic compounds originating in microbial contaminants of indoor spaces, because growth and metabolism of microbial organisms introduce additional variables into the exposure paradigm. Difficulties in measuring microorganisms and their products hinder the accurate determination of human exposure to them. Chapter 3 discusses the methods used and the difficulties in measuring such exposures.

Experimental Data

Because inhalation appears to be an important route of exposure for humans, determining the bioavailability of mycotoxins after inhalation exposure is important for determining the relationship between damp indoor spaces and human health. However, compared with ingestion, relatively few animal experiments have been performed with the inhalation pathway. Some of the studies that have been conducted indicate that acute inhalation exposure, at least of some toxicants, is at least as toxic as exposure by intravenous injection and is more toxic than ingestion or parenteral exposure.

Ueno (1984a) found that inhalation, skin, and parenteral exposure of newborn, young, and older mice to T-2 toxin—a trichothecene mycotoxin produced by *Fusarium* species whose LD_{50} (the lowest dose that kills half the animals that receive it) does not vary much across animal species (Ueno, 1980)—directly affected capillaries, increasing their permeability and leading to intestinal bleeding, diarrhea, and death. However, some ingestion of the toxin might have occurred because of grooming behavior of the animals after it was deposited on their skin. The authors also noted that newborn and young animals were much more susceptible to the mycotoxin than the older mice.

Marrs et al. (1986), using head-only exposures, compared the acute inhalation toxicity of T-2 toxin in guinea pigs, which tend to be sensitive to respiratory irritants, with effects of subcutaneous administration. Respiratory rate and minute volume were measured with whole-body plethysmography. The inhaled dose was estimated by using the concentration of T-2-fluorescein-complexed aerosol collected on a filter at 1.0 L/min. The lethal concentration (LCt_{50}), the air concentration lethal to 50% of the exposed group of animals, was determined by using a range of concentrations and exposure durations. The corresponding dose at which 50% of the exposed group dies (LD_{50}) was estimated from the LCt_{50}. Another group of animals received subcutaneous injections of doses of T-2 toxin ranging from 0.5 to 4.0 mg/kg. The LD_{50} estimated from the inhalation exposure was about twice that of the subcutaneous LD_{50} values, but the authors noted that only about half the inhalation dose was retained. Taking the low retention into account, the lethal dose after inhalation was similar to that by subcutaneous injection. The types of effects on the gastrointestinal tract were similar for the two routes of exposure and are thought to be mediated systemically. The estimated LD_{50}s were also similar to those seen by DeNicola et al. (1978) after oral dosing, and similar gastrointestinal effects have been seen in other oral-exposure studies and appear to be largely independent of route of administration (Ueno, 1984a,b).

Creasia et al. (1987) exposed young adult and mature mice to T-2 toxin by inhalation for 10 min. Tremors, stilted gait, and, in some animals, prostration were observed. Animals in the highest-dose group died 5 h after exposure. After 24 h, the LCt_{50}s were 0.08 ± 0.04 and 0.325 ± 0.010 mg/L of air for young and mature mice, respectively. The corresponding LD_{50}s were 0.24 and 0.94 mg/kg. When those results are compared with results of other studies, inhalation exposure was about 5–10 times more potent than intraperitoneal administration, which had a reported LD_{50} of about 4.5 mg/kg (Bamburg, 1976; Creasia et al., 1987), and at least 10 times more potent than dermal application, which had a reported LD_{50} of at least 10 mg/kg (Schiefer and Hancock, 1984).

Creasia et al. (1990) conducted a nose-only, acute inhalation study of the effects of a 10-min exposure to T-2 toxin in rats and guinea pigs. Respiratory-tract lesions were minimal, and lesions to organs were similar to those described after following systemic administration. LCt_{50}s were 0.02 and 0.21 mg/L of air for rats and guinea pigs, respectively. Deposition dose was measured by extraction of toxin from sacrificed animals, and LD_{50} of 0.05 and 0.4 mg/kg, respectively, were estimated. In that study, inhalation exposure to T-2 was about 20 times as toxic in rats and twice as toxic in guinea pigs as in studies of intraperitoneally administered T-2 toxin (LD_{50}, 1 mg/kg in the rat; and 1–2 mg/kg in the guinea pig.

Coulombe et al. (1991) administered ^{3}H-labeled aflatoxin B_1 (AFB_1) adsorbed to grain dust or in its crystalline form intratracheally to male rats and sampled blood and tissue at selected intervals for 3 weeks to determine the pharmacokinetics of this toxin. After absorption, distribution followed a two-compartment model, with an initial rapid-distribution phase followed by a slower phase. The rate of absorption from the dust-associated dose was much lower for the first 90 min and the time to peak plasma concentration was much longer (12 vs 2 h) than for the crystalline form. Clearance was identical in the two groups. At 3 h, there was a substantially greater amount of AFB_1-DNA adducts in the trachea and lung of the dust group. Retention of dust-associated carcinogens in the lung is an important factor in pulmonary carcinogenesis; it presumably increases the time during which metabolically active cells of the respiratory epithelium capable of transforming procarcinogens to carcinogens are in contact with the carcinogen. In the liver, however, the DNA binding was greater for the crystalline group at 3 h and at 3 days. Zarba et al. (1992) found that nose-only inhalation exposure of rats to aerosolized grain dust that contained AFB_1 resulted in a linear dose-response relationship (correlation coefficient, 0.96) between time of exposure and AFB_1-DNA adducts for 20, 40, 60, and 120 min of exposure. Adduct formation in the lung was not determined.

Dermal absorption of mycotoxins varies. When toxins in excised human skin were tested, the relative penetration rate of toxins dissolved in methanol was T-2 > diacetoxyscirpenol (DAS) > satratoxin H (a trichothecene mycotoxin found in *Stachybotrys*) > AFB_1 (Kemppainen et al., 1988). Systemic toxicity after dermal exposure to a mycotoxin depends on its rate of absorption, relative blood flow to skin area, and the potency of the compound and its metabolites. Of the trichothecenes studied in vivo, relative local and systemic toxicity measured by skin irritation and lethality, respectively, is T-2 > DAS ≈ verrucarin. In vitro studies are fairly consistent with in vivo penetration studies, although the potency of both T-2 and verrucarin is greater than that of DAS (Kemppainen et al., 1988). Kemppainen and colleagues (1988) showed that both aflatoxins and trichothecenes can be absorbed through the skin. Dermal absorption is slow, but increases with the concentration of toxin, with coexposure to solvents (such as DMSO) that enhance penetration, and when the application site is occluded with clothing or wraps. Joffe and Ungar (1969) showed that aflatoxins applied to the skin of rabbits penetrated the stratum corneum and caused changes in the epidermis and dermis. Experiments in newborn, young, and adult mice (Ueno, 1984a,b) and in vitro experiments (Kemppainen et al., 1988) have demonstrated skin penetration of trichothecenes. Kemppainen et al. (1984) showed with ^{3}H-T-2 toxin that T-2 toxin adsorbed onto corn dust can partition and penetrate excised human and guinea pig skin; this indicates that mycotoxin on dust is available for absorption via skin. Those studies indicate that toxins found in damp indoor spaces are bioavailable to people through inhalation and dermal exposure, with the more potent route of exposure depending on the compound. The extent of exposure that occurs in damp indoor spaces, however, has not been studied.

TOXIC EFFECTS OF INDOOR MOLDS AND BACTERIA

Exposure to various mold products—including volatile and semivolatile organic compounds and mycotoxins—and components of and substances produced by bacteria that grow in damp environments has been implicated in a variety of biologic and health effects. This section discusses irritation and inflammation of mucous membranes, respiratory effects, immunotoxicity, neurotoxicity, sensory irritation (irritation of nerve endings of the common chemical sense), dermotoxicity, and carcinogenic effects attributed to such exposure.

Mucous Membrane Irritation and Inflammation

Exposure to microorganisms and their products can irritate mucous membranes, such as those of the eyes and respiratory tract, and lead to inflammation via an immune response. Such immune responses are important in normal host defenses, but chronic or excessive release of inflammatory mediators can cause damage to the lung and other adverse effects (Jussila et al., 2003).

Immune responses triggered by exposure to microorganisms and their products include increased production of inflammatory mediators, such as cytokines (for example, tumor-necrosis factor α [TNFα] and interleukin-6 [IL-6]), reactive oxygen species, and, indirectly, nitric oxide (NO) via the induction of nitric oxide synthetase (iNOS) (Hirvonen et al., 1997a,b; Huttunen et al., 2003; Ruotsolainen et al., 1995). Different bacteria evoke different cellular responses. For example, *Staphylococcus* evokes a response from alveolar macrophages, and *Pseudomonas* evokes a neutrophil response (Rehm et al., 1980). Mold spores and fragments affect the inflammatory response differently (Hirvonen et al., 1999). A number of in vitro, animal, and human studies that have investigated the irritation and inflammation responses to exposure to microorganisms and molds commonly found in damp indoor spaces are discussed below.

In Vitro Experiments

In vitro experiments use animal or human cell lines or primary cell cultures to explore mechanisms of toxicity for specific target tissues or cells. Although toxic exposure of cells and tissues in vitro does not provide information about homeostasis or defenses involved in the responses of an intact animal to exposure by various routes, such studies can avoid some uncertainties of extrapolation from animal to human models, can provide specific, repeatable, precise measures of target-cell effects, and can help to determine their mechanisms (Pitt, 2000).

Hirvonen et al. (1997a) tested the ability of *Streptomyces annulatus* and *S. californicus*—both gram-positive bacteria—and the fungi *Candida*, *Aspergillus*, *Cladosporium*, and *Stachybotrys* to activate the mouse macrophage cell line RAW264.7. All the microorganisms were isolated from moldy houses, and no endotoxin contamination was detected in the cell suspensions. Both bacterial species substantially induced the iNOS enzyme and increased NO, TNFα, and IL-6 production in a dose-dependent manner within 24 h. Only *Stachybotrys* affected cell viability.

Hirvonen et al. (1997b) compared the effect of *Streptomyces* species on macrophages with the macrophage response produced by the gram-positive *Bacillus* sp. and *Micrococcus luteus*, which are common airborne bacteria in normal houses, and the gram-negative bacterium *Pseudomonas fluorescens*, a known activator of macrophages. All *Streptomyces* species tested were able to induce substantial amounts of TNFα and IL-6 and to induce the expression of iNOS and later NO; *Bacillus* sp. and *Micrococcus luteus*, commonly found in houses without dampness problems, did not. None of the bacteria affected cell viability, but endotoxin LPS and *Pseudomonas fluorescens* substantially reduced cell viability within 4 h. For *Streptomyces*, some factor other than NO production seemed to be required to initiate apoptosis, but the induction of proinflammatory mediators may play a role in inflammation related to exposure.

Huttenen et al. (2000) studied inflammatory responses of RAW 264.7

macrophages to three mycobacteria isolated from a moldy building: non-pathogenic *Mycobacterium terrae* and potentially pathogenic *M. avium*-complex and *M. scrofulaceum*. All the bacterial species tested induced time- and dose-dependent production of NO, IL-6, and TNFα, but IL-1 and IL-10 production was not detected. Reactive oxygen species (ROSs) were increased at the highest doses. The level of response differed widely across species. The nonpathogenic *M. terrae* was the most potent inducer, and *M. avium*-complex was the least potent; both pathogenic and nonpathogenic bacteria apparently activate inflammatory processes.

Hirvonen et al. (2001) exposed RAW 264.7 macrophages to *Streptomyces annulatus* spores isolated from a moldy building and then grown on 15 growth media to determine whether growth conditions affected a microorganism's ability to induce inflammatory mediators. After 24 h, bacteria from all growth media induced iNOS in macrophages to some extent; the amount of NO produced ranged from 4.2 to 39.2 μM, depending on the growth medium. ROSs were induced only by the highest dose of *S. annulatus* grown on glycerol-arginine agar. Cytokine production (IL-6 and TNFα) depended on the growth medium. Viability of the RAW 264.7 macrophages varied widely (from 11% to 96%), depending on the growth medium on which the *S. annulatus* was grown.

Murtoniemi et al. (2002) tested the effects of three molds (*Stachybotrys chartarum*, *Aspergillus versicolor*, and *Penicillium spinulosum*) and one gram-positive bacterium (*Streptomyces californicus*) isolated from water-damaged buildings and then grown on different wetted plasterboard cores and liners. Both liners and cores of plasterboard supported microbial growth; all species grew earlier on the core than on the liner material. *Penicillium* grew only on the plasterboard cores. *Aspergillus* and *Streptomyces* grown on those building materials were the most potent of the microorganisms in inducing the production of NO and IL-6 in RAW 264.7 macrophages; *Stachybotrys* spores did not induce NO nor IL-6 but did induce abundant TNFα production. *Aspergillus* also produced high concentrations of TNFα, and both *Aspergillus* and *Stachybotrys* were potently cytotoxic.

Nielsen et al. (2001) examined the cytotoxicity of 20 *Stachybotrys* isolates from water-damaged buildings and their ability to induce inflammatory mediators in RAW 264.7 macrophages. Eleven of the isolates produced satratoxin and were highly cytotoxic to macrophages. Isolates that produced atranone were not cytotoxic but induced inflammatory mediators (ROS, NO, IL-6, and TNFα at doses of 10^6 spores/mL). Pure atranone B and atranone D did not elicit such a response. It should be noted that 30–40% of *Stachybotrys* strains isolated from buildings produce satratoxin (Jarvis et al., 1998).

Animal Experiments

Jussila et al. (2001) compared the mouse inflammatory response to a single intratracheal instillation of one of three doses of *Streptomyces californicus* spore isolates from the indoor air of moldy buildings with the response to 50 µg of lipopolysaccharide (LPS). Effects were assessed daily for 7 days after dosing. Cytokine concentrations were measured in the blood and bronchoalveolar lavage fluid (BALF). Histologic tests were conducted on two mice from each exposure group. *S. californicus* spores induced acute inflammation in mouse lungs, measured in BALF and histologically. The inflammation was still detectable 7 days after exposure. The pattern of cytokine production and the histologic effects were distinguishable from those caused by LPS. Production of the proinflammatory cytokines IL-6 and TNFα was increased in a dose-dependent manner.

Jussila and colleagues have also studied inflammatory and toxic responses to the bacteria *S. californicus* (Jussila et al., 2001, 2003) and *Mycobacterium terrae* (Jussila et al., 2002a) and the fungi *Aspergillus versicolor* (Jussila et al., 2002b) and *Penicillium spinulosum* (Jussila et al., 2002c) after intratracheal instillation in specific-pathogen-free mice. All treatments enhanced TNFα and IL-6 production in BALF after a single dose, but there were marked differences in the time course and magnitude of the response. Details of the responses are provided in Table 4-2. Except for *M. terrae*-treated mice, TNFα concentrations were indistinguishable from those in controls by 3 days after exposure. Both bacterial species induced inflammation at their lowest dose; the fungal spores required higher doses for induction of TNFα response. All the microorganisms increased the total number of inflammatory cells in BALF. Neutrophils were the most typical cells recruited for the acute inflammatory response, and their response peaked at 24 h; the response of macrophages peaked at 3 days, and that of lymphocytes at 7 days.

Repeated dosing with *S. californicus* induced mild to abundant increases in the numbers of mononuclear cells and neutrophils in the alveoli and in the bronchiolar lumen (Jussila et al., 2003). The numbers of peribronchial cells and vascular mononucleated cells also increased. Granuloma-like lesions were seen in one of three mice. Both *M. terrae* and *S. californicus* provoked systemic effects in the lymph nodes and spleen and increased TNFα and IL-6 in the blood.

Respiratory Effects

Microorganisms and their toxins can lead to effects on the tissues and cells of the respiratory system. Some of the effects might be mediated by effects on the immune system.

TABLE 4-2 Summary of Inflammatory and Toxic Responses to Two Bacteria and Two Fungi in Mice

Microorganism	TNFα Response (Lowest Dose to Cause Increase)	IL-6 Response (Lowest Dose to Cause Increase)	iNOS-NO Response	Increased Inflammatory Cells in BALF	Albumin/-LDH Response	Reference
Streptomyces californicus	Intense and rapid (1×10^8 spores)	(2×10^7 spores)	Induced iNOS at 2×10^7 spores within 24 h	Strongest: 6-fold increase at 2×10^7 spores	Increased within 6 h, continuing acute for 24 h; increases in LDH	Jussila et al., 2001
Mycobacterium terrae	83% of that to *S. californicus*; biphasic; intense acute phase followed by sustained phase that lasted more than 14 days (1×10^8 spores)	(1×10^8 spores)	No not increased until 7 days after exposure; iNOS detectable for up to 28 days	Second strongest, but biphasic	Increased within 24 h; peaked at 14 days; acute increases in LDH; strongest and longest dose-dependent response, lasting length of experiment	Jussila et al., 2002a

(continued on next page)

TABLE 4-2 continued

Microorganism	TNFα Response (Lowest Dose to Cause Increase)	IL-6 Response (Lowest Dose to Cause Increase)	iNOS-NO Response	Increased Inflammatory Cells in BALF	Albumin/-LDH Response	Reference
Aspergillus versicolor	Rapid; peaked after 24 h at 7-fold increase (1×10^6 spores)	Massive at 1×10^8 spores (1×10^6 spores)	NA	Dose-dependent increase starting at 1×10^6 spores	(28 days) Slower response; acute increases in LDH	Jussila et al., 2002b
Penicillium spinulosum	8-fold increase within 6 h; disappeared more rapidly than other responses (5×10^6 spores)	Highest level; 20-fold increase; peaked later than response to TNFα; was at control level 24 h after exposure (5×10^6 spores)	NA	Minor cell response even at 5×10^6 spores	Mildest response; no changes in cytotoxicity	Jussila et al., 2002c

Animals and Animal Cells

Pang et al. (1987) studied the effects of a single nebulized dose of T-2 toxin at 9 mg/kg on lung tissues of young pigs (9–11 weeks old). Analyses indicated that 1.8–2.7 mg/kg was retained. Vomiting, cyanosis, anorexia, lethargy, prostration, and death occurred. Those effects are similar to those seen in pigs treated intravenously with the LD_{50} of T-2 toxin, 1.2 mg/kg (Lorenzana et al., 1985a,b). Pang et al. (1987) observed pulmonary and systemic immunologic and morphologic changes of the lung and other organs. Pigs were sacrificed 1, 3, and 7 days after dosing. Morphologic examination of the lungs showed small, dark red foci 2–3 mm in diameter scattered throughout the lobes. Hemorrhages of the gastrointestinal mucous membrane, subendocardial tissue, and subpericardial tissue were also seen. Two pigs that died after 8 h had mild to moderate patchy acute interstitial pneumonia characterized by thickening of the lung septa due to congestion and infiltration of neutrophils and macrophages. There was marked reduction in alveolar macrophage phagocytosis and mitogen-induced blastogenic responses of pulmonary, but not peripheral, lymphocytes at 8, 24, and 72 h, but not 7 days after exposure. Thus, acute exposure to T-2 toxin resulted in mild pulmonary injury and transient impairment of pulmonary immunity (Pang et al., 1987).

Nikulin et al. (1996) isolated spores from two strains of *Stachybotrys atra*, one more toxic[1] and the other less so, from houses with moisture problems. Satratoxins G and H were present in the more toxic strains, and small amounts of stachybotrylactone and stachybotrylactam were found in both strains. A suspension of 10^6 spores of each strain was injected intranasally into a group of four 5-week-old mice. One mouse exposed to the toxic strain died 10 h after dosing, one was moribund at 24 h, and the other two survived the 3-day duration of the experiment. When observed histologically, all treated mice developed inflammatory lung lesions, but the severity and extent of the lesions differed between the two groups. Spores of the more toxic strain induced severe intra-alveolar and interstitial inflammation, and hemorrhagic exudate was found in the alveolar lumina. There was focal aggregation of inflammatory cells (mainly neutrophils and macrophages), especially in the peribronchiolar area. Neutrophilic granulocytes and macrophages containing fungal spores were found in the lung parenchyma. Some lymphocytes were found in the interstitium and necrotic changes were seen. Lungs of mice exposed to the less toxic strain of *S. atra* had much milder inflammatory responses, and no necrotic changes were seen. It is interesting that in this study, in which exposure was to the spores

---

[1]Toxicity was characterized as a function of the amount of crude methanol-extracted solid needed to cause 50% inhibition of growth of feline fetal lung cells.

of *S. atra*, all exposed mice showed pathologic changes; that was not the case in studies in which exposure was to purified T-2 toxins (Creasia et al., 1987, 1990).

Nikulin et al. (1997) examined the effects of intranasal exposure of mice to *S. atra* spores (10^3 and 10^5) twice a week over a 3-week period, using strains and methods similar to those in their earlier work (Nikulin et al., 1996). Five groups of 10 mice (five males and five females) were treated. Mice were evaluated for weight change and blood characteristics over a 3-week period. The sixth and last administration was followed by blood-antibody measurements and hematologic and histologic studies of lung tissue. The severity of changes in lung tissue depended on the concentration and the toxic potency of spores. Treatment with suspensions of 10^5 spores of the more toxic strain caused severe inflammatory changes, with hemorrhagic exudates present in alveolar lumina. Treatment with 10^3 spores of the same strain produced similar but milder changes. Much milder inflammatory changes occurred in the lungs of mice treated with 10^5 spores of the less toxic strain, and no inflammatory changes were seen in mice treated with 10^3 spores of the less toxic strain. In contrast with the earlier acute-exposure testing (Nikulin et al., 1996), necrosis was not found in the lungs of any mice treated repeatedly. The authors attributed that difference to a lower concentration of satratoxins in the spores used in the multidose experiment. Antibodies against *S. atra* were not detected in mice exposed to *S. atra* spores via inhalation, but a separate group of mice intraperitoneally immunized with spores developed IgG antibodies against *S. atra,* as measured with enzyme-linked immunosorbent analysis.

Mason et al. (1998) examined the effect of *Stachybotrys chartarum* conidia and isosatratoxin-F, compared to the negative control fungus *Cladosporium cladosporioides,* on surfactant production in cultures of undifferentiated fetal type II alveolar cells from rabbit lung, and their effects following intratracheal instillation in mice. All concentrations of conidia tested (10^3, 10^4, 10^5, 10^6 conidia/mL) and isosatratoxin-F concentrations of 10^{-9}-10^{-4} M decreased surfactant production, as measured by incorporation of [^{3}H]-choline into surfactant, within 24 h. In mice treated with 50 mg of 10^7 conidia/mL of *S. chartarum* or *Cladosporium cladosporioides* or with 50 µL 10^{-7} M isosatratoxin-F, phospholipid concentrations of the four primary subfractions of surfactant (P10, P60, P100, and S100) measured with lung lavage were changed in a concentration- and time-dependent manner. P10 phospholipid, which is responsible for the surface-tension-lowering properties of the lipid monolayer of alveolar cells, was significantly increased in lungs 12 and 24 h after exposure. *S. chartarum*-treated animals had significantly decreased P60 phospholipid and significantly increased P100 phospholipid 48 h after exposure. In mice, exposure to the toxin resulted in significant increases in P10 phospholipid 12 and 24 h after exposure and in P60 phospholipid 24 and 72 h after exposure. Toxin exposure also significantly decreased S100 phospholipid 24 h and increased P100 phospholipid 72 h after exposure in mice. In *C. cladosporoides*-treated mice, P60 was decreased 24 h after exposure, but no other changes in phospholipids were seen. Thus, mouse surfactant homeostasis can be disrupted by exposure to toxic *S. chartarum* conidia and isosatratoxin-F but not by exposure to *C. cladosporoides* conidia. Such disturbances might lead to disruption of clearance mechanisms and result in increased susceptibility to inhaled infectious organisms, but the exact meaning of the disturbances has not yet been explained.

Rao et al. (2000a) instilled 9.6×10^6 *S. chartarum* spores intratracheally into rats and then performed bronchial alveolar lavage to look at biochemical indicators of injury (albumin, myeloperoxidase [MPO], lactate dehydrogenase [LDH], and hemoglobin) and leukocyte differentials. They observed severe inflammatory changes and interstitial inflammation with hemorrhagic exudate. Exposed animals lost 10% of their body weight within 24 h. The highest inflammatory-cell and polymorphonucleocyte (PMN) count occurred 24 h after exposure. Albumin and LDH concentrations were also significantly increased at that time. Hemoglobin concentrations were different from controls at 72 h. Because *S. chartarum* is not known to cause an immunoglobulin G reaction in rodents or to infect mammalian lungs, the observed injury was thought to be caused by chemical constituents of the spores rather than allergy or infection.

Rao et al. (2000b) extracted spores of a toxic strain of *S. chartarum* with methanol to reduce toxicity, instilled the spores intratracheally into 10-week-old male rats, and then analyzed their BALF for total leukocytes, differential counts of macrophages, PMNs, eosinophils, and lymphocytes 24 h after treatment. Supernatant fluid was analyzed for LDH, MPO, and albumin with spectroscopy. Results were compared with those in rats treated with unextracted spores and with saline. About 0.5 mL of a saline-suspended concentration of 2×10^6, 4×10^6, 1×10^7, and 2×10^7 spores/mL saline was instilled. Physiologic effects of acute pulmonary exposure were examined by measuring body weight, LDH, hemoglobin, blood albumin, and leukocyte, macrophage, lymphocyte, and eosinophil counts. Body weight was decreased in rats treated with non-extracted spores (up to 13% loss of body weight with no loss in saline controls) in the 24 h after exposure. Increased LDH (resulting from cytotoxicity and death), hemoglobin (resulting from erythrocyte infiltration from pulmonary capillary beds), and albumin, a possible early indicator of inflammation, were linearly dose-dependent. The same dose-dependent increases were not seen in methanol-extracted or saline controls. Leukocyte counts were increased, and PMNs were the major contributor to the increases. Total macrophages, lymphocytes, and eosinophils did not increase with increasing instillate concentrations. The authors state that 24 h might not have been enough time so see a rise in macrophages in that they saw such a rise in other rats at 72 h (unpublished data). No assessment of toxin identity or concentration was carried out.

Yike et al. (2002a) developed a model technique for studying pulmonary toxicity in infant rats. They studied the effects of S. *chartarum* spores containing mycotoxins on survival (LD_{50}), growth, lung histopathology, BALF characteristics, and pulmonary function of rat pups treated when 4 days old and observed and weighed for 14 days. Intratracheal instillation of high doses of S. *chartarum* spores—4–8×10^5 spores/g of body weight—led to macroscopically detected hemorrhage that was frequently fatal; 73% and 83% of animals treated with 4×10^5 and 8×10^5 spores/g, respectively, died; the LD_{50} was 2.7×10^5 spores/g. Conidia were present in the alveoli and distal airways of a sample of 75 treated rat pups; the number of spores was greatest 4 days after treatment, at which time they were localized in the macrophages. Acute interstitial or intra-alveolar hemorrhage was observed in 62% of treated animals vs 27% of controls, which received phosphate-buffered saline (PBS); control pups treated with ethanol-extracted spores showed only minimal incidental hemorrhage. The degree of hemorrhage was dose-dependent. Hemoglobin, an indicator of acute alveolar bleeding, was significantly higher in the BALF of treated pups (2.46 ± 0.33 mg/mL of epithelial lining fluid) than PBS pups (1.22 ± 0.17 mg/mL) and pups treated with extracted spores (1.28 ± 0.16 mg/mL). Proinflammatory cytokines and inflammatory cells in lungs were significantly higher after 3 days in pups that received 1×10^5 spores/g than in PBS and ethanol-extracted-spore controls. Those indicators of acute inflammation resolved 8 days after spore instillation. Respiration, measured with whole-body plethysmography, showed statistically significant apnea of more than 3-sec duration in treated pups (28% of exposed pups vs 0% of controls). Minute volume was increased 4 days after treatment, probably because of an increase in tidal volume. Enhanced pause (a noninvasive measure of airway resistance) was also increased. Tidal volume remained elevated at day 7, when pups developed decreased respiratory rate. The experiment of Yike et al. (2002a) differs from those of Nikulin et al. (1996), Nikulin et al. (1997), and Rao et al. (2000a,b) in that trichothecene toxicity was quantified (in toxin equivalents), but total toxin content (that is, phenylspirodrimanes, stachytoxins, cyclosporin, and unknown toxins found in some Cleveland strains) was not characterized. The LD_{50} of 2.7×10^5 spores/g (270 ng of satratoxin G per gram = 0.27 mg/kg) is similar to the values reported for T-2 toxin in other animals: 0.9 mg/kg in adult rats, intravenously; 0.8 mg/kg in monkeys, intramuscularly; 5.2 mg/kg in mice, intravenously (Wannemacher et al., 1991); and 5.2 mg/kg in mice, intraperitoneally (Ueno, 1989).

Yike et al. (2003), using the same infant rat model technique described above, observed that the conidia of S. *chartarum* could germinate in the lungs of infant rats and form hyphae but could not establish an effective infection even in very young rat pups. Germination was observed frequently in the lungs of 4-day-old pups but rarely in 14-day-old pups. However, acute neutrophilic inflammation and intense interstitial pneumonia with poorly formed granulomas observed 3 days after exposure were associated with fungal hyphae and conidia. In 4-day-old pups, pulmonary inflammation with hemorrhagic exudates resulting in about 15% mortality was observed compared with 0% mortality in controls that received PBS.

In related studies that used the Yike et al. technique in juvenile mice, Rand et al. (2002) found that a single intratracheal injection of S. *chartarum* spores or toxins produced marked ultrastructural changes in alveolar type II cells. Both animals that received S. *chartarum* spores and animals treated with isosatratoxin-F demonstrated condensed mitochondria with separated cristae, scattered chromatin and poorly defined nucleoli, cytoplasmic rarefaction, and distended lamellar bodies with irregularly arranged lamellae 48 h after treatment. Point-count stereologic analysis revealed a significant increase ($p = 0.05$) in lamellar body volume density with both treatments.

Rand et al. (2003) compared juvenile mice treated with 50 mL of 1.4×10^6 S. *chartarum* spores/mL saline toxin at ≥ 35 ng/kg of body weight, mice treated with 50 mL of 1.4×10^6 *Cladosporium cladosporioides* spores/mL saline, and mice that received 50 mL of saline solution. Treatment with fungal spores of either species resulted in granuloma formation at the sites of spore impaction, but some lung tissue treated with S. *chartarum* spores exhibited erythrocyte accumulation in the alveolar air space, dilated capillaries engorged with erythrocytes, and hemosiderin accumulation at spore impaction sites. Immunohistochemistry of the granulomas revealed reduced collagen IV distribution in the mice treated with S. *chartarum* but not C. *cladosporioides*. Quantitative analysis of pooled S. *chartarum* and C. *cladosporioides* spore-impacted lungs revealed significant depression of alveolar air space in animals treated with either S. *chartarum* and C. *cladosporioides* relative to that in untreated controls. Significant ($p < 0.05$) alveolar accumulation of erythrocytes was observed: from $1.24 \pm 1.4\%$ in untreated animals to $3.44 \pm 1.5\%$ in the pooled S. *chartarum* mice. It increased significantly over time ($p < 0.001$) from $2.14 \pm 1.7\%$ 12 h, to $5.54 \pm 1.5\%$ 72 h, and remained elevated at $4.94 \pm 1.4\%$ 96 h after treatment. Treatment with S. *chartarum* spores elicited tissue responses significantly different from those associated with pure trichothecene toxin or with a nontoxigenic fungus.

Gregory et al. (2004) used immunocytochemistry to evaluate the distribution of the trichothecene satratoxin G in spores and mycelia of

S. *chartarum* in culture and in the lung tissues of intratracheally exposed mice. Antibodies prepared in rabbits to react against satratoxin G isolated from Cleveland strain 58–17 reacted more to spores than to mycelia; that indicated a higher toxin concentration in the spores. Mice then received 3,000 spores/g of body weight by intratracheal instillation, and the distribution of immune-labeled satratoxin G in tissues was determined. The toxin was observed predominantly in alveolar macrophages, but it was also found in alveolar type II cells; this finding supported other studies that demonstrated that these cells are sensitive to S. *chartarum* spores, isosatratoxin, and satratoxin G exposure, all of which affect surfactant production and composition in BALF (Mason et al., 1998, 2001; Sumarah et al., 1999), regulation and synthesis of pulmonary surfactant (McCrae et al., 2002), and alveolar type II cell microanatomy (Rand et al., 2002, 2003) and function (Mason et al., 1998, 2001).

On the basis of immunochemistry, Gregory et al. (2003) described the localization of stachylysin (which causes lysis of red blood cells in vitro) in *Stachybotrys chartarum* spores and in rat and mice lungs. Stachylysin labeling was greater around 58-06 Cleveland strain spores than the 58-17 strain 72 h after exposure. Granulomatous lesions formed in rat and mouse lungs that contained spores labeled lightly for stachylysin after 24 h and more heavily after 72 h; production of the lesions may be a relatively slow process. The highest stachylysin concentration was found in the inner walls of spores and near the spores, suggesting diffusion of the stachylysin out of the spores. Stachylysin was also localized in alveolar macrophage cytoplasm and mitochondria and in phagolysosomes. The latter suggests that phagolysosomes might be involved in the inactivation and clearance of stachylysin. Satratoxin G was localized in lysosomes, nuclear membranes, heterochromatin, and rough endoplasmic reticulum (RER) of alveolar macrophages (Gregory et al., 2004). Alveolar type II cells showed modest labeling of nuclear heterochromatin and RER.

Flemming et al. (2004) investigated dose-response relationships (30, 300, and 3,000 spores/g of body weight) and time relationships (3, 6, 24, 48, and 96 h after intratracheal instillation) in mice exposed to macrocyclic trichothecene-producing (JS 58-17) and atranone-producing (JS 58-06) S. *chartarum* strains, comparing with results of exposure to C. *cladosporioides* spores. BALF total protein, albumin, proinflammatory cytokine (IL-1β, IL-6, and TNF-α), and LDH concentrations were significantly (p < 0.05) different between fungal species (S. *chartarum* vs C. *cladosporioides*), strains (58-17 vs 58-06), spore doses, and times. Mice exposed to C. *cladosporioides* or S. *chartarum* spores showed no clinical signs of illness or respiratory distress. Total protein in BALF was significantly (p ≤ 0.001) higher after high-dose exposure to S. *chartarum* strain 58-17 than after all other treatments. Albumin concentrations were significantly higher in mouse lungs exposed to high-dose *S. chartarum* strain 58-06 (p ≤ 0.001), and medium-dose (p ≤ 0.01) and high-dose (p ≤ 0.001) *S. chartarum* strain 58-17 than after all other treatments. The changes were similar and dose-dependent. The majority of the increased protein was albumin. The NOAEL for exposure to spores of *S. chartarum* strains JS 58-17 and JS 58-06 was less than 30 spores/g of body weight; for *C. cladosporioides*, it was over 300 spores/g of body weight. Although after moderate and high doses of *S. chartarum* strains the BALF composition reflected differences in strain toxicity, the BALF composition after treatment with either strain at the lowest dose was similar; spore-sequestered factors common to both strains, rather than strain-dependent toxins, might be contributing to lung disease. An important finding was that low doses of the two *S. chartarum* strains (30 spores/g of body weight) still precipitated responses that were significantly higher than those associated with *C. cladosporioides* or saline exposures even though there was no apparent inflammation response in mice to the two *S. chartarum* strains. The concentration of macrocyclic trichothecenes in the 30-spore/g exposure of *S. chartarum* strain JS 58-17 was less than that associated with the NOAEL in in vitro exposures (Sorenson et al., 1987) and may be associated with high concentrations of proteases (stachyrase A) identified by Yike et al. (2002b).

Joki et al. (1993) examined the effect of volatile metabolites of mold (*Trichoderma viride*) and bacteria (two strains of *Actinomycetes*) isolated from moldy houses and *Penicillium* from a dry surface in a nonproblem house on the ciliary beat frequency (CBF) of guinea pig tracheal-tissue explants. The volatile metabolites of the two mold strains and the bacterial strains increased CBF significantly over negative controls (*Actinomycetes*, 19%; *Penicillium*, 25%; *Trichoderma viride*, 30%) after various times. The authors point out that several inflammatory mediators (bradykinin, histamine, and leukotriene D4) increase CBF but that the physiologic meaning of this effect is unexplained.

Di Paolo and co-workers (1993) report a case of acute renal failure (acute tubular necrosis) in a woman farm worker suspected to have inhaled large amounts of ochratoxin A (OTA) because *Aspergillus ochraceous* was found growing on wheat dust in a closed granary and OTA was identified in the wheat. When the worker was examined clinically, she reported that dust was irritating to her lungs; but pulmonary tissues showed only modest effusion 5 days after exposure. Grain dust from the granary was placed in the bottom of a closed chamber that was ventilated with a continuous current of air passed through the wheat. Animals (four rabbits and four guinea pigs) were exposed in the chamber for 8 h to assess toxicity. One rabbit died after 16 h of exposure, one guinea pig after 24 h, and one rabbit after 34 h. An autopsy of the guinea pig revealed renal tubular necrosis. The rest of the animals were sacrificed and examined 5 days after exposure. All the rabbits had fatty liver degeneration and renal tubular necrosis; one rabbit had pulmonary edema. One guinea pig had tubular necrosis, but no anomalies were found in the other three guinea pigs. No quantification of animal exposure to spores or toxins was attempted.

Wilson et al. (1990) examined the action of AFB_1 on cultured airway epithelial cells explanted from species with abundant (rabbit and hamster) and scarce (rat and monkey) distributions of smooth endoplasmic reticulum (SER) in nonciliated tracheal cells. AFB_1 is metabolized by cytochrome P-450 enzymes associated with SER to compounds that are mutagenic and are capable of binding nuclear DNA, as measured by DNA-adduct formation. DNA binding was greatest in rabbit tissue, followed by hamster, monkey, and rat tissue. A plateau of adduct formation was reached in tissues from all species after 12 h in culture. Degenerative changes in the structure of explants, as seen with electron microscopy, were greatest in rabbit and hamster tissue. In rabbit and hamster tissue, binding of AFB_1 and its metabolites—as determined by autoradiography—was greater in nonciliated secretory cells than in ciliated cells, especially in necrotic cells. In rat, tissue binding was evenly distributed between ciliated and nonciliated cells. Population densities of cells, as measured by quantitative microscopy, indicated that the nonciliated secretory cells were the target of AFB_1.

Humans and Human Cells

Pulmonary Hemorrhage in Infants

Chapter 5 details information on and reviews medical and epidemiologic studies of the possible role of *S. chartarum* exposure in a cluster of cases of pulmonary hemorrhage in infants in Cleveland. This has been the subject of a great deal of attention in the scientific community and by the general public. Relevant toxicologic studies are addressed below.

Jarvis and colleagues (1998) examined samples of *Stachybotrys chartarum* (16 isolates) and *Memnoniella echinata* (a fungus closely related to *S. chartarum*; 2 isolates) from the air and surfaces of homes of cases (10 homes) and controls (29 homes) to determine whether these molds isolated in the Cleveland investigation were producing mycotoxins. Isolates were grown in the laboratory, and extracts were tested for cytotoxicity and for specific toxins. The cytotoxicity test was performed in a feline fetal lung cell culture using inhibition of cell proliferation. Cytotoxicity was assayed on crude extracts and on fractions of them. The majority of the cytotoxicity occurred in the fractions that contained most of the macrocyclic and trichoverroid trichothecenes from *S. chartarum*, consisting of satratoxins (macrocyclic trichothecenes) and roridin L-2 and trichoverrol B (trichoverroid trichothecenes). Toxins normally produced by *S. chartarum*, such as trichodermol and verrucarol and their acetates, were not capable of being detected with the analytic method used (high-performance liquid chromatography with ultraviolet detection). Phenylspirodrimanes, which are immune suppressants, were found in all *S. chartarum* and *M. echinata* cultures. No apparent correlation of toxicity between isolates originating from case and control homes was seen. Of the isolates, three (of nine) from case and three (of eleven) from control homes were the most toxic, and three belonging to each group were not toxic. One of the isolates of *S. chartarum* from a case home produced a significantly greater amount of satratoxin F than would be expected from a small culture; thus, there may be significant variation in toxin production among isolates.

Although some macrocyclic trichothecenes are somewhat lipophilic, Sorenson and co-workers (1996) reported that, when extracted with a water wash, they were nearly as toxic as those extracted with methanol. Jarvis and colleagues (1998) found that about 50% of the total trichothecenes produced by *S. chartarum* was found in a water extract. The authors state that the trichothecenes are exported to the fungal-spore surface, where they become water-soluble by being embedded in water-soluble surface polysaccharides. Such toxins might be readily released into the aqueous microenvironment of the lung surface and diffuse to the epithelium, gaining access to capillaries. As previously discussed, trichothecenes cause capillaries to become leaky, and are associated with hemorrhage in tissues (Jarvis, 1995; Ueno, 1984a,b). Alveoli provide ready access to capillaries, which may be a primary target for these toxins.

Cultures of the two isolates of *M. echinata* from one sample from a Cleveland home did not produce macrocyclic trichothecenes but did demonstrate toxicity similar to that of some isolates of *S. chartarum*. The Cleveland isolates of *M. echinata* were moderately cytotoxic and produced the simple trichothecenes trichodermol and trichodermin and substantial amounts of the antifungal agent griseofulvin. *M. echinata* isolates also produced phenylspirodrimanes (Jarvis et al., 1998).

Vesper et al. (1999) explored the toxicity of strains of *S. chartarum* isolated from eight case and eight control homes from the Cleveland outbreak and 12 strains from diverse geographic locations other than Cleveland. The goal of their study was to determine whether the effects of strains from the Cleveland case homes were different from those of strains from control homes and strains isolated elsewhere. The strains were grown on wet wallboard for 8 weeks, and conidia were then subcultured onto sheep's blood agar at 37°C and 23°C. Cultures were examined weekly for evidence of hemolysis. Five Cleveland strains, all from case homes, showed hemolysis at 37°C, and three non-Cleveland strains consistently demonstrated hemolytic activity throughout the 8-week test period. All strains were hemolytic by the end of week 5 at 37° C. None was consistently hemolytic at the lower temperature. All 28 strains of *S. chartarum* showed some toxicity, as measured by inhibition of protein synthesis. Five of the Cleveland strains and two of the non-Cleveland strains were highly toxic, with effects seen above 90 μg of T-2 toxin equivalents per gram wet weight of conidia; one Cleveland strain and three non-Cleveland strains had intermediate toxicity; and the 17 other strains were consistently less toxic than 20 μg of T-2 toxin equivalents per gram wet weight of conidia. Of the 28 strains examined, only three were both highly toxic and consistently hemolytic; all three came from Cleveland homes where infants with bleeding lived. Two of those three strains were significantly different from other highly toxic strains on the basis of random amplified polymorphic DNA (RAPD) analysis. Although the *S. chartarum* strains from Cleveland homes were not more toxic than strains from other locations, the results from this study suggest that a combination of toxicity and hemolytic capability may be characteristic of the Cleveland strains, and this raises the possibility that a combination of toxins and hemolysins induced pulmonary bleeding in the infants. Identification of strains that produce this combination of factors may be possible through RAPD analysis.

A strain of *S. chartarum* isolated from the lung of a child in Texas who had pulmonary bleeding (designated the Houston strain) was studied for hemolytic activity, siderophore production, and relation to five case- and five control-home strains from the Cleveland outbreak (Vesper et al., 2000a). Hemolysin was produced more consistently and in larger amounts in the case strains from Cleveland and in the Houston strain than in most control strains, so it might play a role in pulmonary bleeding. The case strains and the Houston strain also produced more hydroxymate-type siderophores than control strains; this suggests that they may be better able to extract iron from host cells. One control strain, however, was similar to the case strains and the Houston strain in hemolysin production.

Vesper et al. (2001) characterized hemolysin isolated from *S. chartarum*. The hemolysin has been designated as stachylysin, a β-hemolysin. β-Hemolysins are produced by many bacteria and by *Candida albicans* and *Aspergillus fumigatus* and are associated with the virulence of these fungi.

Kordula et al. (2002) isolated an enzyme, named stachyrase A, from *S. chartarum* isolated from the lung of a child with pulmonary hemorrhage. The enzyme is a chymotrypsin-like serine proteinase that cleaves protease inhibitors, peptides, and collagen in the lung. It is possible that the enzyme could provide mycotoxins greater access to capillaries by removing epithelial barriers, but whether that occurred in this case has not been further investigated.

The magnitude of exposure to microbial and other agents, including *Stachybotrys chartarum,* in the Cleveland cluster and other clusters in which *S. chartarum* is thought to be a factor is not known. Some 30–40% of strains of *S. chartarum* isolated from the Cleveland cases contained macro-cyclic trichothecenes associated with hemorrhage via inhalation at high exposures in animal experiments (Jarvis et al., 1998). Phenylsirodrimanes were found in all strains while atranones were found in 60-70% of the strains, and some strains produced hemolysins and enzymes that attack collagen (Kordula et al., 2002; Vesper et al., 2000a, 2001).

The body of research on *S. chartarum*, especially the more recent studies (Andersen et al., 2002; Flemming et al., 2004; Gregory et al., 2003, 2004; Jarvis et al., 1998; Rand et al., 2002, 2003; Rao et al., 2000a,b; Vesper and Vesper, 2002; Vesper et al., 1999, 2000a,b, 2001; Yike et al., 2002a,b), provides a biologically plausible mechanism by which at least some strains of this mold could affect the lungs of young animals. Other potentially toxic molds were isolated in the Cleveland case, but their toxic potency, degree of exposure, and interactions with toxins produced by *S. chartarum* and the closely related fungal species *Memnoniella echinata* (also isolated in the Cleveland case) have not been investigated. No other cluster of similar size pulmonary hemorrhage in infants has been seen since the Cleveland outbreak, although case reports of pulmonary bleeding in infants in whose lungs or environment *S. chartarum* was found to be present have been published (CDC, 1997; Dearborn et al., 2002; Elidemir et al., 1999; Flappan et al., 1999; Knapp et al., 1999; Novotny and Dixit, 2000; Tripi et al., 2000; Weiss and Chidekel, 2002). The role of toxigenic fungal exposure in such cases has yet to be determined.

Other Effects

The effects of mycotoxin-associated fungal spores (152 isolates) from the air of damp domestic environments (503 dwellings) in Scotland were tested in a human embryonic diploid fibroblast lung cell line (MCR-5) (Lewis et al., 1994). At least 37% of the isolates, primarily those of *Penicillium* species, demonstrated cell toxicity when assayed. Of the molds producing mycotoxins, the penicillia were most numerous, accounting for 81.6% of the identified isolates. Of the penicillia identified to the species level, *P. viridicatum, P. expansum,* and several strains of *P. chrysogenum* exhibited the greatest measured toxicity in water-extract trials. Two toxic species of *Aspergillus fumigatus* (an opportunistic pathogen capable of invading the lung) were isolated. Clear variations in cytotoxicity were observed when two additional human cell lines were used: Chang liver cells and Detroit 98 normal human sternal bone marrow cells. Strains of *P. expansum, P. chrysogenum,* and *P. glabrum* collected from dwellings showed no toxicity with the MRC-5 cell line but did with Chang liver cell culture. In addition, a *P. aurentiogriseum* extract resulted in nearly 3 times the mortality of Chang liver cells than of MRC-5 cells. A penicillic acid-producing strain of *P. aurentiogriseum* was more lethal to Detroit 98 cells than to Chang liver cells but showed little toxicity to MRC-5 cells. An additional 23 extracts showed significant toxicity, relative to vehicle control, when dissolved in DMSO.

Amitani et al. (1995) demonstrated in vitro that *Aspergillus fumigatus* produces a number of biologically active molecules, including gliotoxin. Those molecules slow ciliary beating and can damage the respiratory epithelium and so possibly influence colonization of the airway by this mold. Nine clinical isolates of *A. fumigatus* were obtained from the sputum of patients with pulmonary aspergillosis (four cases of chronic necrotizing pulmonary aspergillosis and one case of aspergilloma) and applied to cultured respiratory ciliated epithelium obtained from the nasal mucosa of healthy volunteers. CBF was measured with a photometric technique that used a phase-contrast microscope. Eight of the nine filtrates from cultures of *A. fumigatus* isolates caused a significant decrease in CBF; five of the nine caused at least a 50% decrease. A gliotoxin metabolite coeluted with gliotoxin. An extract similar to but with a slightly different absorbency from laboratory-grade gliotoxin also coeluted; this indicates the presence of other, as yet unidentified ultraviolet-absorbing material in the ciliotoxic fraction. Gliotoxin has been shown to inhibit phagocytosis by rodent macrophages, bactericidal activity of peritoneal macrophages, and the basal rate of hydrogen peroxide production by human neutrophils (Eichner et al., 1986).

Immunotoxicity

Immunotoxicity can result from immunosuppression after exposure to a xenobiotic or from immunoenhancement, autoimmunity, and allergic reactions. Immunosuppression can increase susceptibility to infectious disease and cancer through the loss of immunosurveillance cells (Corrier, 1991; Corrier and Norman 1988). It can result from decreased activity of any of the immune cells, their precursors, or other immune-related cells through inhibition of function, decrease in their population, or other dysregulation. Interpretation of data regarding immunologic end points is extremely difficult, but many mycotoxins have been found to affect alveolar macrophages in in vitro studies. Some of those studies are summarized here.

Alveolar macrophages (called pulmonary macrophages in some studies) are part of the physical defense mechanism of the respiratory system. They help to clear particulate materials, including infectious organisms, from the lower respiratory system by phagocytosis (engulfing, killing, and digesting), or by transporting them out of the respiratory system via the mucociliary escalator. When they and their particulate burden reach the oropharynx, they are swallowed. Alveolar macrophages may also transport particulate material into the lymphatic system and to the regional lymph nodes. Lung-associated lymph nodes contain antibody-forming cells that can be stimulated to form specific antibodies against antigenic material brought to them by alveolar macrophages (Haley, 1993).

Indicators of increased susceptibility to infectious disease are seen in animal field investigations in which flocks of sheep, herds of pigs, or flocks of birds fail to thrive and exhibit reduced immune response to common infectious organisms after exposure to a microbial agent (Corrier, 1991; WHO, 1990). Such responses appear at lower exposures than more overt signs of toxicity, such as vomiting, staggering, or hemorrhage (Corrier, 1991; Smith and Moss, 1985; WHO, 1990). Immunotoxic studies in animals generally examine the effects of short-term exposures. Table 4-3 lists some immunoactive mycotoxins, the fungi that produce them, their potency expressed by various measures, their mechanisms of action (if known), and their effects.

Inhibition or modulation of immune defenses results from exposure to a variety of mycotoxins. According to Pier and McLoughlin (1985), three groups of mycotoxins are predominantly associated with immunosuppressive toxicity: aflatoxins, OTA, and trichothecenes. In all those groups, inhibition of protein synthesis plays a role, although each group acts on a different site of protein formation (Corrier, 1991): aflatoxins bind to DNA and interfere with transcription of RNA from DNA (Hsieh et al., 1977), OTA inhibits the enzyme phenylalanyl-t-RNA synthetase (McLaughlin et al., 1977), and T-2 toxin (representing several other trichothecenes) prevents initiation of translation of mRNA into protein (Hsieh et al., 1977; McLaughlin et al., 1977).

Jakab et al. (1994), using a nose-only exposure chamber, examined inhalation exposure of rats and mice to nebulized AFB_1 (particle mass median diameter, 0.2 μm; time-weighted average concentration, 3.17 μg/L). An initial experiment demonstrated a clear dose-dependent decrease in Fc receptor-mediated alveolar macrophage phagocytosis 3 days after exposure for 20, 60, and 120 min. In a second experiment, alveolar macrophage phagocytosis was measured 2, 4, 7, and 13 days after exposure; a single dose of AFB_1 aerosol suppressed alveolar macrophage phagocytosis for nearly 2 weeks. In a third experiment, 250-μL suspensions of 50 and 150 μg of crystalline AFB_1 were instilled, and alveolar macrophage phagocytosis was assessed 1, 3, and 7 days after exposure by measuring the capacity of alveolar macrophages to produce $TNF\alpha$ elsewhere after LPS stimulation. $TNF\alpha$ production was suppressed in a dose-dependent manner on all days. Similar results were seen in mice into which increasing doses of AFB_1 were instilled intratracheally when alveolar macrophage phagocytosis was mea-

TABLE 4-3 Immunoactive Mycotoxins and Effects

Mycotoxin	Producing Molds	Potency	Mechanism	Effect
Aflatoxin	*Aspergillus flavus, Aspergillus parasiticus*	0.6–10.0 ppm in feed: depressed antibody response in chickens	Inhibits protein synthesis	Immune suppression—various end points
Ochratoxin A (OTA)	*Aspergillus ochraceus, Penicillium verrucosum, Penicillium viridicatum*	5 mg/kg ip in mice: subchronic T-cell suppression	Inhibits φ-alanyl tRNA synthetase	Suppression of antibody production and globulin synthesis
Sterigmatocystin	*Aspergillus versicolor, Aspergillus nidulans*		Inhibits φ-alanyl tRNA synthetase and lipid peroxidation	Suppression of antibody production and globulin synthesis, proliferation of Kupffer cells in liver
Gliotoxin	*Aspergillus fumigatus*		Inhibits MP phagocytosis, induces MP apoptosis, blocks T- and B-cell activation	Infection with A. fumigatus, other microorganisms
Cyclopiazoic acid (CPA)	*Aspergillus* spp., *Penicillium* spp.		Causes lymphoid depletion, inhibits protein synthesis	Immunosuppression

Citrinin	*Penicillium* spp.	0.12–3.9 mg/kg ip mice: lymphocyte blastogenesis	Stimulates mitogen production	Immune modulating
Patulin	*Aspergillus* spp., *Penicillium* spp.	10 mg/kg in rats: decrease IgA	Inhibits protein synthesis, impairs MP activity	Immune modulating
Trichothecenes	*Stachybotrys chartarum; Trichoderma viride*	T-2 toxin: 0.1 µg/kg oral in monkeys: severe immunosuppression Stachybotrytoxin: LD_{50} 0.1 mg/kg ip in mice	Inhibits protein synthesis, inhibits peptidyl synthesis, causes lymphoid necrosis, causes dysregulation of IgA production	Immunosuppression, immune modulating, decrease in host resistance to infection
Cyclosporin A (CsA)	*Stachybotrys chartarum, Streptomyces sutubaensis*	$IC_{50} = 26.8$ ng/mL	Inhibits lymphocyte proliferation, IL-2 and interferon production, protein folding	Immunosuppression
Zearalenone	*Fusarium* spp.	10 ppm oral in mice: decreased resistance to listerosis	Inhibits DNA synthesis, lymphoblastogenesis	Immune modulating
Rapamycin	*Streptomyces hygroscopicus*		Inhibits signal transduction, preventing T-cell activation	Immune modulating

NOTE: Abbreviations: IC_{50}, inhibitory concentration$_{50}$, concentration at which 50% of population shows immune change; ip, intraperitoneal injection; LD_{50}, lethal dose$_{50}$, dose that kills 50% of population; MP, macrophage.

sured 4 days after exposure. As was seen with rats, alveolar macrophage phagocytosis was suppressed 4 days after exposure in a dose-dependent manner.

Jakab et al. (1994) also assessed systemic immune function by measuring primary splenic antibody responses to sheep red cells (SRBCs). Swiss mice immunized with SRBCs a day before or a day after AFB_1 treatment were exposed by inhalation or intratracheally to 75 µg of AFB_1. Splenic antibody response was measured 5 days after immunization by counting the antibody-forming cells. Respiratory tract exposure to AFB_1 significantly suppressed the primary antibody response to SRBCs; the effect was greater when AFB_1 treatment preceded immunization. Neither histologic nor inflammatory alterations occurred as a result of inhalation or intratracheal instillation of AFB_1. Those experiments indicate that suppression of alveolar macrophages could suppress the clearance of particles from the lung, the killing of bacteria, the suppression of tumor cells, and the modulation of inflammatory and immune processes, both through suppression of phagocytosis and through the inhibition of TNFα. Systemic immune function might also be suppressed.

T-2 toxin increased mortality from *Salmonella* administered orally to chickens a week after toxin exposure (Boonchuvit et al., 1975), but neither T-2 toxin nor *Salmonella* caused mortality by itself. Similarly, mice challenged with *Mycobacterium obvis* after T-2 treatment died early and in greater numbers than those not treated with this trichothecene (Otokawa, 1983). Mice treated with T-2 toxin and given intracerebrally injections of Japanese encephalitis virus (JEV) 5 days after cessation of toxin treatment died within 4 days (only one of 10 control mice died). In the early phase of JEV infection, however, the immune response does not seem to play a role in lethality, and the mechanism underlying the effect of T-2 toxin is not known (Otokawa, 1983). Infection experiments show that repeated administration of trichothecenes in animals induces more susceptibility to microbial infection (Otokawa, 1983).

Sumi et al. (1987) exposed (by diet, inhalation, and dermal exposure) 21 male Wistar germ-free rats to *Aspergillus versicolor* in a germ-free isolator for 2 years and compared them with 20 nonexposed controls. Histologic examination found necrosis of the liver (86% of the animals), foamy-cell granuloma of the lung (76%), fibrosis of the pancreas (52%), and nephritic lesions (57%) in the exposed animals. None of the control animals had lesions. Of the exposed rats, 62% exhibited tumors of the pleura, lung, and endocrine organs, compared with 15% of controls. Two exposed rats developed mesothelioma, and another developed squamous-cell carcinoma of the lung. Those results indicated that high-concentration exposure to *A. versicolor* in the absence of other microorganisms induced severe organ damage and some tumors in the rats.

In another study, Sumi et al. (1994) exposed germ-free rats to *A. versicolor* to elucidate the mechanism of lung damage from exposure to *A. versicolor*-contaminated grain dust. They exposed 21 rats for 2 years to a pure culture of *A. versicolor* and evaluated them 1, 2, 3, and 6 months after exposure; the results were compared with those in 21 germ-free controls. After 1 month, alveolar macrophages increased in number and became foamy macrophages as they ingested and digested mold spores. The macrophages expressed IL-1, IgA antigens, and intercellular adhesion molecules intensely bound to lymphocytes. Numerous lymphocytes infiltrated granulomatous lesions consisting of accumulated foamy macrophages and some T lymphocytes, which carried the IL-2 receptor. Granulomatous lesions extended throughout the lung, especially around bronchioles, and were present from the alveolar ducts to alveolar spaces up to 6 months after exposure. The authors concluded that macrophages may be a key effector in producing granulomas of the lung and that inhalation of *A. versicolor* at high concentrations may induce lung damage even in the absence of microbial infection.

Acute and chronic exposures to trichothecene mycotoxins result in depletion of lymphoid tissues, an indication of immune dysfunction, but in vitro experiments indicate that the trichothecenes have both immunosuppressive and immune-enhancing effects (Biagini, 1999). Major trichothecenes whose immunosuppressive activity has been reported are T-2 toxin, DAS, and stachybotryotoxin (Pier and McLoughlin, 1985). Their activity is associated in mice, rats, cattle, turkeys, and guinea pigs with alterations in serum proteins and immunoglobulin profiles, reduced antibody formation, thymic aplasia, reduced cell-mediated immune responses, increased delayed cutaneous hypersensitivity, and impaired bacterial clearance and acquired immunity (Pier and McLoughlin, 1985). Trichothecene immunosuppression also appears to be due to interference with the generation of suppressor cells for the delayed hypersensitivity response (Ueno, 1989) and inhibition of protein synthesis (Hughes et al., 1989). T-2 toxin reduces complement (primarily C_3) formation, diminishes serum immunoglobin (IgA and IgM, but not IgG), and diminishes antibody production (Pier and McLoughlin, 1985). Cells that depend on a high rate of protein synthesis—such as lymphoid cells, those lining the gastrointestinal tract, and hematopoietic cells—seem to be most sensitive to that effect of trichothecene exposure.

OTA inhibits protein synthesis through inhibition of phenylalanyl t-RNA synthetase (McLaughlin et al., 1977). Various effects—including necrosis of the lymph nodes, inhibition of macrophage migration, and reductions in immunoglobulin and antibody production—have been shown to result from that inhibition. The last effect seems most important, and, in contrast with the effects of aflatoxin, there does not seem to be inhibition of complement or cell-mediated immunity (Pier and McLoughlin, 1985). AFB_1

has been shown experimentally to produce thymic aplasia, to reduce T-cell function and number, to diminish antibody response, to suppress phagocytic activity, and to reduce complement (Pestka and Bondy, 1990).

Richard and Thurston (1975) tested the effect of a mixture of aflatoxin (AFB_1, AFB_2, AFG_1, and AFG_2) exposures on phagocytosis of *Aspergillus fumigatus* spores by rabbit alveolar macrophages. Rabbits were exposed orally to daily doses of 0.03, 0.05, and 0.07 AFB_1 equivalents/mL for 2 weeks. *Aspergillus fumigatus* spores were mixed with serum extracted from control and treated rabbits. Rabbits were sacrificed, and macrophages from their lungs were cultured, tested for viability, and inoculated with serum containing the spores. Macrophages from rabbits given any of the doses of aflatoxin in rabbit serum had lower phagocytic activity than controls; the extent of the reduction was dose-dependent. Fresh or frozen, but not heat-treated, rabbit serum was required for any significant phagocytosis by cultured macrophages; this indicated that the reduction of phagocytosis by aflatoxin treatment could be related to a lowering of complement or some other opsonization[2] factor.

Cusumano et al. (1996) used monocytes isolated from healthy human volunteers to study the effects of AFB_1 (0.05, 0.1, 0.2, 0.3, 0.4, 0.5, and 1.0 pg of AFB_1/mL for 2 and 24 h) on phagocytosis, response to microbial activity, superoxide production, and intrinsic antiviral activity. The phagocytic activity of human monocytes was dose-dependent and was significantly lower than controls after 2 hours of pretreatment at 0.5 and 1 pg of AFB_1/mL. Pretreatment for 24 h reduced phagocytic activity significantly at all doses tested. Pretreatment with 1.0 and 0.5 pg/mL for 2 h significantly impaired killing of the yeast *Candida albicans*, and 24-h pretreatment significantly increased the degree of impairment. Production of superoxide anion and antiviral activity were not significantly changed.

Dolimpio et al. (1968) demonstrated an inhibition of mitosis in human leukocytes isolated from three healthy female volunteers after an 8-h exposure to AFB_1 (1–50 µg/mL); greater inhibition was seen after a 48-h exposure. The inhibition was time- and dose-dependent. Chromosomal aberrations—including gaps, breaks, fragments, deletions, and translocations in different chromatids—were also seen.

Pier and McLoughlin (1985) summarized the mechanisms of aflatoxin immunosuppressive actions as follows:

• Most studies suggest that aflatoxin impairs the immune response without affecting antibody formation.

---

[2] The coating of a particle with a substance that helps it to attach to a phagocytic leukocyte.

• Aflatoxin suppresses complement (C_4) and interferon, nonspecific circulating substances related to resistance to infection.

• Aflatoxin suppresses macrophage phagocytosis.

• Aflatoxin causes thymic aplasia.

• Aflatoxin suppresses cell-mediated immunity, especially delayed cutaneous hypersensitivity, lymphoblastogenesis, and leukocyte migration.

Sorenson et al. (1985) studied the toxicity of patulin, which is produced by several species of *Aspergillus* and *Penicillium* and has been shown to have the capability to be mutagenic, teratogenic, and carcinogenic. The researchers looked at cell leakage, energy metabolism, and protein synthesis in alveolar macrophages. Leakage of ^{51}Cr from alveolar macrophage, after exposure to 0.15 mM patulin was time- and concentration-dependent. ATP concentrations were markedly inhibited within 1 h at patulin concentrations of less than 0.05 mM. RNA synthesis and protein synthesis were strongly inhibited with RNA and protein synthesis ED_{50}s (effective doses for 50% inhibition at 1 h) of 0.0016 and 0.019 mM, respectively. Protein synthesis was a much more sensitive end point than RNA synthesis, cell leakage, and ATP concentrations, all of which were more sensitive than cell volume.

Sorensen and Simpson (1986) examined the toxicity of penicillic acid—a toxin similar in chemical nature, molecular size, and toxic end points to patulin that is produced by *Penicillium* species—for rat alveolar macrophages, using similar methods as in the previous experiment. Results were similar to those for patulin in the earlier study, except that patulin is slightly more toxic then penicillic acid.

Neurotoxic Effects

Occupants of damp and moldy buildings have sometimes reported central nervous system symptoms—such as fatigue, headache, memory loss, depression, and mood swings—that they attribute to the indoor environment. However, mycotoxin exposure of those people in their environment has not been identified and measured.

Neurotoxic effects of mycotoxins have been examined in herd animals because consumption of mold-contaminated feed has led to severe neurologic diseases, such as rye grass staggers. Mycotoxins that selectively or specifically target the nervous system have been isolated from species of fungi contaminating grain in incidents of animal toxic response. Table 4-4 lists mycotoxins whose neurotoxic effects have been studied at least in some animals, the genera and species of fungi that produce them, their potency, their mechanism of action (if known or hypothesized), and their neurotoxic end points.

TABLE 4-4 Neurotoxic Mycotoxins and Effects

Mycotoxin	Producing Molds	Potency
Penitrem A	*Penicillium cyclopium, Penicillium verruculosum, Penicillium crustosum*	250 µg/kg ip in mice: tremors, LD_{50} = 1.05 mg/kg; 24–25 µg/kg orally in sheep: lethal; 2.2–4.4 µg/kg iv: tremors
Penitrem E	*Penicillium crustosum*	2.25 mg/kg ip in mice: tremors
Aflatrem	*Aspergillus flavus*	0.5 mg/kg ip in mice: tremors
Roquefortine	*Penicillium commune, Penicillium palitans, Penicillium crustosum*	0.1 potency of penitrem A
Verruculogen	*Penicillium verruculosum* Peyronel *Penicillium simplicissimum Penicillium crustosum Aspergillus caespitosus*	3.0–4.0 µg/kg oral in sheep: tremors; 13.3 µg/kg oral: lethal; LD_{50} = 2.4 mg/kg ip ED_{50} = 0.39 mg/kg ip in mice
Verrucosidin	*Penicillium verruculosum* var. *cyclopium*	4 mg/kg ip in mice
Fumitrem B	*Aspergillus fumigatus*	0.1 potency of Verrucologen
Cyclopiazonic acid	*Penicillium cyclopium, Aspergillus flavus, Penicillium crustosum*	250 µg/kg ip in mice: tremors; 2.5 mg/kg: clonic convulsions, death

Mechanism	Effect	References
Inhibits neurotransmitter release (GABA) in interneurons of anterior horn of spinal cord	Tremors, convulsions	Peterson and Penny, 1982; Selala et al., 1989; Smith, 1997; Wilson et al.,1968
Inhibits neurotransmitter release (GABA) in interneurons of anterior horn of spinal cord	Tremors, convulsions	Kyriakidis et al., 1981
Inhibits neurotransmitter release (GABA) in interneurons of anterior horn of spinal cord	Hyper reactivity to auditory and °tactile stimuli, severe whole-body tremors	Selala et al., 1989; Wilson et al., 1968
	Motor dysfunction in mice, convulsions	Norris et al., 1980; Peterson and Penny, 1982; Selala et al., 1989
	Tremors	Fayos et al., 1974; Peterson et al., 1982
	Paralysis	Hodge et al., 1988
		Peterson and Penny, 1982
		Selala et al., 1989; Wilson et al., 1968

(continued on next page)

TABLE 4-4   continued

Mycotoxin	Producing Molds	Potency
Territrems	*Aspergillus terreus*	
Citreoviridin	*Penicillium citreoviride Biourge, Aspergillus terreus, Penicillium citreonigrum*	$LD_{50} = 7.2$ mg/kg ip in mice
Ochratoxin A (OTA)	*Aspergillus ochraceous, Penicillium verrucosum, Penicillium viridicatum*	
Gliotoxin	*Aspergillus fumigatus*	
Trichothecenes (for example, DON, also known as vomitoxin)	DON *Fusarium* spp., other tricothecenes	TDI in infants 1.5 µg/kg, in adults 3.0 µg/kg
	Stachybotrys chartarum, Trichoderma viride	

NOTE: Abbreviations: DON, deoxynivalenol; ED_{50}, effective dose$_{50}$, amount required to produce specified effect in 50% of population; GABA, gamma-aminobutyric acid; IC_{50}, inhibitory concentration$_{50}$, concentration at which 50% of population shows immune change;

Neurotoxic mycotoxins tend to fall into three general classes: tremorgenic toxins, paralytic toxins, and toxins that interfere with neurotransmitters or receptors either centrally or at the target organ. Many of the toxins are very potent and have immediate effects on animals exposed to a single dose by various routes. Few long-term exposures have been studied, and tests that would evaluate subtle changes in function of animals have not been done. Toxins that exert their effects on the nervous system by interfering with protein, RNA, or DNA synthesis or that exert their effects on membranes have been examined only for short-term exposures. Susceptibility to such toxins varies among animal species. Pigs and sheep seem to be as susceptible as other herd animals and rodents to tremorgens (El-Banna and Leistner, 1988; Peterson and Penny, 1982). Human susceptibility is not well established.

Mechanism	Effect	References
	Tremors, convulsions	
Inhibits mitochondrial ATPases, ATP synthesis, hydrolysis	CNS paralysis, convulsions, respiratory arrest, death, arrest of respiratory center (acute cardiac beriberi in humans)	Franck and Gehrken, 1980; Selala et al., 1989; Sorenson, 1993; Steyn and Vleggaar, 1985; Ueno and Ueno, 1972
Competes with Phe, inhibits protein synthesis, affects neural cell differentiation, neuritic	Microcephaly (mouse pups), sensitive period day 10 of gestation	Bruinick and Sidler, 1997; Miki et al., 1994
Lesions of astrocytes		Chang et al., 1993; Jarvis et al., 1995
Inhibits protein synthesis, increases serotonin	Central nervous system or gastrointestinal effects, probably mediated through vagal receptors because serotonin affects feeding behavior and emesis	Rotter et al., 1996

ip, intraperitoneal injection; iv, intravenous injection; LD$_{50}$, lethal dose$_{50}$, dose that kills 50% of population; Phe, phenylalanine; TDI, tolerable daily intake, amount that can be ingested daily without posing significant health risks.

Tremor

Tremorgenic toxins are produced predominantly by *Aspergillus* and *Penicillium* species (Ciegler et al., 1976; Land et al., 1994). The penitrem-type of mycotoxins produces a neurotoxic syndrome in animals that involves sustained tremors, limb weakness, ataxia, and convulsions (Steyn and Vleggaar, 1985). Tremorgenic toxins generally initiate measurable effects in experimental animals within minutes of exposure. Norris et al. (1980) found that penitrem A (produced primarily by *P. crustosum*) increased spontaneous release of the endogenous neurotransmitters glutamate, γ-aminobutyric acid (GABA), and aspartate by interfering with the neurotransmitter-release mechanisms.

After sublethal doses, animals may suffer effects for hours or days but recover completely from the effects (Knaus et al., 1994; Peterson et al.,

1982). Few long-term exposures have been examined, and tests that would determine subtle changes in function have not been done. Selala et al. (1989) reported that tremorgenic mycotoxins are partial agonists of GABA. Peterson et al. (1982) showed that 5-month-old lambs were more sensitive than 15-month-old sheep, and repeated dosing did not indicate a cumulative effect of verruculogen, a tremorgenic mycotoxin produced by *P. crustosum* and *P. simplicissimum*. *A. clavatus* is toxic to sheep and cattle in pastures; it produces a highly lethal mycotoxicosis that involves neural degeneration and necrosis of the midbrain, medulla, and ventral horns of the spinal cord. The specific toxin involved is not yet known, but it does not seem to be patulin or any other known tremorgen (Kellerman et al., 1976).

Paralysis

Penicillium species also produce neurotoxins that induce paralysis. Citreoviridin, produced by *P. citreo-viride* and *A. terreus*, and verrucosidin, produced by *P. verruculosum* var. *cyclopium*, are examples of such toxins (Franck and Gehrken, 1980; Hodge et al., 1988; Ueno and Ueno, 1972). Those toxins produce a progressive, ascending paralysis and are thought to act at the level of the interneurons and motor neurons of the spinal cord and motor nerve cells of the medulla (Ueno, 1984b). A typical pattern of poisoning begins with paralysis of the hind legs, which is followed by a drop in body temperature and respiratory arrest (Ueno and Ueno, 1972).

The tremorgenic and nontremorgenic mycotoxins from *Aspergillus* and *Penicillium* work at a different functional level of the nervous system from mycotoxins that have more widespread targets for toxicity or work by inhibiting basic cellular functions, such as protein synthesis.

Other Effects

Ochratoxin

OTA is toxic to nephrons and is a known neurotoxicant during prenatal stages (WHO, 1990). It is produced by *Aspergillus* and *Penicillium* species.

In tissue-culture experiments, 5–50 times higher OTA concentrations were required to affect inhibitory (GABA) transmitter levels than to affect markers for neuritic outgrowth and differentiation in both brain and retinal embryonic cell cultures (Bruinink and Sidler, 1997). That indicates that the OTA teratogenic, neurotoxic end point differs from that of the tremorgenic toxins. Bruinink and Sidler (1997) also found neural cells to be more sensitive to OTA than the meningeal fibroblast cultures previously studied by Bruinink et al. (1997); this supports previous suggestions based on in vivo data (Miki et al., 1994) that neural tissues are especially sensitive to OTA. Miki and colleagues (1994) found that the neurosensory and visual cortex of the brain of mice whose dams were treated with OTA during pregnancy had reduced numbers of synapses and that there was a significant deficit in brain, but not body weight, of treated vs age-matched controls.

OTA is a chlorinated dihydroisocoumarin derivative bound to phenylalanine (Phe) through an amide bond. It is an inhibitor of protein synthesis. Previous work in yeast indicated that that might be due to competitive inhibition of the aminoacylation of Phe tRNA by phenylalanyl-tRNA synthetase, and research on rat hepatocytes indicated Phe hydroxylation (Creppy, 1995; Creppy et al., 1983); but these mechanisms do not seem to be involved in brain and retinal cell cultures (Bruinink and Sidler, 1997). The differences, however, might be a function of the various concentrations of OTA used in the several sets of experiments. Bruinink and Sidler (1997) found their effects on neurite formation at concentrations much lower than those at which Creppy and co-workers saw enzyme inhibition. Thus, OTA may affect neural cell differentiation at concentrations much lower than those at which it affects basic cell functions, such as protein synthesis.

OTA has been measured in human maternal and cord blood (Jonsyn et al., 1995a) and in breast milk (Jonsyn et al., 1995b) in Sierra Leone. The consequences of such exposure to the human nervous system, however, have not been studied (Bruinink and Sidler, 1997).

Gliotoxin

Gliotoxin is an epipolythiodioxopiperazine compound that is a potent immunomodulator (Sorenson, 1993). It has been used therapeutically as an immunosuppressive agent for transplantation of organs and tissues. Neurotoxicity has been associated with its use, and data indicate that it can directly affect astrocytes (Chang et al., 1993). Neurotoxic effects might also be indirect through its effects on the immune system. Gliotoxin is commonly produced in cultures of *Aspergillus fumigatus* isolated from tissues of animals that have experimental aspergillosis and in naturally infected tissues (Sorenson, 1993). It might also play a role in the etiology of the disease aspergillosis (Amitani et al., 1995).

Trichothecenes

Trichothecene mycotoxins have a tricyclic trichothane skeleton with an olefinic group at carbon atoms 9 and 10 and an epoxy group at carbon atoms 12 and 13. Macrocyclic trichothecenes have a carbon chain between carbon atoms 4 and 15 that contain an ether or an ester linkage. The 12,13-

epoxide ring, the double bond between carbon atoms 8 and 9, and the presence of various free ester groups are essential to trichothecene toxicity (Bamburg and Strong, 1971). Trichothecene mycotoxins are potent inhibitors of protein, RNA, and DNA synthesis; they act by binding to ribosomes in the cells of eukaryotic organisms (McLaughlin et al., 1977; Ueno, 1980, 1984a).

Because protein synthesis is fundamental to growth and maintenance of cells, inhibition of this fundamental cellular function can have profound effects. Neurotoxic effects in laboratory animals include degeneration of nerve cells in the central nervous system, vomiting, central nervous system-mediated loss of weight and failure to thrive, anorexia, and thirst (Ueno, 1984a).

T-2 toxin, produced by *Fusarium* species, has been used experimentally to study the effects of this class of toxins. T-2 toxin causes neurotoxic effects, including feed refusal, neuromuscular disturbances, and vomiting due to stimulus of the chemoreceptor zone of the medulla (Matsuoka et al., 1979; Weekley et al., 1989). Acute and chronic dosing of female rats with T-2 toxin differentially altered tryptophan, tyrosine, and serotonin concentrations in the cerebellar and brainstem regions; no systemic signs of toxicosis were evident during these neurotransmitter changes (Weekley et al., 1989). The authors suggest that T-2 toxin induces a central neurochemical imbalance that causes an alteration in autonomic function, which can then contribute to the cardiotoxic effects seen with T-2 toxin.

Ueno (1977) reports that T-2 toxin and its metabolites cause depression of the central nervous system that manifests as hyporeflexia, ataxia, and prostration. Bergmann et al. (1988) indicate that T-2 toxin causes cerebral toxicity.

Deoxynevalinol (also called vomitoxin) is a trichothecene produced by *Fusarium* species that is thought to affect 5-hydroxytryptamine receptors in the peripheral nervous system; these receptors are especially prevalent in the gut and are thought to mediate vomiting (Rotter et al., 1996). Effects on the central nervous system may also be mediated through changes in transmitter concentrations in the vomiting center in the medulla (Prelusky, 1993).

Sensory Irritation

The neurotoxic end points that appear to be most affected at low exposures are those which affect the olfactory sense and the "common chemical sense" that responds to pungency (Cometto-Muniz and Cain, 1993; Korpi et al., 1999; Pasanen et al., 1999; Schiffman et al., 2000). It is thought that the common chemical sense resides in the trigeminal, vagus, and glossopharyngeal spinal nerves. Experiments suggest that the sensory nerve endings respond to irritative stimuli, whereas the motor portion responds by smooth muscle contraction, secretion from excretory glands, and central nervous system effects that can include impairment of attention and memory and various fight or flight responses. Perceived pungency can produce reflex constriction of the airways and inflammation and result in nasal stuffiness, headache, malaise, memory loss, and reduced ability to concentrate, depending on the nature of the irritant, its concentration, and individual sensitivity (Cometto-Muniz and Cain, 1993; Kasanen et al., 1998; Lucero and Squires, 1998). In animals, the 50% respiratory dose (RD_{50}, the concentration that causes a 50% decrease in respiratory rate in exposed animals, in this case, in response to trigeminal nerve stimulation by a pungent chemical) varies for different enantiomers of pinene (a terpene produced by some microorganisms); this indicates that sensory irritant receptors respond to the three-dimensional structures of such pungent nonreactive molecules (Korpi et al., 1999).

Microorganisms can produce volatile organic compounds (VOCs). Some microbial VOCs or MVOCs (such as alcohols, aldehydes, and ketones) are products of primary metabolism and are produced throughout an organism's life. Others, which tend to be more complex, have characteristic moldy, musty, or pungent odors. They are produced through secondary metabolism—in *Penicillium* and *Aspergillus*—around the time of sporulation, when mycotoxins also tend to be produced (Fiedler et al., 2001; Larson and Frisvad, 1994). VOCs produced by building materials, paints, solvents, and combustion can irritate the mucous membranes of the eyes and respiratory tract and possibly the nerve endings of the common chemical sense either alone or in concert with other volatile and semivolatile compounds produced by microorganisms (Otto et al., 1990; Schiffman et al., 2000). Miller et al. (1988) measured a total putative MVOC concentration of 2 mg/m^3 in a "moldy" building.

Controlled human experiments indicate that aggregate exposure to non-microbial MVOCs common to new office buildings at a total concentration of 25 mg/m^3 produced subtle changes in some measurable neuropsychologic end points (Hudnell et al., 1992; Otto et al., 1990, 1992). A companion study by Koren et al. (1992) also found increased neutrophils, a sign of inflammation, in 14 volunteers exposed to the VOC mixture, an indication that such VOCs can elicit an inflammatory response. MVOCs include terpenes, sesquiterpenes, and other substances that are highly irritating, but it is unknown whether the concentrations of MVOCs and semivolatile compounds typically found in homes with microbial contamination are sufficient to cause a trigeminal or toxic response (Ammann, 1999; Korpi et al., 1999).

Dermal Toxicity

Indoor surface contamination with molds is common and, because dermal absorption can occur, it is possible that surfaces with large amounts of contamination might provide a means of exposure to occupants or workers who come into contact with such surface contamination.

Simple and macrocyclic trichothecenes are irritating to the skin of animals and humans. Buck and Cote (1991) describe the effects as radiomimetic in potency. A dose as low as 0.5 ng can cause skin reddening in guinea pigs (Ueno, 1984a). In general, type A (such as T-2 toxin) and type D (such as verrucarin A and satratoxins) trichothecenes are highly irritating, and type B (such as deoxynevalinol) trichothecenes less so. All those cause skin reddening in early stages of toxicity, but type D trichothecenes are characterized by edematous damage to skin tissue (Ueno, 1984b). Large volumes of inflammatory exudate containing lower concentrations of sodium and proteins and greater amounts of potassium, calcium, and phosphorus than serum, accumulate in skin tissue; macrocyclic trichothecenes apparently increase the permeability or leakiness of blood vessels (Ueno, 1984a).

Trichothecenes from *Stachybotrys atra* (*S. chartarum*) were isolated from contaminated insulation and ductwork in a house. Workmen handling the material without skin protection suffered painful skin lesions on their hands, armpits, and genitals (Croft et al., 1986; Jarvis, 1990). Hayes and Schiefer (1979) characterized the effect of small doses of T-2 toxin and DAS in the skin of rats and rabbits as an acute inflammatory reaction that involved hyperemia, edema, and neutrophil exudation, with variable amounts of necrosis of the epidermis.

Pang et al. (1987) applied T-2 toxin at 0 and 15 mg/kg in DMSO topically to the skin of SPF juvenile male pigs that had been immunized subcutaneously with sheep red blood cells. Serum samples and whole blood taken periodically were evaluated for clinical pathologic and immunologic changes. Treated pigs displayed anorexia, lethargy, posterior weakness and paresis, persistent high fevers, and reduced weight gain. Neutrophilia, decreased serum glucose, decreased albumin, decreased alkaline phosphatase activity, and increased serum globulin were seen in treated pigs. In addition to severe local dermal injury, this (sublethal) dose of T-2 toxin caused significant systemic effects, including cellular immune responses.

Carcinogenesis

Some bacteria and molds found in indoor environments produce molecules that are known or thought to be carcinogenic in humans and other animals (Table 4-5), and a number of toxins produced by molds are mu-

TABLE 4-5 Carcinogenic Effects of Mycotoxins

Mycotoxin	Producing Molds	Mechanism	Carcinogenic in Humans	Potency or Regulated Levels
Aflatoxin (AFB$_1$)	Aspergillus Flavus, Aspergillus parasiticus, Penicillium puberulum	AFB$_1$ activated to epoxide by cytochrome P-450 enzyme, DNA-adduct formation	Liver, lung cancer cocarcinogen with hepatitis B, Epstein-Barr viruses with other toxins	Canada: VSD (1 × 10^{-5}) = 0.14 ng/kg-day; Swiss standards in milk, cheese
Sterigmatocystin	Aspergillus versicolor, Aspergillus flavus, Penicillium luteum, Bipolaris spp.	Similar to AFB$_1$, but much less potent	Liver tumors, rats, mice; lung cancer, rats (chronic: 2 years, 6+/-4 conidiospores in air)	LD$_{50}$ = 60–800 mg/kg^a due to low solubility in water or gastric juices)
Ochratoxin A (OTA)	Aspergillus ochraceus, Penicillium verrucosum, Aspergillus alutaceus, Penicillium viridicatum, Penicillium cyclopium	Single-strand DNA breaks, unscheduled DNA synthesis, DNA-adduct formation, damages chromosomes	Suspect in kidney, pelvis, urethra, bladder carcinomas in Balkans; IARC possible human carcinogen	Canada: VSD (1 × 10^{-5}) = 0.18 ng/kg-day

(continued on next page)

TABLE 4-5 continued

Mycotoxin	Producing Molds	Mechanism	Carcinogenic in Humans	Potency or Regulated Levels
Zearalenone (ZEN)	Fusarium graminearum	Damages chromosomes, DNA adducts in kidney and liver	Suspected carcinogen (also estrogenic properties)	Canada VSD $(1 \times 10^{-6}) = 0.05$ μg/kg-day
Citrinin	Penicillium citrinum, Penicillium verrucosum, Penicillium vindicatum, Aspergillus terreus	Nephrotoxic, mildly hepatotoxic	NA	LD$_{50}$ = 50 mg/kg in rats; 35–58 mg/kg ip and 110 mg/kg orally in mice[b] 19 mg/kg ip in rabbits
Patulin	Penicillium expansum, Penicillium patulum, Aspergillus terreus		NA	US FDA: 50 μg/kg of body weight per day
Penicillin Acid	Aspergillus ochraceus	Affects heart[c]	Hepatocarcinogen in some animals[c]	
Luteoskyrin	Penicillium islandicum	Hepatotoxic, Nephrotoxic[d]	Carcinogen[d]	UK: 20–50 μg/kg of food

NOTE: Abbreviations: LD$_{50}$, lethal dose$_{50}$, the dose that kills 50% of the population; VSD, virtually safe dose.
[a]Terao, 1983.
[b]Scott, 1977.
[c]Reiss, 1988.
[d]Uraguchi et al., 1972.

tagenic or clastogenic in various species. Others are transformed to carcinogenic chemical species by host metabolism (Wang and Groopman, 1999), such as the epoxide metabolite of AFB_1 that is produced in the liver and lungs via cytochrome P-450 enzyme activity and is considered a possible human carcinogen by the International Agency for Research on Cancer (IARC, 1993). AFB_1 is produced by *Aspergillus flavus* and *A. parasiticus*. *A. flavus* causes a problem in agricultural grains and peanuts grown and stored in hot humid conditions, primarily in tropical and subtropical climates. Contamination with AFB_1 is associated with high rates of hepatocarcinoma in some African countries and appears to potentiate the hepatocarcinogenic properties of hepatitis B virus through its immunotoxic effects (Autrup et al., 1987; Badria et al., 1999; Bechtel, 1989; Groopman et al., 1992).

Exposure to high concentrations of dust from silos, grain, and peanut processing has been associated with liver and lung cancer in a few case studies and epidemiologic studies (Hayes et al., 1984; Olsen et al., 1988; van Nieuwenhuize et al., 1973). Concern about AFB_1-associated grain-dust inhalation by workers led to an intratracheal instillation and inhalation nose-only study of rats exposed to AFB_1 (Zarba et al., 1992). Maximal DNA binding of AFB_1 occurred within 30 min in the livers of the animals and indicated that inhalation exposure results in genotoxic damage to the liver; lung binding of AFB_1 to DNA was not assessed. *A. flavus* is generally not found indoors in northern climates, although it has been isolated from soil of indoor plants. It is found indoors more frequently in warm climates. AFB_1 is activated to a carcinogenic epoxide by human lung microsomes, but the cells that contain the activation enzymes are in low concentration in the human lung compared with human liver (Kelly et al., 1997). However, *A. versicolor*, which produces a precursor of aflatoxin, has been shown to induce tumors in germ-free rats (Sumi et al., 1987).

According to the U.S. National Toxicology Program's 10th report on carcinogens, OTA is reasonably anticipated to be a human carcinogen on the basis of sufficient evidence of carcinogenicity in experimental animals (NTP, 2002). It is produced by *A. ochraceus*, *A. alutaceus*, and *P. viridicatum*, *verrucosum*, and *cyclopium*, which are fairly common contaminants of grain and other foodstuffs. Although molds producing ochratoxins are occasionally found growing indoors, no study of their potential carcinogenic role from indoor exposures has been done. IARC (1993) concluded that there was inadequate evidence of carcinogenicity of OTA in humans but noted that is implicated in high rates of Balkan endemic nephropathy.

Citrinin, produced by *P. aurentiogriseum*, is often found with OTA, but is less potent (Krogh, 1989, 1992). Mayura et al. (1984) have produced experimental evidence of interactions between OTA and citrinin.

Molds that produce carcinogenic mycotoxins have been found among fungal flora indoors, but few studies that have isolated the toxins or looked for biomarkers of exposure have been conducted. All the studies that have implicated inhalation exposures related to cancers were of massive exposures to grain or peanut dust that contained spores with AFB_1 at concentrations hundreds of thousands of times greater in air than those thought to be present in indoor, nonagricultural environments. Therefore, the relevance of such exposures to those due to damp indoor spaces is unknown.

FINDINGS, RECOMMENDATIONS, AND RESEARCH NEEDS

On the basis of its review of the papers, reports, and other information presented in this chapter, the committee has reached several findings and recommendations and has identified several research needs regarding the nonallergic effects of molds and bacteria found in damp indoor environments.

• Molds that can produce mycotoxins under the appropriate environmental and competitive conditions can and do grow indoors. Damp indoor spaces may also facilitate the growth of bacteria that can have toxic and inflammatory effects. Little information exists on the toxic potential of chemical releases resulting from dampness-related degradation of building materials, furniture, and the like.

• In vitro and in vivo studies have established that exposure to microbial toxins can occur via inhalation and dermal exposure and through ingestion of contaminated food. Animal studies provide information on possible target organs, the underlying mechanisms of action, and the potency of many toxins isolated from environmental samples and substrates from damp buildings. The dose required to cause adverse health effects in humans has not been determined.

• In vitro and in vivo studies have demonstrated adverse effects— including immunotoxic, neurologic, respiratory, and dermal responses— after exposure to specific toxins, bacteria, molds, or their products.

• In vitro and in vivo research on *Stachybotrys chartarum* suggests that effects in humans may be biologically plausible; these observations require validation from more extensive research before conclusions can be drawn.

• Information on DNA, RNA, and protein adducts resulting from interactions with toxins is available. However, research is needed to further develop techniques for detecting and quantifying mycotoxins in tissues in order to inform the question of interactions and the determination of exposures resulting in adverse effects.

• Animal studies should be initiated to evaluate the effects of long-term (chronic) exposures to mycotoxins via inhalation. Such studies should establish dose-response, lowest-observed-adverse-effect levels, and no-observed-adverse-effect levels for identified toxicologic endpoints in order to generate information for risk assessment that is not available from studies of acute, high-level exposures.

REFERENCES

Amitani R, Taylor G, Elezis EN, Llewellyn-Jones C, Mitchell J, Kuze F, Cole PJ, Wilson R. 1995. Purification and characterization of factors produced by *Aspergillus fumigatus* which affect human ciliated respiratory epithelium. Infection and Immunity 63(9):3266–3271.

Ammann HM. 1999. Microbial Volatile Organic Compounds. In: Bioaerosols Assessment and Control. J Macher, ed. Cincinnati, OH: American Conference of Governmental Industrial Hygienists, Inc.

Andersen B, Nielsen FG, Jarvis BB. 2002. Characterization of *Stachybotrys* from water-damaged buildings based on morphology, growth, and metabolite production. Mycologia 94(3):392–403.

Andersson MA, Nikulin M, Koljag U, Andersson MC, Rainey F, Reijula K, Hintikkka EL, Salkinoja-Salonen M. 1997. Bacteria, molds and toxins in water-damaged building materials. Applied and Environmental Microbiology 63(2):387–393.

Autrup H, Seremet T, Wakhisi J, Wasunna A. 1987. Aflatoxin exposure measured by urinary excretion of aflatoxin-B_1-guanine adduct and hepatitis B virus infection in areas with different liver cancer incidence in Kenya. Cancer Research 47:3430–3433.

Badria FA, El-Nashur E, Hawas SA. 1999. Mycotoxins and disease in Egypt. Journal of Toxicology. Toxin Reviews 18(3&4):337–353.

Bamburg JR. 1976. Chemical and biochemical studies of the trichothecene mycotoxins. In Mycotoxins and other fungal related food problems. Joseph R. Rodericks, ed. Advances in Chemistry Series 149. Washington, DC: American Chemical Society. pp. 144–162.

Bamburg JR, Strong FM. 1971. 12, 13-Epoxytrichothecenes. In S. Kadis, A. Ciegler, and S. Ajl, eds. Microbial toxins, vol. 7. New York: Academic Press. pp. 207–292.

Bechtel DH. 1989. Molecular dosimetry of hepatic aflatoxin B_1-DNA adduct: linear correlation with hepatic cancer risk. Regulatory Toxicology and Pharmacology 10:74–81.

Bergmann F, Soffer D, Yagen B. 1988. Cerebral toxicity of the trichothecene toxin T-2, of the products of its hydrolysis, and of some related toxins. Toxicon 26(10):923–930.

Betina V. 1989. Mycotoxins Chemical, Biological and Environmental Aspects. New York: Elsevier.

Biagini RE. 1999. From fungal exposure to disease: a biological monitoring conundrum. In: Bioaerosols, Fungi and Mycotoxins: Health Effects, Assessment, Prevention and Control. E Johanning, ed. Albany, NY: Eastern New York Environmental Health Center, Mount Sinai School of Medicine, Department of Community Medicine. pp. 320–329.

Boonchuvit B, Hamilton PB, Burmeister HR. 1975. Interaction of T-2 toxin with *Salmonella* infections of chickens. Poultry Science 54(4):1693–1696.

Bruinink A, Sidler C. 1997. The neurotoxic effects of Ochratoxin A are reduced by protein binding but are not affected by *l*-phenylalanine. Toxicology and Applied Pharmacology 146:173–179.

Bruinick A, Rásonyi T, Sidler C. 1997. Reduction of Ochratoxin A toxicity by heat induced epimerization. In vitro effects of ochratoxins on embryonic chick meningeal and other cell cultures. Toxicology 118(2-3):205–210.

Buck WB, Cote LM. 1991. Trichothecene mycotoxins. In: Handbook of Natural Toxins, Volume 6. Toxicology of plant and fungal compounds. RF Keeler, AT Tu, eds. New York: Marcel Dekker Inc. pp. 523–554.

Buttner MP, Cruz-Perez P, Stetzenbach LD. 2001. Enhanced detection of surface-associated bacteria in indoor environments by quantitative PCR. Applied Environmental Microbiology 67(6):2564–2570.

CDC (Centers for Disease Control and Prevention). 1997. Update: pulmonary hemorrhage/ hemosiderosis among infants—Cleveland, Ohio, 1993–1996. Morbidity and Mortality Weekly Report 46(2):33–35.

Chang FW, Wang SD, Lu KT, Lee EHY. 1993. Differential interactive effects of gliotoxin and MPTP in the substantia nigra and the locus coeruleus in BALB/c mice. Brain Research Bulletin 31:253–266.

Ciegler A, Vesonder RF, Cole RJ. 1976. Tremorgenic Mycotoxins. Advances in Carbohydrate Chemistry and Biochemistry 149:163–177.

Cometto-Muniz JF, Cain WS. 1993. Efficacy of volatile organic compounds in evoking nasal pungency and odor. Archives of Environmental Health 48(5):309–314.

Corrier DE. 1991. Mycotoxicosis: mechanisms of immunosuppression. Veterinary Immunology and Immunopathology 30:73–87.

Corrier DE, Norman JO. 1988. Effects of T-2 mycotoxin on tumor susceptibility in mice. American Journal of Veterinary Research 49(12):2147–2150.

Coulombe RA Jr. 1993. Biological action of mycotoxins. Journal of Dairy Science 76:880–891.

Coulombe RA, Huie JM, Ball RW, Sharma RP, Wilson DW. 1991. Pharmacokinetics of intratracheally instilled aflatoxin B_1. Toxicology and Applied Pharmacology 109(2): 196–206.

Creasia DA, Thurman JD, Jones LJ III, Nealley ML, York CG, Wannemacher RW Jr, Bunner, DL. 1987. Acute inhalation toxicity of T-mycotoxin in mice. Fundamental and Applied Toxicology 8(2)230–235.

Creasia DA, Thurman JD, Wannemacher RW, Bunner DL. 1990. Acute inhalation toxicity of T-2 mycotoxin in the rat and guinea pig. Fundamental and Applied Toxicology 14: 54–59.

Creppy EE. 1995. Ochratoxin A in food: molecular basis of its chronic effects and detoxification. In: Molecular Approaches to Food Safety: Issues involving toxic micro-organisms. M Eklung, JL Richard, K Mise, eds. Fort Collins, CO: Alaken Inc. pp. 145–460.

Creppy EE, Stormer FC, Rosenthaler R, Dirheimer G. 1983. Effects of two metabolites of ochratoxin A (4R)-4hydroxy ochratoxin A and ochratoxin α on immune response in mice. Infection and Immunity 39:1015–1018.

Croft WA, Jarvis BB, Yatawara CS. 1986. Airborne outbreak of trichothecene toxicosis. Atmospheric Environment 20(3):549–552.

Cusumano V, Rossano F, Merendino RA, Arena A, Costa GB, Mancuso G, Baroni A, Losi E. 1996. Immunological activities of mould products: functional impairment of human monocytes exposed to aflatoxin B_1. Research in Microbiology 147:385–391.

Dearborn DG, Smith PG, Dahms BB, Allan TM, Sorenson WG, Montaña E, Etzel RA. 2002. Clinical profile of 30 infants with acute pulmonary hemorrhage in Cleveland. Pediatrics 110(3):627–637.

DeNicola DB, Rebar AH, Carlton WW, Yagen B. 1978. T-2 toxin mycotoxicosis in the guinea-pig. Food and Cosmetics Toxicology 16(6):601–609.

Di Paolo N, Guarnieri A, Loi A, Sacchi G, Mangiarotti AM, Di Paolo M. 1993. Acute renal failure from inhalation of mycotoxins. Nephron 64(4):621–625.

Dolimpio DA, Jacobson C, Legator M. 1968. Effect of aflatoxin on human leukocytes. Proceedings of the Society for Experimental Biology and Medicine 127:559–562.

Eaton DL, Klaassen CD. 2001. Principles of toxicology. In Casarett and Doull's Toxicology The Basic Science of Poisons. 6th edition. Curtis D. Klaassen, ed. New York: McGraw-Hill. pp. 11–34.

Eichner RD, Salami MA, Wood PR, Mullbacher A. 1986. The effects of gliotoxin upon macrophage function. International Journal of Immunopharmacology 8(7):789–797.

El-Banna AA, Leistner L. 1988. Production of penitrem A by *Penicillium crustosum* isolated from foodstuffs. International Journal of Food Microbiology 7:9–17.

Elidemir O, Colasurdo GN, Rossmann SN, Fan LL. 1999. Isolation of *Stachybotrys* from the lung of a child with pulmonary hemosiderosis. Pediatrics 104(4 Pt 1):964–966.

Englehart S, Loock A, Skutlarek D, Sagunski H, Lommel A, Färber H, Exner M. 2002. Occurrence of toxigenic *Aspergillus versicolor* isolates and sterigmatocystin in carpet dust from damp indoor environments. Applied and Environmental Microbiology 68(8): 3886–3890.

Etzel RA. 2002. Mycotoxins. Journal of the American Medical Association 287(4):425–427.

Fayos J, Lokensgard D, Clardy J, Cole RJ, Kirksey JW. 1974. Structure of verruculogen, a tremor producing peroxide from *Penicillium verruculosum*. Journal of the American Chemical Society 96(21):6785–6787. Communications to the Editor. (No. NAS 31).

Fiedler K, Schutz E, Geh S. 2001. Detection of microbial volatile organic compounds (MVOCs) produced by moulds on various materials. International Journal of Hygiene and Environmental Health 204:111–121.

Filtenborg O, Frisvad JC, Svendsen JA. 1983. Simple screening method for molds producing intracellular mycotoxins in pure cultures. Applied and Environmental Microbiology 45(2):581–585.

Fink-Gremmels J. 1999. Mycotoxins: their implications for human and animal health. The Veterinary Quarterly 21(4):115–120.

Flappan SM, Portnoy J, Jones P, Barnes C. 1999. Infant pulmonary hemorrhage in a suburban home with water damage and mold (*Stachybotrys atra*). Environmental Health Perspectives 107(11):927–930.

Flemming J, Hudson B, Rand TG. 2004. Comparison of inflammatory and cytotoxic lung responses in mice after intratracheal exposure to spores of two different *Stachybotrys chartarum* strains. Toxicologic Sciences, January 12 (Epub ahead of print).

Franck B, Gehrken HP. 1980. Citreoviridins from *Aspergillus terreus*. Angewandte Chemie 19(6):461–462.

Górny RL, Reponen T, Willeke K, Schmechel D, Robine E, Boissier M, Grinshpun SA. 2002. Fungal fragments as indoor air biocontaminants. Applied and Environmental Microbiology 68(7):3522–3531.

Gravesen S, Nielsen KF. 1999. Production of mycotoxins on water damaged building materials. In: Bioaerosols, Fungi and Mycotoxins: Health Effects, Assessment, Prevention and Control. E Johanning, ed. Albany, New York: Eastern New York Occupational and Environmental Health Center, Mount Sinai School of Medicine. pp. 423–431.

Gravesen S, Nielsen PA, Iversen R, Nielsen KF. 1999. Microfungal contamination of damp buildings—examples of constructions and risk materials. Environmental Health Perspectives 107(Supplement 3):505–508.

Gregory L, Rand TG, Dearborn D, Yike I, Vesper S. 2003. Immunocytochemical localization of stachylisin in *Stachybotrys chartarum* spores and spore-impacted mouse and rat lung tissues. Mycopathologia 156(2):109–117.

Gregory L, Pestka JJ, Dearborn DG, Rand TG. 2004. Localization of satratoxin-G in *Stachybotrys chartarum* spores and spore-impacted mouse lung using immunocytochemistry. Toxicologic Pathology 32(1):26–34.

Groopman JD, Jiaqi Z, Donahue PR, Pikul A, Lisheng Z, Jun-Shi C, Wogan GN. 1992. Molecular dosimetry of urinary aflatoxin-DNA adducts in people living in Guangxi Autonomous Region, People's Republic of China. Cancer Research 52:45–52.

Haley PJ. 1993. Immunological responses within the lung after inhalation of airborne chemicals. In: Toxicology of the Lung. 2nd Edition. DE Gardner, JD Crapo, RO McClellan, eds. New York: Raven Press, Ltd. pp. 389–416.

Harrach B, Nummi M, Niku-Palova ML, Mirocha CJ, Palyusik M. 1982. Identification of "water-soluble" toxins produced by a *Stachybotrys atra* strain from Finland. Applied and Environmental Microbiology 44(2):494–495.

Hayes MA, Schiefer HB. 1979. Quantitative and morphological aspects of cutaneous irritation by trichothecene mycotoxins. Food and Cosmetic Toxicology 17:611.

Hayes RB, van Nieuwenhuize JP, Raatgever JW, ten Kate FJW. 1984. Aflatoxin exposures in the industrial setting: an epidemiological study of mortality. Food and Chemical Toxicology 22:39–43.

Hirvonen MR, Nevalainen A, Makkkonen N, Savolainen K. 1997a. Induced production of nitric oxide, tumor necrosis factor, and interleukin-6 in RAW 264.7 macrophages by *Streptomycetes* from indoor air of moldy houses. Archives of Environmental Health 52(6):426–432.

Hirvonen MR, Nevalainen A, Makkkonen N, Mönkkönen J, Ruotsolainen M. 1997b. *Streptomyces* spores from mouldy houses induce nitric oxide, TNFα and IL-6 secretion from RAW264.7 macrophage cell line without causing subsequent cell death. Environmental Toxicology and Pharmacology 3:57–63.

Hirvonen MR, Ruotsalainen M, Roponen M, Hyvärinen A, Husman T, Kosma VM, Komulainen H, Savolainen K, Nevalainen A. 1999. Nitric oxide and proinflammatory cytokines in nasal lavage fluid associated with symptoms and exposure to moldy building microbes. American Journal of Respiratory and Critical Care Medicine 160:1943–1946.

Hirvonen MR, Suutari M, Ruotsalainen M, Lignell U, Nevalainen A. 2001. Effect of growth medium on potential of *Streptomyces annulatus* spores to induce inflammatory responses and cytotoxicity in RAW 264.7 macrophages. Inhalation Toxicology 13:55–68.

Hodge RP, Harris CM, Harris TM. 1988. Verrucofortine, a major metabolite of *Penicillium verrucosum* var. *cyclopium*, the fungus that produces the mycotoxin verrucosidin. Journal of Natural Products 51(1):66–73.

Hsieh DPH, Wong ZA, Wong JJ, Michas C, Ruebner BH. 1977. Comparative metabolism of aflatoxin. In: Mycotoxins in Humans and Animal Health. J Rodericks Chesseltine, M Mehlman, eds. Park Forest Sound, IL: Pathotox Publishers. pp. 37–50.

Hudnell HK, Otto DA, House DE, Mølhave L. 1992. Exposure of humans to a volatile organic mixture. II. Sensory. Archives of Environmental Health 47(1):31–38.

Hughes BJ, Hsieh GC, Jarvis BB, Sharma RP. 1989. Effects of macrocyclic trichothecene mycotoxins on the murine immune system. Archives of Environmental and Contamination Toxicology 18(3):388–395.

Huttunen K, Ruotsalainen M, Iivanainen E, Torkko P, Katila ML, Hirvonen MR. 2000. Inflammatory responses in RAW264.7 macrophages caused by mycobacteria isolated from moldy houses. Environmental Toxicology and Pharmacology 8:237–244.

Huttunen K, Hyvärinen A, Nevalainen A, Komulainen H, Hirvonen MR. 2003. Production of proinflammatory mediators by indoor air bacteria and fungal spores in mouse and human cell lines. Environmental Health Perspectives 111(1):85–92.

IARC (International Agency for Research on Cancer). 1993. IARC monographs on the Evaluation of the Carcinogenic Risk of Chemicals to Man. Some Naturally Occurring Substances: Food Items and Constituents, Heterocyclic Aromatic Amines and Mycotoxins. Vol. 56. Lyon, France. pp. 489–520.

Jakab GJ, Hmieleski RR, Hemenway DR, Groopman JD. 1994. Respiratory aflatoxicosis: suppression of pulmonary and systemic host defenses in rats and mice. Toxicology and Applied Pharmacology 125:198–205.

Jarvis BB. 1990. Mycotoxins and indoor air quality. In: Biological Contaminants in Indoor Environments. PR Morey, JC Feeley, JA Otten, eds. Boulder, CO: ASTM Symposium. July 16–19, 1989. pp. 201–214.

Jarvis BB. 1991. Macrocyclic Trichothecenes. In: Mycotoxins and phytoalexins in human and animal health. Sharma RP, Salunkhe DK, eds. Boca Raton, FL: CRC Press. pp. 361–421.

Jarvis BB. 1995. Mycotoxins in the air: keep your buildings dry or the bogeyman will get you. Proceedings of the International Conference: Fungi and Bacteria in Indoor Environments. Health Effects, Detection and Remediation. E Johanning, CS Yang, eds. Saratoga Springs, NY. October 6–7, 1994. pp. 35–44.

Jarvis BB, Salemme J, Morais A. 1995. *Stachybotrys* toxins. 1. Natural Toxins 3:10–16.

Jarvis BB, Sorenson WG, Hintikka EL, Nikulin M, Zhou Y, Jiang J, Wang S, Hinkley S, Etzel R, Dearborn D. 1998. Study of toxin production by isolates of *Stachybotrys chartarum* and *Memnoniella echinata* isolated during a study of pulmonary hemosiderosis in infants. Applied and Environmental Microbiology 64(10):3620–3625.

Joffe AZ, Ungar H. 1969. Cutaneous lesions produced by topical application of aflatoxin to rabbit skin. Journal of Investigative Dermatology 52(6):504–507.

Joki S, Saano V, Reponen T, Nevalainen A. 1993. Effect of indoor microbial metabolites on ciliary function of respiratory airways. Proceedings of Indoor Air '93 Helsinki, 1:259–263.

Jonsyn FE, Maxwell SM, Hendrickse RG. 1995a. Human fetal exposure to ochratoxin A and aflatoxins. Annals of Tropical Paediatrics 15(1):3–9.

Jonsyn FE, Maxwell SM, Hendrickse RG. 1995b. Ochratoxin A and aflatoxins in breast milk samples from Sierra Leone. Mycopathologia 131(2):121–126.

Jussila J, Komulainen H, Huttunen K, Roponen M, Hälinen A, Hyvärinen MR, Pelkonen J, Hirvonen MR. 2001. Inflammatory responses in mice after intratracheal instillation of spores from *Streptomyces californicus* isolated from indoor air of a moldy building. Toxicology and Applied Pharmacology 171:61–69.

Jussila J, Komulainen H, Huttunen K, Roponen M, Iivanainen E, Torkko P, Kosma VM, Pelkonen J, Hirvonen MR. 2002a. *Mycobacterium terrae* isolated from indoor air of a moisture-damaged building induces sustained biphasic inflammatory response in mouse lungs. Environmental Health Perspectives 110(11):1119–1125.

Jussila J, Komulainen H, Kosma VM, Nevalainen A, Pelkonen J, Hirvonen MR. 2002b. Spores of *Aspergillus versicolor* isolated from indoor air of a moisture-damaged building provoke acute inflammation in mouse lungs. Inhalation Toxicology 144:1261–1277.

Jussila J, Komulainen H, Kosma VM, Pelkonen J, Hyvärinen MR. 2002c. Inflammatory potential of the spores of *Penicillium spinulosum* isolated from indoor air of a moisture-damaged building in mouse lungs. Environmental Toxicology and Pharmacology 12:137–145.

Jussila J, Pelkonen J, Kosma VM, Mäki-Paakkanen J, Komulainen H, Hirvonen MR. 2003. Systemic immunoresponses in mice after repeated exposure of lungs to spores of *Streptomyces californicus*. Clinical and Diagnostic Laboratory Immunology 10(1):30–37.

Kasanen JP, Pasanen AL, Pasanen P, Liesvuori J, Kosma VM, Alarie Y. 1998. Stereospecificity of the sensory irritation receptor for nonreactive chemicals illustrated by pinene enantiomers. Archives of Toxicology 72(8):514–523.

Kellerman TS, Pienaar JG, van der Westhuizen GC, Anderson GC, Naude TW. 1976. A highly fatal tremorgenic mycotoxicosis of cattle caused by *Aspergillus clavatus*. Onderstepoort Journal of Veterinary Research 43(3):147–154.

Kelly JD, Eaton DL, Guengerich FP, Coulombe RA Jr. 1997. Aflatoxin B_1 activation in human lung. Toxicology and Applied Pharmacology 144(1):88–95.

Kemppainen BW, Riley RT, Pace JG. 1984. Penetration of [^{3}H]T-2 toxin through excised human and guinea-pig skin during exposure to [^{3}H]T-2 toxin adsorbed to corn dust. Food and Chemical Toxicology 22(11):893–896.

Kemppainen BW, Riley RT, Pace JG. 1988. Skin absorption as a route of exposure for aflatoxin and trichothecenes. Journal of Toxicology. Toxin Reviews 7(2):95–120.

Knapp JF, Michael JG, Hegenbarth MA, Jones PE, Black PG. 1999. Case records of the Children's Mercy Hospital, Case 02-1999: a 1-month-old infant with respiratory distress and shock. Pediatric Emergency Care 15(4):288–293.

Knaus HG, McManus OB, Lee SH, Schmalhofer WA, Garcia-Calvo M, Helms LMH, Sanchez M, Giangiacomo K, Reuben JP, Smith AB III, Kaczorowski GJ, Garcia ML. 1994. Tremorgenic indole alkaloids potently inhibit smooth muscle high-conductance calcium-activated potassium channels. Biochemistry 33:5819–5828.

Kordula T, Banbula A, Macomson J, Travis J. 2002. Isolation and properties of stachyrase A, a chymotrypsin-like serine protease from *Stachybotrys chartarum*. Infection and Immunity 70(1):419–421.

Koren HS, Graham DE, Devlin RB. 1992. Exposure of humans to a volatile organic mixture. III. Inflammatory response. Archives of Environmental Health 47(1):39–44.

Korpi A, Kasanen JP, Alarie Y, Kosma VM, Pasanen AL. 1999. Sensory irritating potency of some microbial volatile organic compounds (MVOCs) and a mixture of five MVOCs. Archives of Environmental Health 54(5):347–352.

Krogh P. 1989. The role of mycotoxins in disease of animals and man. Journal of Applied Bacteriology Symposium Supplement 79:99S–204S.

Krogh P. 1992. Role of ochratoxin in disease causation. Food and Chemical Toxicology 30(3):213–224.

Kuiper-Goodman T, Scott PM, Watanabe H. 1987. Risk assessment of the mycotoxin zearaleonone. Regulatory Toxicology and Pharmacology 7:253–306.

Kyriakidis N, Waight ES, Day JB, Mantle PG. 1981. Novel metabolites from *Penicillium crustosum*, including penitrem E, a tremorgenic mycotoxin. Applied and Environmental Microbiology 42(1):61–62.

Land CJ, Rask-Anderssen A, Werner S, Bardage S. 1994. Tremorgenic mycotoxins in conidia of *Aspergillus fumigatus*. In: Health Implications of Fungi in Indoor Environments. Air Quality Monograph, Vol. 2. RA Samson, B Flannigan, ME Flannigan, AP Verhoeff, ACG Adan, ES Hoekstra, eds. New York: Elsevier.

Larsen TO, Frisvad JC. 1994. Production of volatiles and presence of mycotoxins in conidia of common indoor *Penicillia* and *Aspergilli*. In: Health Implications of Fungi in Indoor Environments. Air Quality Monograph, Vol. 2. RA Samson, B Flannigan, ME Flannigan, et al., eds. New York: Elsevier. pp. 251–279.

Larsen TO, Svendsen A, Smedsgaard J. 2001. Biochemical characterization of Ochratoxin A-producing strains of the genus *Penicillium*. Applied and Environmental Microbiology 67(8):3630–3635.

Lewis CW, Smith JE, Anderson JG, Murad YM. 1994. The presence of mycotoxin-associated fungal spores isolated from the indoor air of the damp domestic environment and cytotoxic to human cell lines. Indoor Environment 3:323–330.

Lorenzana RM, Beasley VR, Buck WB, Ghent AW. 1985a. Experimental T-2 toxicosis in swine. II. Effect of intravascular T-2 toxin on serum enzymes and biochemistry, blood coagulation, and hematology. Fundamental and Applied Toxicology 5(5):893–901.

Lorenzana RM, Beasley VR, Buck WB, Ghent AW, Lundeen GR, Poppenga RH. 1985b. Experimental T-2 toxicosis in swine. I. Changes in cardiac output, aortic mean pressure, catecholamines, 6-keto-PGF1 alpha, thromboxane B2, and acid-base parameters. Fundamental and Applied Toxicology 5(5):879–892.

Lucero MT, Squires A. 1998. Catecholamine concentrations in rat nasal mucus are modulated by trigeminal stimulation of the nasal cavity. Brain Research 807(1-2):234–236.

Malmstrom J, Christophersen C, Frisvad JC. 2000. Secondary metabolites characteristic of *Penicillium citrinum*, *Penicillium steckii* and related species. Phytochemistry 54:301–309.

Marrs TC, Edginton JAG, Price PN, Upshall DG. 1986. Acute toxicity of T-2 mycotoxin to the guinea-pig by inhalation and subcutaneous routes. British Journal of Experimental Pathology 67(2):259–268.

Mason CD, Rand TG, Oulton M, MacDonald JM, Scott JE. 1998. Effects of *Stachybotrys chartarum (atra) conidia* and isolated toxin on lung surfactant production and homeostasis. Natural Toxins 6(1):27–33.

Mason CD, Rand TG, Oulton M, MacDonald J, Anthes M. 2001. Effects of *Stachybotrys chartarum* on surface convertase activity in juvenile mice. Toxicology and Applied Pharmacology 172(1):21–28.

Matsuoka Y, Kubota K, Ueno Y. 1979. General pharmacological studies of fusarenon-x. A trichothecene mycotoxin from *Fusarium* species. Toxicology and Applied Pharmacology 50:87–94.

Mayura K, Parker R, Berndt WO, Phillips TD. 1984. Effect of simultaneous prenatal exposure to ochratoxin A and citrinin in the rat. Journal of Toxicology and Environmental Health 13(4-6):553–561.

McCrae KC, Rand T, Shaw RA, Mason C, Oulton MR, Hastings C, Cherlet T, Thliveris JA, Mantsch HH, MacDonald S, Scott JE. 2002. Analysis of pulmonary surfactant by Fourier–transform infrared spectroscopy following exposure to *Stachybotrys chartarum (atra)* spores. Chemistry and Physics of Lipids 110(1):1–10.

McLaughlin CS, Vaughan MH, Campbell IM, Wei CM, Stafford ME, Hansen BS. 1977. Inhibition of protein synthesis by trichothecenes. In: Mycotoxins in Human and Animal Health. JS Rodrick, CM Hesseltine, MA Mehlman, eds. Park Forest Sound, IL: Pathotox Publishers. pp. 263–273.

Miki T, Fukui Y, Uemura N, Takeuchi Y. 1994. Regional difference in the neurotoxicity of ochratoxin A on the developing cerebral cortex in mice. Developmental Brain Research 82:259–264.

Miller JD, LaFlamme AM, Sobol Y, LaFontaine P, Greenhalgh R. 1988. Fungi and fungal products in some Canadian homes. International Biodeterioration 24:103–120.

Miller RV, Martinez-Miller C, Bolin V. 2001. Application of a novel risk assessment model to evaluate exposure to molds and mycotoxins in indoor environments. Proceedings, Second NSF International Conference on Indoor Air Health. January 2001, Miami Beach, FL.

Morgan KT, Gross EA, Bonnefoi M. 1993. Nasal structure, function, and toxicology. In: Toxicology of the Lung. 2nd Edition. DE Gardner, JD Crapo, RO McClellan, eds. New York: Raven Press, Ltd.

Murtoniemi T, Nevalainen A, Suutari M, Hirvonen MR. 2002. Effect of liner and core materials of plasterboard on microbial growth, spore-induced inflammatory responses, and cytotoxicity in macrophages. Inhalation Toxicology 14:1087–1101.

Nielsen KF, Thrane U, Larsen TO, Nielsen PA, Gravesen S. 1998. Production of mycotoxins on artificially inoculated building materials. International Biodeterioration and Biodegradation 42:9–16.

Nielsen KF, Huttunen K, Hyvärinen A, Andersen B, Jarvis BB, Hirvonen MR. 2001. Metabolite profiles of *Stachybotrys* isolates from water-damaged buildings and their induction of inflammatory mediators and cytotoxicity in macrophages. Mycopathologia 154:201–205.

Nieminen SM, Kärki R, Auriola S, Toivola M, Laatsch H, Laatikainen R, Hyvärinen A, von Wright A. 2002. Isolation and identification of *Aspergillus fumigatus* mycotoxins on growth medium and some building materials. Applied and Environmental Microbiology 68(10):4871–4875.

Nikulin M, Reijula K, Jarvis BB, Hintikka EL. 1996. Experimental lung mycotoxicosis in mice induced by *Stachybotrys atra*. International Journal of Experimental Pathology 77:213–218.

Nikulin M, Rejula K, Jarvis BB, Veijalainen P, Hintikka EL. 1997. Effects of intranasal exposure to spores of *Stachybotrys atra* in mice. Fundamental and Applied Toxicology 35(2):182–188.

Nishie K, Cole RJ, Dorner JW. 1988. Toxicity of citreoviridin. Research Communications in Chemical Pathology and Pharmacology 59(1):31–52.

Norris PJ, Smith CC, De Bellerroche J, Bradford HF, Mantle PG, Thomas AJ, Penny RH. 1980. Actions of tremorgenic fungal toxins on neurotransmitter release. Journal of Neurochemistry 34(1):33–42.

Novotny WE, Dixit A. 2000. Pulmonary hemorrhage in an infant following 2 weeks of fungal exposure. Archives of Pediatrics and Adolescent Medicine 154(3):271–275.

NTP (National Toxicology Program). 2002. Report on Carcinogens, Tenth Edition; U.S. Department of Health and Human Services, Public Health Service, National Toxicology Program, December 2002.

Olsen JH, Dragsted L, Autrup H. 1988. Cancer risk and occupational exposure to aflatoxins in Denmark. British Journal of Cancer 58:392–396.

Otokawa M. 1983. Immunological disorders. In: Trichothecenes: chemical, biological and toxicological aspects. Y Ueno, ed. New York: Elsevier. pp. 163–170.

Otto D, Mølhave L, Rose G, Hudnell HK, House D. 1990. Neurobehavioral and sensory irritant effects of controlled exposure to a complex mixture of volatile organic compounds. Neurotoxicology and Teratology 12:649–652.

Otto DA, Hudnell HK, House DE, Mølhave L, Counts W. 1992. Exposure of humans to a volatile organic mixture. I. Behavioral assessment. Archives of Environmental Health 47(1):23–30.

Pang VF, Lambert RJ, Felsburg PJ, Beasley VR, Buck WB, Haschek WM. 1987. Experimental T-2 toxicosis in swine following inhalation exposure: effects on pulmonary and systemic immunity, and morphologic changes. Toxicologic Pathology 15(3):308–319.

Pasanen AL, Korpi A, Kasaen JP, Pasanen P. 1999. Can microbial volatile metabolites cause irritation at indoor air concentrations? In: Bioaerosols, Fungi and Mycotoxins: Health Effects, Assessment, Prevention and Control. E Johanning, ed. Albany, NY: Eastern New York Occupational and Environmental Health Center. pp. 60–65.

Peltola J, Andersson MA, Mikkola R, Mussalo-Rauhamaa H, Salkinoja-Salonen M. 1999. In: Bioaerosols, Fungi and Mycotoxins: Health Effects, Assessment, Prevention and Control. E Johanning, ed. Albany, NY: Eastern New York Environmental Health Center. pp. 432–443.

Peltola J, Niessen L, Jarvis BB, Andersen B, Salkinoja-Salonen M, Moller EM. 2002. Toxigenic diversity of two different RAPD groups of *Stachybotrys chartarum* isolates analyzed by potential for trichothecene production and for boar sperm cell motility inhibition. Canadian Journal of Microbiology 48(11):1017–1029.

Pestka JJ, Bondy GS. 1990. Alteration of immune function following dietary mycotoxin exposure. Canadian Journal of Physiology and Pharmacology 68:1009–1016.

Peterson DW, Penny RH. 1982. A comparative study of sheep and pigs given the tremorgenic mycotoxins verruculogen and penitrem A. Research in Veterinary Science 33:183–187.

Pier AC, McLoughlin ME. 1985. Mycotoxin suppression of immunity. Trichothecenes and other mycotoxins. JC Lacey, ed. New York: John Wiley and Sons.

Pitt JI. 2000. Toxigenic fungi and mycotoxins. British Medical Bulletin 56(1):184–192.

Prelusky DB. 1993. The effect of low-level deoxynivalenol on neurotransmitter levels measured in pig cerebral spinal fluid. Journal of Environment, Science, and Health B 28: 731–761.

Rand TG, Mahoney M, White K, Oulton M. 2002. Microanatomical changes in alveolar type II cells in juvenile mice intratracheally exposed to *Stachybotrys chartarum* spores and toxin. Toxicologic Science 65(2):239–245.

Rand TG, White K, Logan A, Gregory L. 2003. Histological, immunohistochemical and morphometric changes in lung tissue in juvenile mice experimentally exposed to *Stachybotrys chartarum* spores. Mycopathologia 156(2):119–131.

Rao CY, Burge HA, Brain JD. 2000a. The time course of responses to intratracheally instilled toxic *Stachybotrys chartarum* spores in rats. Mycopathologia 149:27–34.

Rao CY, Brain JD, Burge HA. 2000b. Reduction of pulmonary toxicity of *Stachybotrys chartarum* spores by methanol extraction of mycotoxins. Applied Environmental Microbiology 66(7):2817–2821.

Rehm SR, Gross GN, Pierce AK. 1980. Early bacterial clearance from murine lungs. Species-dependent phagocyte response. Journal of Clinical Investigation 66:194–199.

Reiss J. 1988. Study on the formation of penicillic acid by moulds on bread. XVIII. Mycotoxins in foodstuffs. Deutsche Lebensmittel.-Rundschau 84:318–320.

Richard JL, Thurston JR. 1975. Effect of ochratoxin and aflatoxin on serum proteins, complement activity, and antibody production to *Brucella abortus* in guinea pigs. Applied Microbiology 29:27–29.

Rotter BA, Prelusky DB, Pestka JJ. 1996. Toxicology of deoxynivalenol (vomitoxin). Journal of Toxicology and Environmental Health 48(1):1–34.

Rozman KK, Klaassen CD. 1996. Absorption, distribution, and excretion of toxicants. In: Casarett and Doull's Toxicology. The Basic Science of Poisons. 5th edition. CD Klaassen, ed. New York: McGraw-Hill.

Ruotsolainen M, Hirvonen MR, Hyvärinen A, Meklin T, Savolainen K, Nevalainen A. 1995. Cytotoxicity, production of reactive oxygen species and cytokines induced by different strains of *Stachybotrys* sp. From moldy buildings in RAW 264.7 macrophages. Environmental Toxicology and Pharmacology 6:193–199.

Russell FE. 1996. Toxic effects of animal toxins. In: Cassarett and Doull's Toxicology. The Basic Science of Poisons. 5th edition. CD Klaasen, MO Amdur, J Doull, eds. New York: McGraw-Hill. p. 802.

Schiefer HB, Hancock DS. 1984. Systemic effects of topical application of T-2 toxin in mice. Toxicology and Applied Pharmacology 76(3):464–472.

Schiffman SS, Walker JM, Dalton P, Lorig TX, Raymer JH, Schusterman D, Williams CM. 2000. Potential health effects of odor from animal operations, wastewater treatment, and re-cycling by-products. Journal of Agromedicine 7(1):7–81.

Scott PM 1977. Penicillium Mycotoxins. In Mycotoxic Fungi, Mycotoxins, Mycotoxicoses. An Encyclopedic Handbook. TD Wyllie, LG Morehouse, eds. New York: Marcel Dekker. pp. 283–356.

Selala MI, Daelmans F, Schepens PJC. 1989. Fungal tremorgens: the mechanism of action of single nitrogen containing toxins—a hypothesis. Drug and Chemical Toxicology 12(3&4):237–257.

Smith BL. 1997. Effect of the mycotoxins penitrem, paxilline, and lolitrem B on the elctromyographic activity of skeletal and gastrointestinal smooth muscle of sheep. Research in Veterinary Science 62:111–116.

Smith JE, Moss MO. 1985. Mycotoxins Formation, Analysis, and Significance. New York: John Wiley and Sons.

Sorenson WG. 1993. Mycotoxins: toxic metabolites of fungi. In: Fungal Infections and Immune Response. JW Murphy et al., eds. New York: Plenum Press. pp. 469–491.

Sorenson WG, Simpson J. 1986. Toxicity of penicillic acid for rat alveolar macrophages *in vitro*. Environmental Research 41:505–513.

Sorenson WG, Simpson J, Castranova V. 1985. Toxicity of the mycotoxin patulin for rat alveolar macrophages. Environmental Research 38:407–416.

Sorenson WG, Gerberick GF, Lewis DM, Castranova V. 1986. Toxicity of mycotoxins for the rat pulmonary macrophage *in vitro*. Environmental Health Perspectives 66:45–53.

Sorenson WG, Frazer DG, Jarvis BB, Simpson J, Robinson VA. 1987. Trichothecene mycotoxins in aerosolized conidia of *Stachybotrys atra*. Applied and Environmental Microbiology 53(6):1370–1375.

Sorenson, WB, Jarvis BB, Zhou Y, Jiang J, Wang S, Hintikka E-L, Nikulin M. 1996. Toxine im zusammenhang mit *Stachybotrys* und *Memnoniella* in häusern mit wasserschaden. (Toxins associated with *Stachybotrys* and *Memnoniella* in houses with water damage). M. Gareis and R. Sheuer (eds). Myktoxin Workshop. Institut für Mikobiologie und Toxikologie der Bundesanstalt für Fleischforschung, Kulmbach, Germany. pp. 207–214.

Steyn PS, Vleggaar R. 1985. Tremorgenic mycotoxins. Fortschritte der Chemie Organischer Naturstoffe. (Progress in the Chemistry of Organic Natural Products) 48:1–80.

Sumarah MW, Rand TG, Mason CD, Oulton M, MacDonald J, Anthes M. 1999. Effects of *Stachybotrys chartarum* spores and toxin on mouse lung surfactant phospholipid composition. Proceedings of the Third International Conference on Bioaerosols, Fungi and Mycotoxins. Health Effects, Assessment, Prevention and Control. Mount Sinai Medical School and Eastern NY Center for Occupational Health. Saratoga Springs, NY. pp. 444–452.

Sumi Y, Hamasaki T, Miyakawa M. 1987. Tumors and other lesions induced in germ-free rats exposed to *Aspergillus versicolor* alone. Japanese Journal of Cancer Research 78(5):480–486.

Sumi Y, Nagura H, Takeuchi M, Miyakawa M. 1994. Granulomatous lesions in the lung induced by inhalation of mold spores. Virchows Archiv 424(6):661–668.

Terao K. 1983. The target organella of trichothecenes in rodents and poultry. In Ueno Y, ed. Trichothecenes—Chemical, Biological, and Toxicological Aspects. Developments in Food Sciences 4. New York: Elsevier.

Tripi PA, Modlin S, Sorensen WG, Dearborn DG. 2000. Acute pulmonary haemorrhage in an infant during induction of general anesthesia. Paediatric Anaesthia 10(1):92–94.

Ueno Y. 1977. Mode of action of trichothecenes. Annales de la Nutrition et de l'Alimentation 31(4-6):885–900.

Ueno Y. 1980. Trichothecene mycotoxins—mycology, chemistry, and toxicology. In Advances in Nutritional Research, Volume 3. HH Draper, ed. New York: Plenum Publishing Corp. pp. 301–353.

Ueno Y. 1984a. Toxicological features of T-2 toxin and related trichothecenes. Fundamental and Applied Toxicology 4:S124.

Ueno Y. 1984b. The toxicology of mycotoxins. CRC Critical Reviews in Toxicology 14:99–152.

Ueno Y. 1989. Trichothecene mycotoxins: mycology, chemistry, and toxicology. Advances in Nutritional Research 3:301–353.

Ueno Y, Ueno I. 1972. Isolation and acute toxicity of citreoviridin, a neurotoxic mycotoxin of *Penicillium citreo-viride* Biourge. Japanese Journal of Experimental Medicine 42(2): 91–105.

Uraguchi K, Saito M, Noguchi Y, Takahashi K, Enomoto M. 1972. Chronic toxicity and carcinogenicity in mice of the purified mycotoxins, luteoskyrin and cyclochlorotine. Food and Cosmetics Toxicology 10(2):193–207.

van Nieuwenhuize JP, Herber RFM, De Bruin A, Meyer PB, Duba WC. 1973. Aflatoxinen: Epidemiologisch onderzoek naar carcinogeniteit bij langdurige "low level" expositie van een fabriekspopulatie [Aflatoxin: epidemiological study on the carcinogenicity of prolonged exposure to low levels among plant workers]. Tijdschrift voor sociale geneeskunde 51:706–710; 717; 754.

van Walbeek W, Scott PM, Harwig J, Lawrence JW. 1969. *Penicillium viridicatum* Westling: a new source of ochratoxin A. Canadian Journal of Microbiology 15:1281–1285.

Vesper S, Vesper MJ. 2002. Stachylysin may be a cause of hemorrhaging in humans exposed to *Stachybotrys chartarum*. Infection and Immunity 70(4):2065–2069.

Vesper S, Dearborn DG, Yike I, Sorenson WG, Haugland RA. 1999. Hemolysin, toxicity, and randomly amplified polymorphic DNA analysis of *Stachybotrys chartarum* strains. Applied and Environmental Microbiology 65(7):3175–3181.

Vesper SJ, Dearborn DG, Elidemir O, Haugland RA. 2000a. Quantification of siderophore and hemolysin from *Stachybotrys chartarum* strains, including a strain isolated from the lung of a child with pulmonary hemorrhage and hemosiderosis. Applied and Environmental Microbiology 66(6):2678–2681.

Vesper S, Dearborn DG, Yike I, Allan T, Sobolewski J, Hinckley SF, Jarvis BB, Haugland RA. 2000b. Evaluation of *Stachybotrys chartarum* in the house of an infant with pulmonary hemorrhage: quantitative assessment before, during, and after remediation. Journal of Urban Health 77(1):68–85.

Vesper S, Magnuson ML, Dearborn DG, Yike I, Haugland RA. 2001. Initial characterization of the hemolysis from *Stachybotrys chartarum*. Infection and Immunity 69(2):912–916.

Wainman T, Zhang J, Weschler CJ, Lioy PJ. 2000. Ozone and limonene in indoor air: a source of submicron particle exposure. Environmental Health Perspectives 108(12): 1139–1145.

Wang JS, Groopman JD. 1999. DNA damage by mycotoxins. Mutation Research 424:167–181.

Wannemacher RW Jr, Bunner DL, Neufeld HA. 1991. Toxicity of trichothecenes and other related mycotoxins in laboratory animals: In JE smith, RS Henderson, eds. Mycotoxins and Animal Foods. Boca Raton: CRC Press. pp. 449–452.

Waring P, Beaver J. 1996. Gliotoxin and related epipolythiodioxopiperazines. General Pharmacology 27(8):1311–1316.

Weekley LB, O'Rear CE, Kimbrough TD, Llewellyn GC. 1989. Acute effects of the trichothecene mycotoxin T-2 on rat brain regional concentrations of serotonin, tryptophan, and tyrosine. Veterinary and Human Toxicology 31(3):221–224.

Weiss A, Chidekel AS. 2002. Acute pulmonary hemorrhage in a Delaware infant after exposure to *Stachybotrys atra*. Delaware Medical Journal 74(9):363–368.

WHO (World Health Organization). 1990. Selected Mycotoxins: Ochratoxins, Trichothecenes, Ergot. Environmental Health Criteria 105. World Health Organization. Geneva.

Wilson BJ, Wilson CH, Hayes AW. 1968. Tremorgenic toxin from penicillium cyclopium grown on food materials. Nature 220:77–78.

Wilson DW, Ball RW, Coulombe A Jr. 1990. Comparative action of aflatoxin B_1 in mammalian airways. Cancer Research 50:2493–2498.

Yike I, Miller MJ, Sorenson WG, Walenga R, Tomashefski JF Jr, Dearborn DG. 2002a. Infant animal model of pulmonary mycotoxicosis induced by *Stachybotrys chartarum*. Mycopathologia 154(3):139–152.

Yike I, Rand TG, Dearborn D. 2002b. Proteases from the spores of toxigenic fungus *Stachybotrys chartarum*. American Journal of Respiratory and Critical Care Medicine. Supplement American Thoracic Society Abstracts 165:A537.

Yike I, Vesper S, Tomashefski JF Jr, Dearborn DG. 2003. Germination, viability and clearance of *Stachybotrys chartarum* in the lungs of infant rats. Mycopathologia 156(2):67–75.

Zarba A, Hmielski R, Hemenway DR, Jakab GJ, Groopman DR. 1992. Aflatoxin B_1-DNA adduct formation in rat liver following exposure by aerosol inhalation. Carcinogenesis 13(6):1031–1033.

5

Human Health Effects Associated with Damp Indoor Environments

INTRODUCTION

Various human health effects have been attributed to damp or moldy indoor environments. Respiratory symptoms are most often researched, but other symptoms and clinical outcomes have also been examined in studies and anecdotal reports.

Previous chapters of this report have addressed the scientific literature regarding the biologic and chemical agents encountered in damp indoor environments: the factors influencing their presence or release, actions that can be taken to prevent or remediate contamination by them, the means available to characterize human exposure to them, and their toxic properties. This chapter evaluates the strength of the scientific evidence concerning the possible association between the agents and adverse health outcomes. Although it does not review all such literature[1]—an undertaking beyond the scope of this report—it attempts to cover the most recent studies and other work that the committee believed to be influential in shaping scientific understanding at the time it completed its task in late 2003. The chapter is organized by health outcome. Each major section describes the characteristics or symptoms of the health outcome, discusses the evidence

---

[1]Several surveys and reviews of the literature regarding damp indoor spaces and health or of specific exposures related to damp indoor spaces have been published in recent years, including Bornehag et al. (2001), Fung and Hughson (2003), Kolstad et al. (2002), Kuhn and Ghannoum (2003), Peat et al. (1998), Piecková and Jesenská (1999), and Robbins et al. (2000).

of possible association between the outcome and the presence of dampness or dampness-related microbial agents, and presents the committee's conclusions regarding the evidence of the association. (Chapter 1 describes the methodologic considerations underlying the evaluation of the evidence and definitions of the categories used to summarize the committee's findings.) Because there are great differences between specific health outcomes in the amount and type of information available, the sections vary in their depth and focus.

This chapter, like other parts of the report, focuses on studies that examine the health effects of dampness or of fungi and bacteria associated with damp indoor spaces. Other exposures that may be found in such environments—notably, house dust mites, viruses, and environmental tobacco smoke (ETS)—are not addressed here, although their presence may have important effects on occupants. The health effects of those agents are covered in detail in the Institute of Medicine reports *Clearing the Air* (IOM, 2000), as related to asthma, and *Indoor Allergens* (IOM, 1993), as related more generally to allergic responses. Smoking and ETS in particular are established confounding factors in studies of respiratory health outcomes and serve as sensitizing agents and potentiators of effect (IOM, 2000; Scientific Committee on Tobacco and Health, 1998; U.S. EPA, 1992). Larger organisms, such as cockroaches, also inhabit damp spaces and may be responsible for some of the health problems attributed to these spaces and are addressed in the previously cited IOM reports. Studies of such microbial infections as tinea pedis (athlete's foot) that are associated with moisture but not the damp indoor conditions addressed in this report are excluded.

An extensive literature examines the influence on occupants' health of various agents found indoors—such as pesticides (Lewis, 2000), nitrogen dioxide (NO_2) from gas appliances (Neas et al., 1991), and volatile organic compounds and formaldehyde outgassing from furnishings or construction materials (Norbäck et al., 1995; U.S. EPA, 1989)—or characteristics of indoor environments, including ventilation rate, temperature, and the use of circulated air and sealing measures to improve energy efficiency (Engvall et al., 2003; Seppänen and Fisk, 2002). When reading this chapter, one should remember that many of the health effects attributed to the presence of mold or other dampness-related agents in the papers cited here have also been attributed to other factors. Not all papers that address damp indoor spaces control for those other factors, just as dampness-related agents are not always examined as possible factors in studies of the health effects of indoor spaces. This weakness in the literature underlines the importance of the committee's recommendations for research on improved methods of exposure assessment.

Indoor environments are complex. They subject occupants to multiple exposures that may interact physically or chemically with one another and with the other characteristics of the environment, such as humidity, temperature, and ventilation. Synergistic effects—interactions among agents that result in combined effects greater than the sums of the individual effects—may take place. Information on the combined effects of multiple factors and on synergist effects among agents is cited wherever possible. However, as was noted in *Clearing the Air*, little information is available on this topic and it remains one of active research interest.

Finally, some factors may influence people's exposure to indoor agents, their ability to respond to circumstances in which indoor exposure may increase the risk of adverse health outcomes, and their health in general. Notable among those is socioeconomic status (SES). Low SES may be a contributory or independent factor in some of the health outcomes addressed below, affecting their incidence of severity.

Thus, when the committee draws conclusions about the association between damp indoor environments and health outcomes, it is not imposing the assumption—and readers should not presume—that these outcomes are necessarily associated with exposure to a specific microbial agent or to microbial agents in general. When an association between a particular indoor dampness-related agent and a particular health outcome is addressed, it is specified in the text. However, even in those cases, it is likely that people are being exposed to multiple agents.

The following sections draw conclusions about the state of the scientific literature regarding association of health outcomes with two circumstances: exposure to a damp indoor environment and the presence of mold or other microbial agents in a damp indoor environment. As noted in Chapter 2, the term dampness has been applied to a variety of moisture problems in buildings that include high relative humidity, condensation, and signs of excess moisture or microbial growth. Most of the studies considered by the committee did not specify which microbial agents were present in the buildings occupied by subjects, and this likely varied between and even within study populations. The conclusions presented here qualify the term mold with quotation marks to indicate the uncertainty regarding the agents that may be involved.

To fulfill their charge to evaluate the effect of damp indoor spaces on health, the committee conducted a review of epidemiologic studies. The committee began their evaluation presuming neither the presence nor the absence of association. They sought to understand the strengths and limitations of the available evidence. These judgments have both quantitative and qualitative aspects. They reflect the nature of the exposures, health outcomes, and populations exposed; the characteristics of the evidence examined; and the approach taken by the study's authors to evaluate this evidence. Because of the great differences among the studies reviewed, the committee concluded it was inappropriate to use quantitative summary techniques such as meta-analysis. Instead, as detailed in Chapter 1, the committee summarized their judgment of the association between dampness or mold and particular health outcomes by using a common format to categorize the strength of the scientific evidence. Fungi and other microbial agents are omnipresent in the environment, and the committee restricted its evaluation to circumstances that could be reasonably associated with damp indoor environments. Studies regarding homes, schools, and office buildings were considered; such other indoor environments as barns, silos, and factories—which may subject people to high occupational exposures to organic dusts and other microbial contaminants—were not.

EVALUATING HEALTH EFFECTS

Most of the research about the health effects resulting from damp indoor spaces is the result of epidemiologic investigations of associations between self-reported symptoms or clinical outcomes and the presence of dampness (however it might be defined or termed) or "mold," either reported or measured. The studies examined in this report primarily addressed dampness or mold in the home, reflecting the focus of researchers working in this field. A small number of studies of office or school environments were also evaluated. As detailed in Chapters 2 and 3, many of the studies use reports of current or past signs of dampness and mold or general measures of it as a proxy for the agents of interest. A few have considered dampness as a risk factor separate from the presence of microbial agents indoors.

There are thought to be more than one million species of fungi, but humans are routinely exposed to only about 200 (IOM, 2000), and fewer than 50 are commonly identified and described in epidemiologic studies of indoor environments (Asero and Bottazzi, 2000). Many health studies that evaluate the presence of mold do not formally identify species. Instead, they use "mold" as a generic term to describe microbial growth. From a practical standpoint, that means that fungi—perhaps several species—are grouped with fungus-like bacteria (such as thermophilic actinomycetes) when the health consequences of microbiologic agents are being investigated. Epidemiologic studies that examine particular mold species or strains often fail to factor or minimize the possible influence of other mold species and bacteria and other agents associated with damp indoor environments. Only a handful of researchers have explicitly examined chemical emissions from water-damaged materials. Their studies are discussed in Chapter 2.

Clinical studies and case reports are additional sources of information on some health outcomes, but they are often limited by the small number of subjects examined. Clinical studies may involve exposure scenarios (such as intentional installation) that are not encountered outside the laboratory, and case reports often address unusual or unusually high potential exposures that are not representative of those experienced in homes, schools, or office buildings. Some clinical studies and case reports are cited below; their results were considered by the committee with the understanding of their inherent limitations.

Anecdotal reports of health problems attributed to mold indoors often dominate mass-media attention, but they are not a source of reliable information. Good epidemiologic and clinical practice in investigations of potential environmental health problems requires—to the extent possible—the evaluation of all suspect environmental agents, valid measures of exposures and health outcomes, and thorough consideration of alternative explanations for observed signs, symptoms, and diseases. Those criteria can be difficult to completely fulfill in scientific studies, and they are seldom met in outlets where information is not subject to rigorous scientific standards.

Epidemiologists most commonly use questionnaires to collect information about symptoms, signs, and diseases. Exposures are often characterized through self-reports or expert-reports of the presence of dampness or visible mold. While self-reports are often the only way to gather information from large numbers of subjects in a cost-efficient manner, they have disadvantages that must be considered when evaluating studies that use them. A self-report of dampness or visible mold, for example, may indicate rather a wide range of potential exposures: particular molds, endotoxin, gram-negative bacteria, microbial VOCs, house dust mites, and dampness-related chemical releases from building materials, among others. Except in cases where studies carefully separate dampness-related exposures or where specific biomarkers of exposure exist, it can be difficult to identify the responsible agent and even then the identification of the agent may be problematic. It is not always possible to determine whether a specific health outcome examined in a study is caused by an allergic reaction versus an infectious agent, an irritant stimulus, a toxic agent, or some other cause. The clinical literature and to a lesser extent, toxicological studies, inform the interpretation of some epidemiological findings—especially those studies that are carried out under carefully controlled conditions. However, confident attribution of an outcome to a particular pathological mechanism is often limited by the observational (rather than experimental) nature of epidemiological studies and more than one mechanism may be responsible for the results in a particular study. Studies reviewed by the committee examined populations from across the United States and numerous foreign countries including Canada, Australia, New Zealand, and the nations of Europe. Differences in such factors as climate, predominant mycoflora, building practices, the genetic make up of subjects, and cultural traditions may affect results. Despite these limitations, epidemiological studies provide useful information for studying patterns of disease in populations and drawing conclusions about possible environmental influences.

Clinical measures are sometimes used in smaller-scale studies. For respiratory disease outcomes, these include lung-function testing based on spirometry or peak expiratory flow measures. Challenge testing with inhalation of methacholine, histamine, or other substances designed to induce bronchospasm in susceptible people has been used to measure the extent of bronchial hyperresponsiveness in clinical settings and epidemiologic studies. The thorax has been imaged radiographically with chest x-rays and computed tomographic (CT) scans to evaluate individual patients in clinical studies. Lung biopsy may be indicated to confirm or rule out the diagnosis of particular diseases, such as hypersensitivity pneumonitis. Direct objective means of measuring nasal function have not been widely applied to the evaluation of complaints related to damp indoor spaces. Similarly, objective clinical measures have not been widely used to investigate gastrointestinal, dermatologic, rheumatologic, or neurologic complaints.

A variety of biologic markers of inflammation are increasingly being applied to measure the effects of exposure to dampness and dampness-related agents in indoor environments (Purokivi et al., 2001; Roponen et al., 2001a; Trout et al., 2001; Wålinder et al., 2001). Circulating immunoglobulin G (IgG) antibodies to microbial agents that can cause hypersensitivity pneumonitis have been shown to have limited prognostic significance as markers of the chronic form of this disorder (Cormier and Bélanger, 1989; Guernsey et al., 1989; Marx et al., 1990), but it has been asserted that these markers are more useful as indicators of recent high-level exposure to specific molds and thermophilic actinomycete antigens (Lacasse et al., 2003). A 2003 study suggests that stachylysin—a proteinaceous hemolysin—may be a useful indicator of human exposure to *Stachybotrys chartarum* (Van Emon et al., 2003). Immunologic markers that have been examined in relation to indoor environmental exposures include cytokines, other mediators of inflammation, and antibodies to mycotoxins measured in nasal lavage fluids and in serum. Exhaled nitric oxide (NO) is a biomarker of respiratory tract inflammation that is elevated in some inflammatory lung conditions but not in others (Robbins et al., 1996). Measurement of substances in induced sputum samples and exhaled-breath condensate samples has not yet been applied to dampness or mold in indoor spaces but might be used to investigate them (Mutli et al., 2001). Variability in individual susceptibility as mediated by genetic risk factors is beginning to be explored by investigators. Chapter 3 addresses the use of biomarkers in exposure assessment.

Although this report focuses on health effects associated with excessive indoor dampness, excessive dryness may also be a problem. An indoor environment is typically considered to be dry if the relative humidity level falls below 30% (Nagda and Hodgson, 2001). Low indoor relative humidity conditions are more likely in winter when cold outdoor air, which is less able to hold moisture, is drawn indoors and warmed. Health complaints associated with indoor dryness include skin irritation, drying of the lining (mucous membranes) of the nose, mouth, and throat, nosebleeds, eye irritation, sore throat, and minor respiratory difficulties (Arundel et al., 1986; Berglund, 1998). A 2003 study by Reinikainen and Jaakkola found that, in dry conditions, increasing the humidity level alleviated some of these symptoms.

RESPIRATORY SYMPTOMS

Respiratory symptoms—possible indications of disease rather than disease itself—have been ascribed to numerous agents found in and characteristics of indoor environments. This section divides them between upper respiratory tract (URT) and lower respiratory tract (LRT) symptoms. Studies reviewed vary in which symptoms and sets of symptoms they examine.

Upper Respiratory Tract Effects

The URT comprises the nose, mouth, sinuses, and throat. The committee identified numerous studies that examine either individual URT symptoms (such as nasal congestion or sore throat) or groups of symptoms. Rhinitis, an inflammatory condition that involves the nasal mucosa, constitutes one such group: nasal congestion, sneezing, and runny or itchy nose (Jaakkola et al., 1993; Koskinen, 1999). Allergic rhinitis ("hay fever") is rhinitis caused by IgE-mediated inflammation. Sinusitis symptoms are similar to those of the common cold; they result from the inflammation of the paranasal sinuses. Ear and eye symptoms related to URT infections are sometimes grouped with them. Sinus disease related specifically to *Aspergillus* is discussed later in this chapter.

Overview of the Evidence

Table 5-1 summarizes results of studies that address URT symptoms. Because these symptoms often occur together, the table includes papers that address several different outcomes. Among the studies summarized in the table is an investigation by Rylander and Mégevand (2000) of risk factors for respiratory infections. The investigation, which examined 304 Swiss children 4–5 years old, reported associations between humidity in the home and colds (odds ratio [OR], 2.71; 95% confidence interval, 1.07–6.91), sore throats (3.02; 1.14–7.98), and ear infections (2.78; 1.13–6.80) after adjusting for mother's age and allergy status. The ORs for the association

TABLE 5-1 Selected Epidemiologic Studies—Upper Respiratory Tract Symptoms and Exposure to Damp Indoor Environment or Presence of Mold or Other Agents in Damp Indoor Environments

Reference	Subjects	Dampness or Mold Measure	Risk Estimate
Cross-sectional studies			
Engvall et al., 2002	3,241 adults in multifamily buildings	Self-reported moldy odor and water leakage in preceding 5 years	1.92 (1.78–2.07) for nasal symptoms; 4.42 (4.09–4.77) for throat irritation
Kilpeläinen et al., 2001	10,667 students (18–25 years)	Self-reported visible mold	1.29 (1.01–1.66) for allergic rhinitis; 1.48 (1.17–1.88) for common cold ≥ 4 times per year
		Self-reported visible mold or damp stains or water damage	1.30 (1.12–1.51) for allergic rhinitis 1.28 (1.09–1.47) for common cold ≥ 4 times per year
Rylander and Mégevand, 2000	304 children (4–5 years old)	Self-reported humidity	2.71 (1.07–6.91) for cold; 3.02 (1.14–7.98) for sore throat
		Self-reported mold at home	2.27 (.082–6.33) for cold; 2.57 (0.86–7.71) for sore throat
Zacharasiewicz et al., 2000	2,849 children (6–9 years old)	Self-reported dampness at home	1.51 (1.31–1.74) for nasal symptoms

Dales and Miller, 1999	403 elementary school children	Self-reported mold or mildew	1.81(1.02–3.24) for itchy eyes, rash or itch, nose irritation
Wan and Li, 1999	1,113 workers in 19 office buildings	Self-reported mold	0.94 (0.50–1.77) for nasal congestion or runny nose
		Self-reported flooding	1.55 (0.79–3.06) for nasal congestion or runny nose
Wieslander et al., 1999	95 staff members in 4 hospitals	Measured dampness in concrete floor	1.10 (1.15–1.45) for irritated, stuffy, or runny nose; 1.29 (1.02–1.18) for itching, burning, or irritated eyes
Koskinen et al., 1999a	699 adults (16+ years) in 310 households	Surveyor-assessed moisture	1.06 (0.71–1.59) for rhinitis; 1.92 (1.11–3.30) for sinusitis; 1.46 (1.03–2.08) for sore throat
		Self-reported mold	1.89 (1.15–3.11) for rhinitis; 1.36 (0.78–2.39) for sinusitis; 2.40 (1.56–3.69) for sore throat

(continued on next page)

TABLE 5-1 continued

Reference	Subjects	Dampness or Mold Measure	Risk Estimate
Koskinen et al., 1999b	204 children (7–15 years old) in 310 households	Surveyor-assessed moisture	4.31 (1.80–10.34) for rhinitis; 0.75 (0.19–2.98) for sinusitis; 2.34 (1.13–4.86) for sore throat
Thorn and Rylander, 1998a	129 adults (18–83 years old)	Measured airborne glucan (> 2–4 ng/m^3)	1.23 (0.85–1.77) for irritation in nose
Pirhonen et al., 1996	1,460 adults (25–64 years old)	Self-reported damp or mold problem	1.68 (0.97–2.89) for dry or sore throat
Jaakkola et al., 1993	2,568 preschool children	Self-reported mold odor in preceding year	2.39 (1.15–4.98) for nasal congestion; 2.38 (1.13–4.99) for nasal excretion
		Self-reported water damage >1 year ago	4.60 (2.57–8.22) for nasal congestion; 3.19 (1.64–6.19) for nasal excretion
Brunekreef, 1992	2,685 adult parents of children 6–12 years old	Self-reported damp stains or mold growth (last 2 years)	1.03 (0.79–1.35) among women (for allergy); 1.24 (0.95–1.73) among men (for allergy)
Brunekreef et al., 1989	4,625 children (7–11 years old)	Self-reported molds (ever)	1.57 (1.31–1.87) for hay fever
		Self-reported dampness (ever)	1.26 (1.06–1.50) for hay fever

between visible mold in the home and colds, bronchitis, and sore throats were also greater than 1.0 but did not achieve statistical significance.

Koskinen and colleagues (1999a) studied 699 adults in Finland and found that those who reported moisture in their homes were more likely to have common colds (1.62; 1.08–2.41), hoarseness (1.44; 0.99–2.10), sore throat (2.40; 1.56–3.68), rhinitis (1.89; 1.15–3.11), and eye irritation (1.43; 0.84–1.83) in the preceding 12 months, after adjusting for smoking, age, sex, allergy, indoor pets, and atopic predisposition. A companion study of children in the same residences (Koskinen et al., 1999b) yielded ORs greater that 1.0 for some URT symptoms (cold, hoarseness, sore throat, rhinitis, and eye irritation) but not others (sinusitis and otitis). However, the number of observations was relatively small, and the confidence intervals were wide. Kilpeläinen et al. (2001) surveyed over 10,000 college students and observed an association between incidence of common colds (≥ 4 per year) and visible mold by itself (1.48; 1.17–1.88) and "visible mold or damp stains or water damage" (1.28; 1.09–1.47). An association was also found with allergic rhinitis and visible mold (1.29; 1.01–1.66) and "visible mold or damp stains or water damage" (1.30; 1.12–1.51). No association was observed between allergic conjunctivitis and either exposure surrogate. The analyses controlled for parental education, smoking, presence of second-hand smoke, pets, wall-to-wall carpeting, place of residence (farm, rural nonfarm, or urban), and type of residence (apartment vs other building types).

Wieslander et al. (1999) found that damp concrete floors were associated with an increased risk of irritated, stuffy, or runny nose (1.10; 1.02–1.18) and itching, burning, or irritated eyes (1.29; 1.15–1.45). The researchers, who were investigating the influence of building characteristics on URT symptom incidence, hypothesized that the health outcomes were the result of exposure to the emission of 2-ethyl-1-hexanol due to alkaline degradation of octylphthalates in floor materials. The concentrations of total and viable molds and bacteria were low in the buildings evaluated in the study.

A study of 4,625 children by Brunekreef et al. (1989) found significant relationships between rates of hay fever and "mold" or "dampness" in homes (ORs, 1.57 and 1.26, respectively), although neither finding correlated with fungal counts (Su et al., 1989, 1990).

Some other studies reviewed by the committee examined less common symptoms or focused on particular dampness-related agents. Subjects in moldy environments sometimes report an impaired sense of smell (Koskinen et al., 1999b; Nevalainen et al., 2001). The ability to smell can be tested and measured, but the method has not been applied in studies of dampness or mold in indoor environments, so interpretation of results is problematic. A 2000 study showed an association between nasal polyposis and skin reactivity to *Candida albicans* in a study of 15 patients but did not specifically examine whether the exposure had indoor sources (Asero and Bottazzi, 2000). Of the studies summarized in Table 5-1, only Thorn and Rylander (1998a) measured a specific dampness-related agent—airborne $(1\rightarrow3)$-β-D-glucan concentrations.

Conclusions

Several epidemiologic studies address the association between one or more URT symptoms—nasal congestion, sneezing, runny or itchy nose, and throat irritation—and indoor dampness or microbial contamination. Studies are uniform in showing an increased risk of those symptoms; some, although not all, of the studies report statistically significant associations. The committee did not conduct an empirical investigation of the possible effect of publication or respondent bias; as indicated in Chapter 1, it does not believe either to be the determining factor in the results. It concludes as follows:

- There is sufficient evidence of an association between exposure to a damp indoor environment and upper respiratory tract symptoms.
- There is sufficient evidence of an association between the presence of "mold" (otherwise unspecified) in a damp indoor environment and upper respiratory tract symptoms.

Lower Respiratory Tract Effects

The major passages and structures of the lower respiratory tract (LRT) include the windpipe (trachea) and within the lungs, the bronchi, bronchioles, and alveoli. Although a number of LRT symptoms are reported when people are exposed to agents or particular environmental conditions in the home and other indoor spaces, limitations of study design sometimes preclude conclusions to be drawn regarding whether symptoms reported by participants indicate of the presence of defined disease entities. LRT symptoms include cough with or without production of phlegm, wheeze, chest tightness, and shortness of breath (dyspnea).

Cough

Cough can be triggered by a variety of means, including exposure to allergens or irritants. It may either be a nonspecific complaint or be associated with a clinical syndrome.

Overview of the Evidence   Among the more recent studies reviewed in Table 5-2 is a 2003 effort by Belanger and colleagues that prospectively examined

TABLE 5-2 Selected Epidemiologic Studies—Cough and Exposure to Damp Indoor Environment or Presence of Mold or Other Agents in Damp Indoor Environments

Reference	Subjects	Dampness or Mold Measure	Risk Estimate
Cross-sectional studies			
Belanger et al., 2003[a]	593 infants (1–12 months old) with asthmatic sibling	Reported persistent mold or mildew in previous 12 months	1.53 (1.01–2.30) for persistent cough
		Measured fungal colonies (species not specified)	0.99 (0.89–1.10) for persistent cough
	256 infants with asthmatic sibling + asthmatic mother	Reported persistent mold or mildew in previous 12 months	1.91 (1.07–3.42) for persistent cough
		Measured fungal colonies (species not specified)	1.04 (0.87–1.24) for persistent cough
Gunnbjörnsdottir et al., 2003	1,853 young adults	Self-reported water damage or visible mold in home (n = 74)	2.23 (1.24–4.00) for long-term cough
Gent et al., 2002	880 infants (1–12 months old) with asthmatic sibling	Measured *Penicillium* low (1–499 CFU/m^3)	1.01 (0.80–1.28)
		Measured *Penicillium* medium (500–999 CFU/m^3)	1.62 (0.93–2.82)
		Measured *Penicillium* high ≥ 1,000 CFU/m^3)	2.06 (1.31–3.24)
		Measured *Cladosporium* low (1–499 CFU/m^3)	1.03 (0.79–1.35)
		Measured *Cladosporium* medium (500–999 CFU/m^3)	1.45 (0.99–2.12)
		Measured *Cladosporium* high ≥ 1,000 CFU/m^3)	0.72 (0.42–1.24)
		"Other mold" low (1–499 CFU/m^3)	1.05 (0.83–1.33)

(continued on next page)

TABLE 5-2 continued

Reference	Subjects	Dampness or Mold Measure	Risk Estimate
		"Other mold" medium (500–999 CFU/m^3)	0.78 (0.42–1.45)
		"Other mold" high (≥ 1,000 CFU/m^3)	1.18 (0.63–2.21)
		Water leaks	1.17 (0.91–1.49)
		Humidifier use	1.26 (1.01–1.56)
Engvall et al., 2001	3,241 adults in multifamily buildings	Self-reported moldy odor and signs of high humidity	3.97 (3.74–4.22)
		Self-reported moldy odor and major water leakage	3.78 (3.46–4.12)
Dales and Miller, 1999	403 elementary-school children	Self-reported mold or mildew	1.28 (0.74–2.23) for nocturnal cough or wheeze
Koskinen et al., 1999a	699 adults (16+ years old) in 310 households	Surveyor-assessed moisture	2.11(1.21–4.98) for nocturnal cough; 1.42 (0.92–2.19) for cough without phlegm; 1.15 (0.78–1.69) for cough with phlegm
		Self-reported mold	2.30 (1.32–4.01) for nocturnal cough; 1.60 (1.01–2.53) for cough without phlegm; 1.44 (0.95–2.19) for cough with phlegm
Koskinen et al., 1999b	204 children (≤ 15 years old) in 310 households	Surveyor-assessed moisture	5.72 (1.22–26.83) for nocturnal cough; 3.23 (1.43–7.31) for

Reference	Population	Exposure	Results
			cough without phlegm; 0.94 (0.47–1.87) for cough with phlegm
Taskinen et al., 1999	622 children (7–13 years old), 101 cases total	Self-reported dampness (school)	2.3 (1.3–4.1)
		Self-reported dampness (school, home)	4.7 (2.1–10.8)
Andriessen et al., 1998	1,614 children (5–13 years old)	Self reported moisture stains	1.01 (0.89–1.16)
Jedrychowski and Flak, 1998	1,129 children (age 9)	Self reported molds	1.01(0.87–1.18)
		Self-reported molds or dampness	1.13 (0.64–2.02)
Rylander et al., 1998	347 children in 2 schools (1 with previous mold problem)	Problem school vs control school	26.1% vs 10.3% ($p = 0.024$) in nonatopics, for dry cough; 54.5% vs 7.1% ($p = 0.003$) in atopics, for dry cough; 23.1% vs 8.2% ($p = 0.012$) in nonatopics, for dry cough at night without cold 58.3% vs 6.5% ($p < 0.001$) in atopics, for dry cough at night without cold; 17.0% vs 7.5% ($p = $ NS) in nonatopics,

(continued on next page)

TABLE 5-2 continued

Reference	Subjects	Dampness or Mold Measure	Risk Estimate
			for cough with phlegm; 40.0% vs 7.1% (p = 0.031) in atopics, for cough with phlegm
Thom and Rylander, 1998a	129 adults (18–83 years old)	Measured airborne glucan (>2–4 ng/m^3) Measured airborne glucan (>4 ng/m^3)	1.05 (0.72–1.52) for dry cough 1.08 (0.74–1.56) for dry cough
Austin and Russell, 1997	1,537 children (12–14 years old)	Self-reported damp	1.62 (1.06–2.48)
		Self-reported mold	1.78 (1.10–2.89)
Pirhonen et al., 1996	1,460 adults (25–64 years old)	Self-reported damp or mold problem	1.37 (0.99–1.88)
Jaakkola et al., 1993	2,568 preschool children	Self-reported mold odor in past year Self-reported water damage >1 year ago	3.88 (1.88–8.01) for persistent cough 2.54 (1.16–5.57) for persistent cough
Brunekreef, 1992	2,685 adult parents of children 6–12 years old	Self-reported damp stains or mold growth (last 2 years)	1.75 (1.30–2.36) among women 2.56 (1.94–3.38) among men
Dales et al., 1991a	13,495 children (5–8 years old)	Self-reported dampness or mold Self-reported flood Self-reported moisture Self-reported mold site	1.89 (1.63–2.20) 1.38 (1.16–1.65) 1.91 (1.60–2.27) 1.61 (1.36–1.89)

Study	Population	Exposure	Result
		Self-reported mold sites (2 vs 0)	2.26 (1.80–2.83)
Dijkstra et al., 1990	775 children (6–12 years old)	Self-reported damp stains or mold	0.57 (0.13–2.56)
		Self-reported damp stains, mold	3.62 (1.57–8.36)
Bruneksef et al., 1989	4,625 children (7–11 years old)	Self-reported molds (ever)	2.12 (1.64–2.73)
		Self-reported dampness (ever)	2.16 (1.64–2.84)
Waegemaekers et al., 1989	164 men	Self-reported damp homes	1.35 (nonsignificant; no CI reported)
	164 women	Self-reported damp homes	3.48 (nonsignificant; no CI reported)
	190 children	Self-reported damp homes	2.99 (1.28–6.97) for morning cough; 1.54 (0.77–3.10) for day or night cough
		Author-measured airborne fungi	1.98 (no CI reported)
Case-control studies			
Verhoeffet al., 1995	Children (6–12 years old): 84 chronic cough cases, 247 controls	Self-reported dampness	1.70 (0.94–3.09)
		Self-reported mold	1.90 (1.02–3.52)
		Surveyor-observed dampness	1.18 (0.70–1.99)
		Surveyor-observed mold	1.26 (0.70–2.25)

[a]Same population studied by Gent et al., 2002.

persistent cough and wheeze in a cohort of 849 infants (up to 1 year old) who had at least one sibling with physician-diagnosed asthma. Telephone interviews were used to ascertain symptoms and home characteristics; indoor allergens (house dust mites, cockroaches, cats, and dogs), airborne fungal spores, and NO_2 were measured. In models that controlled for allergen concentrations, the presence of a gas or wood stove, maternal education, ethnicity, the sex of the child, and smoking in the home, measured mold or mildew was associated with persistent cough both in infants whose mothers had asthma (1.91; 1.07–3.42) and in those whose mothers did not have asthma (1.53; 1.01–2.30). A study of the same cohort by Gent et al. (2002) classified airborne concentrations of *Penicillium*, *Cladosporium*, and "other mold"[2] as low, medium, and high. It found an association between high *Penicillium* and greater incidence of persistent cough (p < 0.05 for trend) in an analysis that accounted for the influence of socioeconomic factors and housing characteristics; neither *Cladosporium* nor "other mold" showed such a relationship. The authors concluded that susceptible infants in homes with high *Penicillium* were at greater risk for cough but noted that the study was limited by the fact that a single airborne sample was used to represent exposure and that samples were taken at different times of the year and some molds exhibit seasonality.

Gunnbjörnsdottir et al. (2003) found an association between "self-reported mold and water damage" and long-term cough (2.23; 1.24–4.00) in a study of young adults that controlled for age, sex, smoking history, environmental tobacco smoke, total serum IgE, and sensitization to a number of environmental agents. However, no relationship was found for mold or water damage alone or for nocturnal cough. The Engvall et al. (2001) study of 3,241 people in multifamily buildings found a significantly increased risk of cough where there was either self-reported moldy odor and signs of high humidity (3.97; 3.74–4.22) or moldy odor and major water leakage (3.78; 3.46–4.12). Their analyses accounted for subject age, sex, current smoking, number of subjects per room, and type of ventilation.

Two large studies of children are included in Table 5-2. Dales et al. (1991a) used questionnaires to gather data on the health and home characteristics of over 13,000 children 5–8 years old in 30 communities across Canada. Cough was associated with parent-reported basement flooding, water damage, or leaks in the preceding year (1.38; 1.16–1.65); wet or damp spots in rooms other than the basement (1.91; 1.60–2.27); having mold in two or more sites in the home (2.26; 1.8–2.83); or having at least one of the dampness-mold indicators (1.89; 1.63–2.20). Those estimates

---

[2]No other mold was found in a sufficient number of households to permit separate analysis. The "other mold" spore count was determined by subtracting *Penicillium* and *Cladosporium* from the total.

were not adjusted for confounders, but the authors stated that analyses that adjusted for age, sex, race, parental education, presence of environmental tobacco smoke (ETS), presence of gas appliances, and hobbies that generate airborne contaminants yielded similar results. Questionnaires were also used by Brunekreef et al. (1989) to obtain information on 4,624 children 7–11 years old living in homes that were part of the Harvard Six Cities Study.

Similar ORs were estimated for the relationship between cough and the presence of mold (2.12; 1.64–2.73) or dampness (2.16; 1.64–2.84) in subjects' homes. Estimates were slightly higher for nonasthmatic wheezers (OR = 1.73) and nonwheezers (1.59) than for asthmatics (1.50); no confidence intervals were reported, but the relationships were all statistically significant (p < 0.05).

Conclusions Despite the variations in methods used to collect information, studies report a remarkably consistent association between cough and damp indoor conditions. Statistically significant associations between cough and visible signs of dampness or mold have been described by a number of investigators, with ORs of 1.3 to over 5.0. The committee concludes

- There is sufficient evidence of an association between exposure to a damp indoor environment and cough.
- There is sufficient evidence of an association between the presence of "mold" (otherwise unspecified) in a damp indoor environment and cough.

Wheeze

Wheeze is a musical or whistling sound, typically accompanied by labored breathing, produced when a person exhales; it may be accompanied by a feeling of tightening in the chest. It is a subjective finding that may be a sign or symptom of asthma but can also occur in persons who are not considered to be asthmatic. Before the age of about 3 years, children may exhibit wheeze or other symptoms that are characteristic of asthma, but they might not exhibit persistent asthmatic symptoms or other related conditions, such as bronchial reactivity or allergy, later in life. Wheeze in these children may thus signify a non-allergic inflammatory process. In adults and older children, wheeze in the presence of dampness or mold more likely signifies an allergic response, although high level exposure to microbial agents may trigger an irritant response. Several agents found indoors and several indoor characteristics have been associated with increased likelihood of wheeze, including high relative humidity, low temperatures, gas heating, and the presence of smokers (Ross et al., 1990).

Overview of the Evidence   Studies of wheeze that examined direct or indirect measures of the presence of dampness or mold are summarized in Table 5-3. Many of the studies cited in the table also examined the LRT health outcomes reviewed above. Among the others is the Zock et al. (2002) meta-analysis of 18,873 adult subjects in a wide-ranging multicenter asthma study. The researchers found increased ORs for the associations between self-reported wheeze apart from colds in the last year (characterized as an asthma symptom in the study) and home water damage in the last year (1.23; 1.06–1.44), water on the basement floor (1.26; 0.81–1.98), and visible mold or mildew in the last year (1.44; 1.30–1.60). The analyses controlled for sex, age, and smoking status.

Maier et al. (1997) surveyed parents of 925 children 5–9 years old in Seattle to obtain information on home characteristics and wheezing in the preceding 12 months in children without a previous diagnosis of asthma. Data on four measures of household dampness were analyzed: household water damage, visible mold, water in the basement, and wall or window dampness. Reported household water damage was associated with an OR greater than 1.0 for current wheezing (1.7; 0.9–3.0) in an analysis that adjusted for sex, ethnicity, allergy history, SES, and parental asthma; a combined "any wetness or no water damage" measure was not (1.0; 0.6–1.8). The authors suggested that household water damage might be an indicator of poorer housing quality and thus a surrogate for lower SES, noting that they found a stronger nonadjusted risk of current wheezing among children in lower SES categories. Alternatively, they proposed that household water damage might simply have been more easily recognized than other forms of household dampness by study participants. When data were separated by subjects' race, household water damage was associated with current wheezing among both blacks (prevalence ratio, 3.2; 1.0–9.9; nonadjusted) and other nonwhites (2.5; 0.7–8.3).

Jędrychowski and Flak (1998) examined the influence of outdoor and indoor air quality on respiratory-health outcomes in 1,129 children 9 years old in Cracow, Poland. Surveys of parents were used to obtain information on the children's health and home characteristics; questionnaire data and measurements of suspended particulate matter and sulfur dioxide (SO_2) were used to construct an outdoor air-pollution index score. After adjustment for outdoor air pollution, sex, parental education, type of home heating system, ETS, and the presence of a physician-diagnosed allergy, home mold or dampness was found to be associated with wheezing (1.63; 1.07–2.48). When subjects were separated by allergic status, nonallergic children of nonallergic parents were found to be at increased risk for two or more respiratory symptoms in the presence of home mold or dampness (2.27; 1.07–4.82).

TABLE 5-3 Selected Epidemiologic Studies—Wheeze and Exposure to Damp Indoor Environment or Presence of Mold or Other Agents in Damp Indoor Environments

Reference	Subjects	Dampness or Mold Measure	Risk Estimate
Adolescents and adults			
Cross-sectional studies			
Gunnbjörnsdottir et al., 2003	1,853 young adults	Self-reported water damage or visible mold in home	1.4 (0.78–2.52)
Zock et al., 2002	18,872 adults in 38 centers of European Community Respiratory Health Survey	Self-reported water damage in last year	1.16 (1.00–1.34) for wheeze, breathlessness 1.23 (1.06–1.44) for wheeze apart from colds
		Self-reported water on basement floor	1.46 (1.07–2.01) for wheeze, breathlessness 1.26 (0.81–1.98) for wheeze apart from colds
		Self-reported mold or mildew in last year	1.34 (1.18–1.51) for wheeze, breathlessness 1.44 (1.30–1.60) for wheeze apart from colds
Nicolai et al., 1998	155 adolescents (mean age, 13.5 years)	Self-reported dampness	14.3% exposed vs. 5.3% nonexposed (p = 0.06).
		Self-reported dampness (adjusted for mite allergen concentrations)	5.77 (1.17–28.44)
Brunekreef, 1992	2,685 parents of children 6–12 years old	Self-reported damp stains or mold growth (last 2 years)	1.43 (1.15–1.77) in women; 1.63 (1.30–2.06) in men

(continued on next page)

TABLE 5-3 continued

Reference	Subjects	Dampness or Mold Measure	Risk Estimate
Waegemaekers et al., 1989	164 men	Self-reported damp homes	4.06 (p < 0.01; no CI reported)
	164 women	Self-reported damp homes	4.79 (p < 0.01; no CI reported)
Case-control studies			
Norbäck et al., 1999	98 cases and 357 controls nested among Swedish adult (20–45 years old) cohort	Self-reported water damage or flooding	1.6 (1.03–2.6)
		Dampness in floor	2.8 (1.4–5.5)
		Visible mold on indoor surfaces	2.4 (1.4–4.3)
		Moldy odor	1.5 (0.74–3.1)
		At least one sign of dampness	2.2 (1.5–3.2)
Infants and children			
Cross-sectional studies			
Belanger et al., 2003[a]	849 infants with asthmatic siblings	Reported persistent mold or mildew in previous 12 months	2.51 (1.37–4.62) mother with asthma 1.22 (0.80–1.88) mother without asthma
		Measured fungal colonies (species not specified)	1.23 (1.01–1.49) mother with asthma 1.10 (0.99–1.23) mother without asthma
Gent et al., 2002	880 infants (1–12 months old) with asthmatic siblings	Measured *Penicillium* low	1.11(0.87–1.42)
		Measured *Penicilliurn* medium	1.29 (0.65–1.48)
		Measured *Penicillium* high	2.15 (1.34–3.46)

Study	Population	Exposure	Result
Park et al., 2001	499 infants with at least one parent with asthma or allergy	Measured *Cladosporium* low	0.92 (0.69–1.22)
		Measured *Cladosporium* medium	0.95 (0.61–1.49)
		Measured *Cladosporium* high	0.91 (0.53–1.56)
		Other mold low	0.97 (0.75–1.26)
		Other mold medium	0.91 (0.49–1.68)
		Other mold high	1.02 (0.49–2.11)
		Water leaks	1.18 (0.90–1.55)
		Humidifier use	1.41 (1.11–1.79)
		Family-room dust endotoxin ≥ 100 EU/mg	1.33 (0.99–1.79) for any wheeze; 1.55 (1.00–2.42) for repeated wheeze
Taskinen et al., 1999	622 children (7–13 years old); 76 cases total	Self-reported dampness (school)	3.8 (1.8–8.3)
		Self-reported dampness (home)	3.4 (0.8–14.2)
		Self-reported dampness (school and home)	3.8 (1.3–11.3)
Jedrychowski and Flak, 1998	1,129 children (9 years old)	Self-reported molds or dampness	1.63 (1.07–2.48)
Rylander et al., 1998	347 children in 2 schools (1 with previous mold problem)	Problem school vs control school	13.5% vs 2.8% (p = 0.014) in nonatopics; 36.4% vs 13.3% (p = NS) in atopics

(continued on next page)

TABLE 5-3 continued

Reference	Subjects	Dampness or Mold Measure	Risk Estimate
Slezak et al., 1998	1,085 children in Head Start programs (3–5 years old)	Self-reported dampness or mold	2.01 (1.38–2.93)
Maier et al., 1997	925 children (5–9 years old)	Self-reported mold	1.2 (0.7–1.9)
		Self-reported water damage	1.7 (1.0–2.8)
		Self-reported basement water	1.0 (0.6–1.7)
		Self-reported water condensation	1.3 (0.8–2.1)
		Any of above	1.1 (0.7–1.8)
Jaakkola et al., 1993	2,568 preschool children	Self-reported moldy odor in preceding year	4.31 (1.61–11.6) for persistent wheeze
		Self-reported water damage >1 year ago	8.67 (3.87–19.4) for persistent wheeze
Dales et al., 1991a	13,495 children (5–8 years old)	Self-reported dampness or mold	1.58 (1.42–1.76)
		Self-reported flood	1.25 (1.10–1.41)
		Self-reported moisture	1.74 (1.53–1.98)
		Self-reported mold site	1.42 (1.26–1.59)
		Self-reported mold sites (2 vs 0)	1.73 (1.45–2.06)

Dijkstra et al., 1990	775 children (6–12 years old)	Self-reported damp stains or mold	1.13 (0.45–2.88)
		Self-reported damp stains and mold	1.54 (0.59–4.00)
Strachan et al., 1990	1,000 children (7 years old)	Self-reported mold	3.70 (2.22–6.15)
Brunekreefet al., 1989	4,625 children (7–11 years old)	Self-reported molds (ever)	1.79 (1.44–2.32)
		Self-reported dampness (ever)	1.23 (1.10–1.39)
Waegemaekers et al., 1989	190 children	Self-reported damp homes	2.80 (1.18–6.64)
		Author-measured airborne fungi	1.28 (includes shortness of breath, asthma)

[a]Same population studied by Gent et al., 2002.

Conclusions Studies demonstrate a consistent association between wheeze and various indications of indoor dampness, although the association of wheeze with exposure to indoor allergens (notably, house dust mite) in damp environments somewhat complicates the evaluation. Studies addressing infants and children and those addressing adolescents and adults yield similar relative risk estimates. The committee concludes on the basis of its review that

• There is sufficient evidence of an association between exposure to a damp indoor environment and wheeze.
• There is sufficient evidence of an association between the presence of "mold" (otherwise unspecified) in a damp indoor environment and wheeze.

Shortness of breath (Dyspnea)

Dyspnea is the medical term for shortness of breath. It is a common complaint in persons suffering from a variety of respiratory illnesses and can also be a symptom in persons suffering from cardiac disease. Acute inhalation of high concentrations of endotoxin may also cause dyspnea (Jagielo et al., 1996; Reed and Milton, 2001).

Overview of the Evidence Clinicians and some investigators routinely perform lung-function testing as an objective measure of respiratory physiology in persons with dyspnea. Results from the epidemiologic studies that address dyspnea are summarized in Table 5-4.

Conclusions Available studies consistently report an association between exposure to dampness or the presence of mold (otherwise unspecified) and dyspnea. However, the small number of studies lessens the confidence with which conclusions can be drawn. Taken together,

• There is limited or suggestive evidence of an association between exposure to a damp indoor environment and episodes of dyspnea (shortness of breath).
• There is inadequate or insufficient evidence to determine whether an association exists between the presence of mold or other agents in damp indoor environments and episodes of dyspnea (shortness of breath).

RESPIRATORY TRACT DISORDERS

Sinusitis

Sinusitis is a common clinical problem that consists of an inflammation of the paranasal sinuses of the face (Braunwald et al., 2001). It is usually

TABLE 5-4 Selected Epidemiologic Studies—Dyspnea and Exposure to Damp Indoor Environment or Presence of Mold or Other Agents in Damp Indoor Environments

Reference	Subjects	Dampness or Mold Measure	Risk Estimate
Cross-sectional studies			
Koskinen et al., 1999a	699 adults (16+ years old) in 310 households	Surveyor-assessed moisture	2.33 (1.09–4.98) for nocturnal dyspnea
		Self-reported mold	1.58 (0.74–3.39) for nocturnal dyspnea
Norbäck et al., 1999	98 cases and 357 controls nested among Swedish adult (20–45 years old) cohort	Self-reported water damage or flooding	2.2 (1.4–3.7) daytime 2.2 (1.4–3.5) nocturnal
		Dampness in floor	3.1 (1.5–6.2) daytime 2.7 (1.3–5.4) nocturnal
		Visible mold on indoor surfaces	2.2 (1.2–4.0) daytime 2.5 (1.4–4.5) nocturnal
Jedrychowski and Flak, 1998	1,129 children (9 years old)	Self-reported molds or dampness	2.01 (1.24–3.28)
Waegemaekers et al., 1989	164 men	Self-reported damp homes	9.38 (nonsignificant; no CI reported)
	164 women	Self-reported damp homes	2.25 (nonsignificant; no CI reported)
	190 children	Self-reported damp homes	0.92 (0.32–2.61)

caused by infectious organisms, including viruses, bacteria, and, less often, fungi. It may also be caused by the inhalation of irritant substances. The inflammatory process that is common to the various etiologies of sinusitis leads to the presence of increased amounts of mucus and mucosal edema that prevent drainage of mucus through the ostia of the sinuses. This obstructive process helps to cause or perpetuate microbial infection of the sinuses. Sinusitis can be an acute process that resolves spontaneously, can lead to a serious infection of the soft tissues surrounding the sinuses, and can become chronic and result in mucosal thickening and nasal polyps. The immune status of the patient helps to determine the course of the disease. Signs and symptoms of sinusitis are similar to those of allergic rhinitis and viral URT infections, and this can make diagnosis difficult.

Overview of the Evidence

Fungal sinusitis can present in a variety of ways (Schubert, 2000). The clinical syndromes that result from the presence of fungi in the paranasal sinuses include acute invasive fungal sinusitis, chronic invasive fungal sinusitis, mycetoma, and allergic fungal sinusitis.

Fungi are commonly isolated from the nasal secretions of patients with chronic rhinosinusitis and from healthy people. The presence of the fungi is indicative of colonization or noninvasive infection in most cases. In one study, 91% of healthy volunteers had positive fungal cultures; separately, 91% of patients with chronic rhinosinusitis were also found to have positive cultures (Braun et al., 2003). Some 33 genera of fungi were isolated, with a mean of 3.2 species per subject. Persons with chronic rhinosinusitis had eosinophilic mucin, which was not present in the normal control subjects. The presence of such allergic mucin is characteristic of the disorder termed allergic fungal rhinosinusitis (AFS). Marple (2002) states that fungal exposure alone is thought to be insufficient to initiate AFS and that instead it is a multifactorial event that results from exposure to specific fungi, an IgE-mediated atopy, specific T-cell HLA receptor expression, and aberration of local mucosal defense mechanisms.

Invasive fungal sinusitis almost always occurs in persons who are immunocompromised by diabetes, hematologic malignancies, or immunosuppressive treatments after transplantation or chronic glucocorticoid therapy (Malani and Kaufman, 2002). It may present as an acute, fulminant process that ends in death. If the host is immunocompetent, it is more likely to be a subacute disorder.

Conclusions

It is not known whether the organisms that cause the various clinical syndromes of fungal sinusitis come from the indoor or the outdoor environment. The committee did not identify any studies that associated the condition specifically with damp or moldy indoor spaces.

As noted above, fungal colonization is often found in the sinuses of both healthy persons and those who have sinusitis or nasal symptoms. Available information does not indicate that exposure to a damp indoor environment or the presence of agents associated with them places otherwise-healthy people at risk for the various forms of sinusitis.

Airflow Obstruction

Overview of the Evidence

Airflow obstruction can be easily measured with a spirometer and can be seen in the context of asthma, chronic obstructive pulmonary disease (COPD), and other lung disorders that are more uncommon and less clearly linked to environmental exposures.

Relatively little research has examined whether indoor dampness-related agents are associated with changes in FEV_1 or FEV_1:FVC,[3] the most reliable measures of obstructed airflow. Strachan et al. (1990) noted that total mold counts were nonsignificantly higher in the homes of children who exhibited a 10% or greater decline in FEV_1 after exercise. The Thorn and Rylander (1998b) experimental study of the response of healthy adult volunteers to 40 μg of inhaled lipopolysaccharide (endotoxin)—a rather large amount—under controlled conditions found relatively small but statistically significant decrements in both measures; the FEV_1:FVC ratio remained almost unchanged.

Norbäck et al. (1999) noted that FEV_1 was lower and peak expiratory flow (PEF) variability higher in subjects whose dwellings had floor dampness. Sunyer et al. (2000) found—after adjusting for bronchial hyperresponsiveness, respiratory symptoms, and smoking—that specific immunoresponse to *Alternaria alternata* was associated with a lower baseline FEV_1. In contrast, Gunnbjörnsdottir et al. (2003) did not observe any difference in FEV_1 or FVC measurements between subjects who reported water damage or visible mold and those who did not. An association between a reduction in FEF_{25-75}[4] and living in a damp or mold-contaminated home was identified in one study in children (Brunekreef et al., 1989).

---

[3]Forced vital capacity (FVC) is the total volume of air that can be expired after deep inhalation. Forced expiratory volume in (FEV_1) is the volume of air that can be expired in the first second of FVC; the normal value is at least 80% of FVC (an FEV_1:FVC ratio of 0.8).

[4]Forced expiratory flow 25%–75% (FEF_{25-75}) is the average forced expiratory flow at the middle part of the FVC maneuver. It is used as an indication of the state of the lower airways. Normal values vary with sex, age, and height.

Bronchial hyperresponsiveness (BHR), also known as bronchial hyper-reactivity, is a clinical finding defined by a fall in FEV_1 of more than 20% in response to inhalation of methacholine, histamine, or other substances known to cause bronchospasm. Such substances include common air pollutants, such as sulfur dioxide and ozone, and inhaled allergens (Tilles and Bardana, 1997). The excessive responsiveness presents as cough and wheeze. Bronchial hyperresponsiveness is seen not only in persons with asthma and other chronic inflammatory diseases of the airways but also in persons with allergic rhinitis and in many otherwise healthy people (O'Byrne and Inman, 2000). Measurement of BHR is an important research tool in epidemiologic studies and clinical evaluations. BHR has been associated with exposure to organic dust in occupational environments (Carvalheiro et al., 1995), and some researchers have reported an association with sensitization to molds in the indoor environment or the presence of visible mold in the home (Chinn et al., 1998; Dharmage et al., 2001; Nelson et al., 1999; Zock et al., 2002).

Conclusions

Airflow obstruction is a prominent clinical finding of symptom exacerbation in people with asthma or COPD. The following conclusions apply to persons who do not suffer from those diseases, which are addressed separately in this chapter:

• There is inadequate or insufficient evidence to determine whether an association exists between exposure to a damp indoor environment and airflow obstruction in otherwise healthy people.
• There is inadequate or insufficient evidence to determine whether an association exists between the presence of mold or other agents in damp indoor environments and airflow obstruction in otherwise healthy people.

Mucous Membrane Irritation Syndrome

Overview of the Evidence

Mucous membrane irritation syndrome (MMI) consists of symptoms such as rhinorrhea (runny nose), nasal congestion, and sore throat that are secondary to irritation of the nose, eyes, and throat (Richerson, 1990). The symptoms often occur with cough and other LRT complaints.

MMI is most commonly associated with exposures in agricultural environments (Von Essen and Romberger, 2003) but studies have shown that the symptoms also occur in people exposed to damp buildings with or without visible mold growth. Rudblad et al. (2001, 2002) found persistent mucosal hyperreactivity in adult subjects exposed to a damp indoor environment.

The mechanisms by which these effects might occur in humans include release of proinflammatory cytokines—including tumor necrosis factor alpha (TNF-α), interleukin-1 (IL-1), and IL-6—and nitric oxide (NO) (Hirvonen et al., 1999; Purokivi et al., 2001). Mold spores have been shown to stimulate the release of proinflammatory cytokines from macrophages (Hirvonen et al., 1997). Studies indicate that indoor-air fungi activate leukocytes to produce oxidative stress (Ruotsalainen et al., 1995) and that molds can be directly toxic to macrophages (Murtoniemi et al., 2001). The findings from studies done in humans are supported by studies of laboratory animals (Jussila et al., 2002a,b; Shahan et al., 1998). Chapter 4 discusses the results of several experiments performed by exposing animals to fungal spores or mycotoxins; they suggest that such exposures could cause MMI symptoms and other respiratory and nonrespiratory symptoms. Difficulties in performing the necessary exposure assessment have prevented a more direct evaluation of whether mycotoxins from indoor microbial agents might have that effect on humans. Interestingly, one study found that high-level nasal exposure to fungal spores in an occupational setting was not associated with high NO or proinflammatory cytokines in nasal lavage fluids; this suggests that microbial exposure does not always result in inflammatory changes detectable by this method (Roponen et al., 2002), although such changes were detected in people who worked in damp school and office environments (Hirvonen et al., 1999; Roponen et al., 2001b).

Conclusions

Mucous membrane irritation syndrome is one of the respiratory problems associated with high-level exposure to organic dust. Little attention has been paid to it as a health outcome possibly associated with nonagricultural indoor exposure to microbial agents. The committee concludes that there is inadequate or insufficient information to determine whether an association exists between indoor dampness or dampness-related agents and mucous membrane irritation syndrome.

Chronic Obstructive Pulmonary Disease

The Global Initiative for Chronic Obstructive Lung Disease (GOLD) defines COPD as ". . . a disease state characterized by airflow limitation that is not fully reversible. The airflow limitation is usually both progressive and associated with an abnormal inflammatory response of the lungs to noxious particles or gases" (Pauwels et al., 2001). Smoking is regarded as the primary cause of COPD; other agents that have been implicated include

ETS, air pollutants, and a variety of organic and inorganic dusts (Blanc et al., 2003). Symptoms include cough that is often productive of sputum, persistent dyspnea, wheeze, and diminished exercise tolerance. Spirometry often shows chronic airflow limitation, usually with limited reversibility by bronchodilator medication. Little is known about the heritability of COPD, although it is being actively investigated (Wilk et al., 2003).

Overview of the Evidence

There is some evidence that persons with COPD symptoms are more likely than those without to experience an exacerbation of their symptoms when exposed to damp indoor spaces (Brunekreef, 1992). There is also a recognized association between bronchial hyperresponsiveness, atopy, and COPD (IOM, 1993).

Several studies (reviewed above) address COPD symptoms, but relatively little research examines an association between dampness-related agents and diagnosed COPD. Thorn and Rylander (1998a) measured airborne $(1\rightarrow3)\beta$-D-glucan in the homes of 129 adults and reported an increased risk of chronic bronchitis at both lower concentrations (>2–4 ng/m^3: OR, 7.99; CI, 0.65–98.05) and higher concentrations (>4 ng/m^3: 2.51; 0.23–27.83). (Studies of exposure to glucans and other health outcomes are discussed below in the section titled "Severe Respiratory Infections.") Although some of the homes surveyed had experienced water damage, they were, according to the researchers, "not particularly damp" when the study was conducted. In a case study of two critically ill patients who were receiving high-dose corticosteroid medication for COPD symptoms, Kistemann et al. (2002) suggest that they may have acquired aspergillosis because of exposure to airborne *Aspergillus* conidia (>10^2 CFU/m^3) released during reconstruction activities near the intensive-care unit where they were housed. Other studies have also reported *Aspergillus* infections in COPD patients who were undergoing chronic corticosteroid therapy (Bulpa et al., 2001).

Conclusions

Studies identified by the committee indicate that immunocompromised people with COPD are at increased risk for fungus-related illness. However, there is inadequate or insufficient information to determine whether an association exists between indoor dampness or dampness-related agents and chronic obstructive pulmonary disease itself. Symptoms experienced by those diagnosed with COPD are addressed in the section titled "Respiratory Symptoms" earlier in the chapter.

Asthma

Definition of Disorder

Asthma, like COPD, is a disorder of airflow obstruction. As was noted in the 2000 Institute of Medicine report *Clearing the Air,* finding a widely accepted definition for this disease has proved problematic. The definition adopted by the committee responsible for that report states that (pp. 23–24)

> . . . asthma is understood to be a chronic disease of the airways characterized by an inflammatory response involving many cell types. Both genetic and environmental factors appear to play important roles in the initiation and continuation of the inflammation. Although the inflammatory response may vary from one patient to another, the symptoms are often episodic and usually include wheezing, breathlessness, chest tightness, and coughing. Symptoms may occur at any time of the day but are more commonly seen at night. These symptoms are associated with widespread airflow obstruction that is at least partially reversible with pharmacologic agents or time. Many persons with asthma also have varying degrees of bronchial hyperresponsiveness. Research has shown that after long periods of time this inflammation may cause a gradual alteration or remodeling of the architecture of the lungs that cannot be reversed with therapy.

Objective information about the presence and severity of asthma can be obtained by demonstrating airflow obstruction with spirometry. Alternatively, evidence of airflow obstruction can be obtained by measuring peak expiratory flow. Asthma is distinguished from other airway disorders in part by reversibility of the airflow obstruction, which may be spontaneous or induced by bronchodilator medications.

Role of Sensitization

Asthma may be allergic or nonallergic. Allergic asthma is IgE-mediated.[5] A number of cell types play an important role in asthma, including eosinophils, macrophages, and T lymphocytes. Basophils and neutrophils are also present in increased numbers in the inflammatory infiltrates of asthma. The interactions of those cell types with one another and the substances they release are complex; they are addressed in Chapter 4 of *Clearing the Air* (IOM, 2000) and elsewhere (Oh et al., 2002; Wills-Karp

---

[5]Douwes and Pearce (2003) note that "molds are known to produce immunoglobulin E-inducing allergens, and some studies have shown a higher prevalence of mold sensitization among subjects living in damp buildings (Norbäck et al., 1999) and among severe asthmatics (Black et al., 2000)."

and Chiaramonte, 2003). It is now recognized that there are at least two distinct variants of atopic diseases: an extrinsic allergic variant that occurs in the context of sensitization to environmental allergens and an intrinsic, nonallergic variant with no detectable sensitization and with low IgE concentrations. A third form, a combination of the first two, has also been described (Novak and Bieber, 2003).

Nonallergic asthma may be mediated by irritant responses. Mechanisms of irritant asthma are largely unknown, but there is evidence that a localized airway response and activation of neural pathways are involved (Balmes, 2002). It has been proposed that neutrophilic airway inflammation may also play a role in causing asthma (Douwes et al., 2002).

Other Asthma Issues

Allergic responses to fungi have been well documented (Halonen et al., 1997; Norbäck et al., 1999). However, the difficulty of performing population-based surveys to assess subjects for allergies to molds is compounded by the relative lack of standardized mold antigens for use in skin-prick testing. *Clearing the Air* (IOM, 2000) notes that about 6–10% of the population are sensitized to fungal allergens; Mari et al. (2003) found that 19% of atopics reacted to at least one of seven fungal extracts—*Alternaria, Aspergillus, Candida, Cladosporium, Penicillium, Saccharomyces,* and *Trichophyton*—administered via skin-prick test. Skin-test surveys most often focus on broad panels of allergens, including fungi as a single representative extract (usually *Alternaria*), two or three extracts, or mixtures of several fungi. However, Galant et al. (1998), in a survey of California allergy patients, revealed that nine fungal extracts were necessary to detect 90% of mold-allergic patients. Most studies that focus specifically on fungal sensitivity also use only a few extracts, *Alternaria* again being the dominant type. A 2002 study of IgG and IgE antibodies in Finnish children who attended a water-damaged school found that the number of positive IgG antibodies did not correlate with respiratory symptoms or illnesses although the mean number of positive IgG findings was higher in the exposed group (Savilahti et al., 2002).

Fungal skin-sensitivity rates increase with age (Erel et al., 1998). Production of IgG antibodies as a result of allergen exposure may block the skin-test response even in patients who clearly are experiencing symptoms on exposure (Witteman et al., 1996). Fungal allergens can produce a strong IgG response and possibly make reported incidences of skin reactivity underestimates.

Asthma has an important genetic component (Burke et al., 2003). Current evidence indicates that the clinical syndrome of asthma is a diverse group of related conditions that require multiple genes to be expressed for the clinical syndrome to become apparent (Blumenthal, 2002; Cookson, 2002). In addition to the presence of genes that predispose persons to asthma, some exposures may be required for the symptoms to appear. The nature of the gene-environment interactions relevant to damp indoor environments is a topic of a great deal of research and is not completely understood. For example, studies about genetic susceptibility to the inflammatory effects of endotoxin are yielding interesting findings, but no firm conclusions can yet be drawn that can be applied directly to the issue of endotoxin exposure in damp indoor spaces (Vercelli, 2003). Substances in the damp indoor environment other than endotoxin could be identified as important in the future.

As discussed in Chapter 4, animal studies have shown that exposure to fungal spores causes a profound inflammatory response in the LRT (Jussila et al., 2001, 2002a,b; Shahan et al., 1998). Intratracheal exposure of mice to *Streptomyces californicus* resulted in release of TNFα and IL-6 and in inflammatory-cell recruitment to the airways (Jussila et al., 2001). A similar inflammatory response was seen when mice were challenged with other fungi commonly found in damp indoor spaces: *Aspergillus versicolor* and *Penicillium spinulosum* (Jussila et al., 2002a,b). Inflammatory-cell recruitment to the airways and release of TNFα and IL-6 were also seen in mice experimentally exposed to the nonfungal contaminant *Mycobacterium terrae* isolated from a moisture-damaged building (Jussila et al., 2002c). It must be determined whether other mediators of inflammation are also involved in the inflammatory response of the respiratory tract after exposure to fungi and other microorganisms that are present in large numbers in damp indoor spaces.

High indoor humidity has been associated with increased endotoxin concentrations, and this might help to explain why damp indoor spaces cause symptoms (Park et al., 2001). Contaminated humidifiers can spray spores, fragments, fungal components (including endotoxins) and dissolved allergens into the air (Baur et al., 1988; Burge et al., 1980; McConnell et al., 2002; Tyndall et al., 1995) and may contribute to exposure in some instances. Exposure to high concentrations of endotoxin has been associated with a lower FEV_1 in asthmatic people (Michel et al., 1996). There is evidence from animal studies that long-term endotoxin exposure results in chronic lung inflammation (Vernooy et al., 2002), but it is not known whether that occurs in humans. Some studies (Böttcher et al., 2003; Braun-Fahrländer et al., 2002; Gehring et al., 2002) suggest that indoor exposure to endotoxin protects against allergic sensitization and allergic wheeze, but the notion remains controversial (Maziak et al., 2003; Weiss, 2002).

Exacerbation of Asthma

Overview of the Evidence   Studies of asthma can be divided into those dealing with factors that lead to the development of asthma and those dealing with factors that lead to the onset or worsening of symptoms—some combination of shortness of breath, cough, wheeze, and chest tightness—in someone who has already developed asthma.

Multiple indoor environmental factors are thought to exacerbate asthma. Table 5-5 summarizes the conclusions reached in *Clearing the Air* on the state of the scientific evidence (as of the middle of 1999) regarding the association between various indoor agents and asthma exacerbation. Conclusions regarding microbial agents that are addressed in the present report are omitted from the table.

Some investigators have examined the association between self-reports of asthma symptoms in people who identify themselves as asthmatic and their presence in damp or moldy indoor environments. The health outcome that is evaluated varies but typically involves two components: a self-report of physician-diagnosed asthma as evidence of the disorder and a self-report of one or more of the asthma symptoms listed above or use of asthma medication. The outcome is variously labeled *current asthma, asthma symptoms, symptomatic asthma,* or *exacerbations.* Exacerbation is used here as an umbrella term for consistency with the 2000 IOM report *Clearing the Air,* but it should be noted that the studies reviewed do not in general ask whether the subject experienced a worsening of symptoms or of symptom frequency in the presence of dampness or mold—a more customary definition of exacerbation. It is thus not always clear whether subjects are experiencing their typical level of symptoms or an adverse reaction in response to a trigger. The discussion and conclusions in this section should be interpreted with that caveat in mind.

The committee did not review the literature regarding outdoor fungal levels. It notes that some (Dales et al., 2003, 2004; Delfino et al., 1997) but not all (Lierl and Hornung, 2003) studies of hospital admissions suggest an association between increased levels and asthma exacerbations.

Table 5-6 summarizes the results of studies that addressed asthma symptoms in asthmatic people and exposure to a damp indoor environment or the presence of mold or other agents in damp indoor environments.

The studies reviewed in the table include a 2002 paper by Zock et al. that drew data from nearly 19,000 adult subjects in Europe, Australia, New Zealand, India, and the United States who participated in a multicenter asthma study. A meta-analysis of data from the centers—adjusted within study centers for sex, age group (20–29, 30–39, 40–45 years), and smoking status—yielded a statistically significant relationship between "current asthma" (defined as asthma symptoms, medication use, or both within

TABLE 5-5   Selected Findings from *Clearing the Air* (IOM, 2000) Regarding Association Between Biologic and Chemical Exposures and Exacerbation of Asthma in Sensitive Persons

Biologic Agents	Chemical Agents
Sufficient Evidence of a Causal Relationship	
Cat	ETS (in preschool-age children)
Cockroach	
House dust mite	
Sufficient Evidence of an Association	
Dog	NO_2, NO_x (high-level exposures[a])
Rhinovirus	
Limited or Suggestive Evidence of an Association	
Domestic birds	ETS in school-age and older children and in adults
Chlamydia pneumoniae	Formaldehyde
Mycoplasma pneumoniae	Fragrances
Respiratory syncytial virus (RSV)	
Inadequate or Insufficient Evidence to Determine Whether an Association Exists	
Cow and horse	Pesticides
Rodents (as pets or feral animals)	Plasticizers
Chlamydia trachomatis	
Houseplants	
Pollen exposure in indoor environments	
Insects other than cockroaches	
Limited or Suggestive Evidence of No Association	
(no agents met this definition)	(no agents met this definition)

[a]At concentrations that may occur only when gas appliances are used in poorly ventilated kitchens.

preceding 12 months in persons who reported that a physician had diagnosed asthma) and mold or mildew in the home in the preceding year (1.28; 1.13–1.46). The increased ORs for current asthma and self-reported water damage in the last year (1.13; 0.94–1.35) and water on the basement floor (1.54; 0.84–2.82) were nonsignificant. Those observations were homogeneous across centers and stronger in subjects sensitized to *Cladosporium* species.

A study by Engvall et al. (2001) used a questionnaire to gather information on 3,241 people living in randomly selected units in 231 multifamily buildings in Stockholm. Their analyses—which controlled for age, sex, current smoking, number of subjects per room, and type of ventilation—found relationships between self-reported "asthma symptoms" and a number

TABLE 5-6 Selected Epidemiologic Studies—Asthma Symptoms in Asthmatic People and Exposure to Damp Indoor Environment or Presence of Mold or Other Agents in Damp Indoor Environments

Reference	Subjects	Dampness or Mold Measure	Risk Estimate
Adults			
Cross-sectional studies			
Zock et al., 2002	18,872 adults in 38 study centers of European Community Respiratory Health Survey	Self-reported water damage in last year	1.13 (0.95–1.35) for "current asthma"
		Self-reported water on basement floor	1.54 (0.84–2.82) for "current asthma"
		Self-reported mold or mildew in last year	1.28 (1.13–1.46) for "current asthma"
Kilpeläinen et al., 2001	10,667 college students (18–25 years old)	Self-reported visible mold	2.21 (1.48–3.28) for "current asthma symptoms"
		Self-reported visible mold, damp stains, water damage	1.66 (1.25–2.19) for "current asthma symptoms"
Engvall et at., 2001	3,241 persons randomly sampled in multi-family buildings	At least one sign of dampness	2.28 (2.19–2.37) for "asthma symptoms"
		At least one dampness-related odor	2.38 (2.30–2.47) for "asthma symptoms"
		Reports of at least one odor and structural building dampness	3.59 (3.37–3.82) for "asthma symptoms"
Hu et al., 1997	2,041 young adults (20–22 years old)	Self-reported water leaking	1.6 (0.7–3.5), physician-reported 1.6 (0.7–3.8), self-reported

Study	Population	Exposure measure	Result
Pirhonen et al., 1996	1,460 adults (25–64 years old)	Self-reported indoor dampness	1.2 (0.8–1.9), physician-reported 1.3 (0.7–2.2), self-reported
		Self-reported visible mold	1.5 (1.0–2.4), physician-reported 2.0 (1.2–3.2), self-reported
Brunekreef, 1992	2,685 adult parents of children 6–12 years old	Self-reported damp or mold problem	1.02 (0.60–1.72)
		Self-reported damp stains or mold growth (last 2 years)	1.25 (0.94–1.66) among women 1.29 (0.92–1.81) among men
Waegemaekers et al., 1989	164 men	Self-reported damp homes	1.15 (nonsignificant)
	164 women	Self-reported damp homes	4.16 ($p<0.05$)
Case-Control Studies			
Norbäck et al., 1999	98 cases and 357 controls nested among Swedish adult (20–45 years old) cohort	Self-reported water damage or flooding	1.9 (1.2–2.9) for at least one asthma symptom
		Dampness in floor	3.3 (1.6–6.8) for at least one asthma symptom
		Visible mold on indoor surfaces	2.9 (1.6–5.3) for at least one asthma symptom
		Moldy odor	1.8 (0.9–3.8) for at least one asthma symptom
		At least one sign of dampness	2.2 (1.5–3.2) for at least one asthma symptom

(continued on next page)

TABLE 5-6 continued

Reference	Subjects	Dampness or Mold Measure	Risk Estimate
Williamson et al., 1997	102 asthmatics; 196 matched controls (5–44 years old)	Self-reported dampness or condensation in present home	1.93(1.14–3.28), physician-diagnosed asthma
		Self-reported dampness in previous home	2.55 (1.49–4.37), physician-diagnosed asthma
		Observed severe dampness	2.36 (1.34–4.01), physician-diagnosed asthma
		Observed significant mold	1.70 (0.78–3.71)
Children *Cross-sectional studies*			
Wever-Hess et al., 2000	113 infants (0–1 years old)	Self-reported damp housing	7.6 (2.0–28.6) 3.8 (1.1–12.8) for "recurrent exacerbations"
Taskinen et al., 1999	622 children (7–13 years old); 28 cases total	Self-reported dampness (school)	1.0 (0.4–2.3) for symptomatic asthma
		Self-reported dampness (home)	1.9 (0.4–10.4) for symptomatic asthma
		Self-reported dampness (school and home)	1.1 (0.2–5.5) for symptomatic asthma
Jędrychowski and Flak, 1998	1,129 children (9 years old)	Self-reported molds or dampness	2.65 (0.96–7.13) for diagnosed asthma
Nicolai et al., 1998	155 adolescents (mean age, 13.5 years)	Past or present self-reported Dampness in the home	61.5% exposed vs 37.7% non-exposed (p<0.05) for ≥5 asthma attacks in previous year

Study	Population	Exposure	Odds Ratio (95% CI)
Rönmark et al., 1999	3,431 children (7–8 years old) in northern Sweden	Parent-reported dampness at home	1.40 (0.8 1–2.42) for atopic asthma 1.78 (1.10–2.89) for nonatopic asthma
Yang et al., 1997	4,164 primary-school children (3–15 years old)	Self-reported home dampness	1.73 (1.20–2.49), physician confirmed asthma
Jaakkola et al., 1993	2,568 preschool children	Self-reported water damage >1 year ago Self-reported mold odor in last year	2.52 (0.93–6.870) 1.46 (0.34–6.29)
Dales et al., 1991a	13,495 children (5–8 years old)	Self-reported flood Self-reported moisture Self-reported dampness or mold Self-reported mold site Self-reported mold sites (2 vs 0)	1.29 (1.06–1.56) 1.58 (1.29–1.94) 1.45 (1.23–1.71) 1.40 (1.16–1.68) 1.67 (1.27–2.19)
Dijkstra et al., 1990	775 children (6–12 years old)	Self-reported damp stains or mold Self-reported damp stains and mold	1.16 (0.38–3.52) 1.56 (0.50–4.87)
Brunekreefet al., 1989	4,625 children (7–11 years old)	Self-reported dampness (ever)	1.42 (1.04–1.94)
Waegemaekers et al., 1989	190 children	Self-reported damp homes	2.80 (0.39–20.02)

(continued on next page)

TABLE 5-6 continued

Reference	Subjects	Dampness or Mold Measure	Risk Estimate
Case-control studies			
Yazicioglu et al., 1998	597 controls, 85 asthmatics	Self-reported home dampness	2.62 (1.13–6.81)
Dekker et al., 1991	13,495 children (5–8 years old)	Self-reported dampness or visible mold	1.46 (1.22–1.74)
Mohamed et al., 1995	77 child cases and 77 controls (9–11 years old)	Author-observed damp damage in child's bedroom	4.9 (2.0–11.7)

of dampness indicators. Where both damp odor and structural building dampness were reported, the adjusted OR was 3.59 (3.37–3.82).

Kilpeläinen et al. (2001) analyzed the results of 10,667 surveys returned by university students 18–25 years old. The researchers observed an OR of 2.21 (1.48–3.28) between "current asthma" (defined as self-reported, physician-diagnosed asthma with symptoms during the preceding year) and visible mold and an OR of 1.66 (1.25–2.19) between "current asthma" and visible mold, damp stains, or water damage. The analyses controlled for parental education, smoking, ETS, pets, wall-to-wall carpeting, place of residence (farm, rural nonfarm, or urban), and type of residence (apartment vs other building types). After controlling for the effects of genetic predisposition to asthma or allergies, this relationship was significant only in those whose parents were asthmatic or had atopic disease, and the p value for the interaction between atopic heredity and dampness was statistically significant (0.033).

Norbäck et al. (2000) reported an association between current asthma symptoms (bronchial hyperresponsiveness and either wheezing or dyspnea in the preceding 12 months) and building dampness at work (8.6; 1.3–56.7) in a study of 87 personnel working in four geriatric hospitals during the winter months. The authors suggested that the result might be a consequence of exposure to the emission of 2-ethyl-1-hexanol due to alkaline degradation of octylphthalates in damp floor materials.

Dharmage et al. (2001) measured home allergen concentrations and performed skin-prick and lung-function tests in 485 adults in Melbourne, Australia. Dust samples in subjects' bedrooms and beds were evaluated for dust mites (Der p 1) and cat allergen (Fel d 1); ergosterol was used as a proxy of fungal biomass. Air samples were screened for five allergenic fungi—*Cladosporium, Alternaria, Epicoccum, Penicillium, and Aspergillus*—and for Der p 1 and Fel d 1, and questionnaires were used to gather personal and sociodemographic characteristics of the cohort. The authors reported that high ergosterol was associated with an increased risk of current asthma after adjustment for potential confounders, but the relationship was not statistically significant (data not provided).

Among studies of children, the Taskinen et al. (1999) analysis of 622 children (7–13 years old) found no relationship between physician-diagnosed, parent-reported asthma and moisture problems in the home (1.9; 0.4–10.4), school (1.0; 0.4–2.3), or both combined environments (1.1; 0.2–5.5). Dales et al. (1991a), however, did identify a relationship between home dampness or mold and physician-diagnosed asthma (1.45; 1.23–1.71) in a survey of the parents of 13,495 children 5–8 years old.

Conclusions Numerous studies of adults and children uniformly report odds ratios over 1 for the association between exposure to dampness or the presence of mold or other agents in damp indoor environments and self-reports of symptoms in people with physician-diagnosed asthma. Most of the observed associations are statistically significant.

From the reviewed body of evidence, the committee concludes that

• There is sufficient evidence of an association between exposure to a damp indoor environment and asthma symptoms in sensitized asthmatic people.

• There is sufficient evidence of an association between the presence of "mold" (otherwise unspecified) in a damp indoor environment and asthma symptoms in sensitized asthmatic people.

Studies used to draw this conclusion use self-reports or parent reports of physician-diagnosed asthma in combination with self-reports of asthma symptoms or medication use as their measure of health outcome.

Development of Asthma

Overview of the Evidence   The other asthma outcome reviewed by the committee was development—that is, the initial onset of the illness. Asthma is defined by the manifestation of a set of symptoms rather than by any one objective test. With asthma symptoms ranging from clearly episodic to nearly continuous, from mild to severe, and from coughing without other respiratory symptoms to a loud wheeze, the initial diagnosis of the illness can be complicated and subject to controversy. It is thus difficult to study the determinants of and influences on asthma development. An additional complication arises in interpreting studies of infants and younger children. Before the age of about 3 years, children may exhibit symptoms that are characteristic of asthma, but they might not exhibit persistent asthmatic symptoms or other related conditions, such as bronchial reactivity or allergy, later in life.

Table 5-7, excerpted from *Clearing the Air*, shows the conclusions drawn on the state of the scientific evidence (as of the middle of 1999) regarding the association between the development of asthma and various agents indoors. Conclusions regarding microbial agents that are addressed in the present report are omitted from the table.

Studies of damp indoor environments or dampness-related agents and asthma development are summarized in Table 5-8. The table includes studies of LRT illnesses in infants and young children thought to be at risk for allergic sensitization to microbial agents and asthma development.

A 2002 study examined the possible relationship between the presence of indoor molds and development of asthma in adulthood with a population-based indirect case-control design (Jaakkola et al., 2002a,b). The authors

TABLE 5-7   Selected Findings from *Clearing the Air* (IOM, 2000) Regarding Association Between Indoor Biologic and Chemical Exposures and Development of Asthma

Biologic Agents	Chemical Agents
Sufficient Evidence of a Causal Relationship	
House dust mite	(no agents met this definition)
Sufficient Evidence of an Association	
(No agents met this definition)	ETS (in preschool-age children)
Limited or Suggestive Evidence of an Association	
Cockroach (in preschool-age children)	(no agents met this definition)
Respiratory syncytial virus (RSV)	
Inadequate or Insufficient Evidence to Determine Whether an Association Exists	
Cat	NO_2, NO_x
Cow and horse	Pesticides
Dog	Plasticizers
Domestic birds	Formaldehyde
Rodents	Fragrances
Cockroaches (except for preschool-age children)	ETS (in school-age and older children, and in adults)
Chlamydia pneumoniae	
Chlamydia trachomatis	
Mycoplasma pneumoniae	
Houseplants	
Pollen	
Limited or Suggestive Evidence of No Association	
Rhinovirus (adults)	(no agents met this definition)

concluded that there was an association between visible mold or mold odor in the workplace and adult-onset asthma (1.54; 1.01–2.32) in an analysis that controlled for the possible confounding influences of sex, age, parental atopy or asthma, education, smoking, ETS, pets, work indoors, self-reported occupational exposures, and signs of dampness or water damage (Jaakkola et al., 2002a). An earlier case-control study of adult asthma and self-reported dampness-related agents in the home (Thorn et al., 2001) reached like conclusions, finding an association with visible mold (2.2; 1.4–3.5) but not with visible dampness (1.3; 0.9–2.0). Similar results for males and females were obtained when the data were analyzed separately. That study, which examined 174 adult cases and 870 referents, controlled for age, sex, smoking habits, and atopy.

Longitudinal study of birth cohorts is the best research design to examine the onset of asthma in relation to environmental conditions. There is

TABLE 5-8 Selected Epidemiologic Studies—Asthma Development and Exposure to Damp Indoor Environment or Presence of Mold or Other Agents in Damp Indoor Environments

Reference	Subjects	Dampness or Mold Measure	Risk Estimate
Adults			
Population-based nested case-control studies			
Jaakkola et al., 2002a	521 newly diagnosed adult cases; 932 controls	Visible mold or odor (work)	1.54 (1.01–2.32)
		Damp stains or paint peeling (work)	0.84 (0.56–1.25)
		Water damage (work)	0.91 (0.60–1.39)
		Visible mold or odor (home)	0.98 (0.68–1.40)
		Damp stains or paint peeling (home)	1.02 (0.73–1.41)
		Water damage (home)	0.90 (0.61–1.34)
Thom et al., 2001	174 adult (20–50 old) asthma cases diagnosed in last 15 years; 870 referents	Self-reported visible dampness	1.3 (0.9–2.0)
		Self-reported visible mold growth	2.2 (1.4–3.5)
		Self-reported dampness or visible mold growth	1.8 (1.1–3.1) prevalence
Children			
Nested and incident case-control studies			
Øie et al., 1999	172 children <2 years old with bronchial obstruction; 172 matched controls from population-based sample of 3,754 newboms (same population as Nafstad et al., 1998)	Surveyor-verified	2.4 (1.25–4.44) for bronchial obstruction
Nafstad et al., 1998	251 children <2 years old with bronchial obstruction and 251 controls (0–2 years old) from population-based sample of 3,754 newboms	Parent-reported dampness	2.5 (1.1–5.5) for bronchial obstruction
		Surveyor-verified dampness	3.8 (2.0–7.2) for bronchial obstruction
		Inspector-observed dampness	3.8 (2.0–7.2) for bronchial obstruction first diagnosed

TABLE 5-8   continued

Reference	Subjects	Dampness or Mold Measure	Risk Estimate
Yang et al., 1998	86 cases with first-time diagnosis of asthma and 86 controls (3–15 years old)	Parent-reported home dampness	1.77 (1.24–2.53)
Infante-Rivard, 1993	457 newly diagnosed infant cases and 457 controls (3–4 years old)	Parent-reported humidifier use	1.89 (1.30–2.74)
Cohort Studies			
Slezak et al., 1998	1,085 children in Head Start programs (3–5 years old)	Self-reported dampness or mold	1.94 (1.23–3.04) self-reported, physician-diagnosed asthma
Maier et al., 1997	925 children (5–9 years old) followed prospectively for 1 year	Water damage	1.7 (1.0–2.8) for physician-diagnosed asthma
		Other wetness or no water damage	1.1 (0.6–1.8) for physician-diagnosed asthma

evidence that some environmental factors, such as ETS and allergens from house dust mites and cockroaches, cause allergic disease in genetically predisposed subjects (IOM, 2000). In a 2-year-long birth-cohort study of 3,754 children born in Oslo (Nafstad et al., 1998), exposure to indoor dampness problems increased the risk of development of clinically confirmed bronchial-obstruction signs and symptoms (3.8; 2.0–7.2) in children 0–2 years old, even when the presence of house dust mites was controlled for. The association was observed regardless of whether the parents or trained inspectors reported the dampness, but it was enhanced if both parents and trained inspectors reported it. A later paper on the same cohort (Øie et al., 1999) found that the adjusted OR for bronchial-obstruction signs and symptoms was higher in low-ventilation-rate homes (9.6; 1.05–87.45) than in high-ventilation-rate homes (2.4; 1.25–4.44); low-ventilation-rate homes had total air change rates of less than 0.5/hr while high were anything above that. The authors observed that the results were consistent with the hypothesis that low ventilation rates strengthen the effects of indoor air pollutants.

Yang et al. (1998) examined the influence of indoor environmental factors on the development of asthma in children 3–15 years old. They drew 86 cases and 86 controls from patients at a teaching hospital in southern Taiwan. Cases had a first-time physician diagnosis of asthma. The study found a statistically significant association between parent-reported home dampness and asthma (1.77; 1.24–2.53) when they controlled for age, sex, parental education, parental asthma, physician-confirmed allergy to food, breast-feeding, and even the presence of mold or dust.

Infante-Rivard observed (1993) significant associations between first-time diagnosed asthma and the parent-reported use of a humidifier in the child's bedroom (1.89; 1.30–2.74) after adjusting for maternal smoking and presence of an electric home-heating system in a case-control study of 914 children 3–4 years old.

A 2001 study found an association between age and IgE sensitization to *Alternaria alternata, Aspergillus fumigatus, Cladosporium herbarum,* and *Penicillium chrysogenum* in atopic children 0–15 years old, with maximal sensitization prevalence at about 8 (Nolles et al., 2001). A second study showed that children who attended a water-damaged school and had increased respiratory complaints were more likely to have high IgE values to a variety of common indoor allergens than were children who attended a reference school (Savilahti et al., 2001). However, Taskinen et al. (2002) did not find an association between asthma and IgG antibodies to molds in the schools of 126 screened children.

In a 1997 study, Taskinen et al. compared 133 children in two schools— one with "moisture problems," the other without—and observed that eight of nine asthma-diagnosed children attended the moisture-problem school. No data were provided on the date of diagnosis, but skin-prick tests revealed positive mold tests in six of the eight asthmatic children in the moisture-problem school.

Two studies that were previously summarized in the discussions of wheeze and cough above also provide information related to asthma development. Gent et al. (2002) and Belanger et al. (2003) evaluated a cohort of 849 infants (<1 year old) who had at least one sibling with physician-diagnosed asthma. Belanger et al. (2003) used models that controlled for allergen concentrations, the presence of a gas or wood stove, maternal education, ethnicity, the sex of the child, smoking in the home, and respiratory illness; measured mold or mildew was associated with wheeze (2.51; 1.37–4.62) and persistent cough (1.91; 1.07–3.42) in infants whose mothers had asthma. Gent et al. examined airborne concentrations of *Penicillium, Cladosporium,* and "other mold" in the cohort. (No other mold was found in enough households to allow separate analysis; the "other mold" spore count was determined by subtracting *Penicillium* and *Cladosporium* from the total.) They found an association between higher *Penicillium* concentrations and greater incidence of wheeze and persistent cough (p < 0.05 for trend) in an analysis that accounted for the influence of socioeconomic factors and housing characteristics. Neither *Cladosporium* nor "other mold" showed such a relationship. Persistent cough and wheeze in infancy is associated with a greater risk of asthma development.

Research by Stark et al. (2003)—also cited below under "Respiratory Infections"—is related to this issue as well. Those investigators examined the incidence of croup, pneumonia, bronchitis, and bronchiolitis in a cohort of 499 infants genetically predisposed to asthma. They found associations between the illnesses and high concentrations of some microbial agents commonly found indoors in the United States (*Penicillium*, *Cladosporium*, *Zygomycetes*, and *Alternaria*). An independent association was found with measures of indoor dampness. There is an association between infection with one respiratory virus, respiratory syncytial virus, and later development of asthma (IOM, 2000). Other respiratory viruses cause exacerbations of acute severe wheeze in persons with established asthma but are not associated with the development of incident asthma (Gern et al., 1999).

Conclusions In summary, studies reviewed by the committee indicate that

• There is limited or suggestive evidence of an association between exposure to a damp indoor environment and the development of asthma. It is not clear whether this association reflects exposure to fungi, bacteria or their constituents and emissions, such other agents related to damp indoor environments as house dust mites and cockroaches, or some combination thereof. The responsible factors may vary among individuals.

• There is inadequate or insufficient evidence to determine whether an association exists between the presence of mold or other agents in damp indoor environments and the development of asthma. The exposure-assessment problems in the papers examined and the small number of longitudinal studies performed limit the confidence that can be placed in their results.

Hypersensitivity Pneumonitis

Overview of the Evidence

Hypersensitivity pneumonitis (HP), also called extrinsic allergic alveolitis, is a granulomatous lung disease that is the result of exposure and sensitization to antigens inhaled with a variety of organic dusts (Terho, 1982). Symptoms of acute HP include dry cough, dyspnea, and fever experienced several hours after the causative exposure (Fraser et al., 1999). Acute bronchospasm may also be present. A subacute form of the disease has been described, and there is a chronic form of HP that may present with pulmonary fibrosis (Yi, 2002). More detail on the diagnosis and etiology of HP are available in literature reviews on the disorder (Lacasse et al., 2003; Wild and Lopez, 2001).

HP is diagnosed by collecting information from multiple sources. There is no single characteristic finding; clinicians need several pieces of evidence that make HP the most likely diagnosis before subjects are described as having it. These include a careful clinical and exposure history, environmental sampling for the presence of microorganisms known to cause HP, chest x-ray pictures and high-resolution CT scanning of the thorax, lung-function and serologic tests, and lung biopsy specimens. Circulating immunologic antibodies have been shown to have little value as markers for chronic HP (Cormier and Bélanger, 1989; Guernsey et al., 1989; Marx et al., 1990), but it is generally agreed that they are better used as indicators of recent high exposure to specific molds and thermophilic actinomycete antigens (Lacasse et al., 2003).

HP has been described in case reports after environmental exposures to buildings contaminated with fungi in a variety of settings, including showers and home ultrasonic humidifiers (Hogan et al., 1996; Lee et al., 2000; Suda et al., 1995). Bacteria contaminating an ultrasonic cold-air home humidifier have been implicated as a causative agent (Kane et al., 1993), as have bacteria (thermophilic actinomycetes) that are able to tolerate very warm environments. A seasonal form of building-related HP caused by the fungus *Trichosporon cutaneum* is found mainly in Japan (Ando et al., 1995).

Management of building-related HP includes standard medical therapy and removing sources of fungal contamination from the environment. In some cases, efforts to remove mold from a building are unsuccessful in relieving symptoms, and moving to another home or office may be necessary (Apostolakos et al., 2001).

It is not uncommon for building-related illness to occur when, despite careful investigation, it is not clear whether the respiratory illness observed is HP or another clinical syndrome that is not yet defined (Trout et al., 2001). The possible role of mycotoxins in causing this type of respiratory illness was raised by Trout and colleagues. They found IgG antibodies to roridin-A, a tricothecene mycotoxin, in a person who had an illness suspected of being HP but whose clinical findings were not classic for this disorder. However, the test was not useful for distinguishing persons who were probably exposed to mold indoors from those who were not. Their paper raises the question of whether mycotoxins may have a role in acute and chronic lung injury in humans. Further exploration of the subject is needed, including research to find biomarkers that can be used to assess exposure to mycotoxins reliably.

Exposure to disease-mediating organic material clearly is not the sole factor in determining whether a person will develop HP. Only a small minority of exposed people manifest the disease, and those who do are not necessarily the most highly exposed. Genetic factors also have a role, and research suggests that some polymorphisms may modify a person's risk of the disease (Schuyler, 2001).

Conclusions

HP is a relatively rare immune-mediated condition, and only susceptible people exposed to a sensitizing antigen develop clinically significant disease. It has thus been studied with relation to specific agents rather than dampness in general. Studies reviewed by the committee indicate that there is sufficient evidence of an association between the presence of mold and bacteria in damp indoor environments and hypersensitivity pneumonitis in such people. Others are not at risk for this disease.

Inhalation Fevers

Inhalation fever is the general name given to any one of a number of influenza-like, self-limited syndromes caused by a heterogeneous group of stimuli (Blanc, 1997). Two that have been potentially associated with damp indoor environments are briefly addressed here.

Humidifier fever is an illness that consists of a febrile reaction accompanied by respiratory tract symptoms and fatigue. It does not manifest the radiographic or laboratory abnormalities consistent with HP and it is thought to be a nonimmunologic reaction (Baur et al., 1988).

Organic dust toxic syndrome (ODTS) is a self-limiting noninfectious febrile illness that occurs after heavy organic-dust exposure by inhalation (Emanuel et al., 1975; Marx et al., 1981; Von Essen et al., 1990). Common symptoms include malaise, myalgia, headache, nonproductive cough, fever and nausea—symptoms that resemble those of acute HP. However, unlike HP, prior sensitization is not required in ODTS, serum precipitin antibodies against fungi are negative, the chest x-ray picture usually does not show infiltrates, there is no hypoxemia, there is no restriction or low CO diffusing capacity on lung-function testing. ODTS shares with HP the laboratory finding of leukocytosis with a predominance of neutrophils and a left shift during the acute phase.

Overview of the Evidence

Humidifier fever has been reported most commonly in industrial settings where workers are exposed to microorganisms—notably, thermophilic actinomycetes but also including other bacteria and protozoa—growing in humidification systems. However, humidifier fever has also been reported in an office building where a culture of the water from the humidifier yielded *Pseudomonas* spp. (Forsgren et al., 1984) and may occur in offices where a humidifier in the air-conditioning system is in operation (Robertson et al., 1985). It is a short-term ailment, and removal of the individual from the environment or the contaminant from the air-handling system is effective. Little research has been published on humidifier fever in recent years, perhaps because increased attention to the health effects of contamination in humidifying systems has resulted in decreased incidence and decreased use of such systems.

ODTS is well established as an ailment associated with occupational exposures in agricultural environments (Seifert et al., 2003a). Although it has been referred to as pulmonary mycotoxicosis (May et al., 1986), the relevant exposure is a complex mixture of bacteria, fungi, their byproducts, and other contaminants; the components responsible for the syndrome are not known (Blanc, 1997).

As this chapter notes, some studies describe subjects who experience headache, nausea, or fatigue after exposure to damp indoor environments (Dales et al., 1991a,b; Hyndman, 1990; Koskinen et al., 1999a,b; Norbäck and Edling, 1991; Thorn and Rylander, 1998a,b; Wan and Li, 1999). ODTS has been identified as posing a risk for workers performing renovation work on building materials that are contaminated with fungi (NYCDOH, 2000) but has not, to the committee's knowledge, been explored as a possible explanation of symptoms experienced by some occupants of highly contaminated indoor environments. A paper by Kolmodin-Hedman et al. (1986) did describe a case of ODTS in a museum worker who had high airborne exposure from moldy books. Evidence indicates that ODTS is underrecognized as a diagnosis (Seifert et al., 2003b). Although concentrations of organic dust consistent with the development of ODTS are very unlikely to be found in homes or public buildings, clinicians should consider the syndrome as a possible explanation of symptoms experienced by some occupants of highly contaminated indoor environments.

Conclusions

Inhalation fevers are associated with occupational exposure to high concentrations of organic materials, including bacteria, fungi, and their associated constituents and emissions. Humidifier fever, which may have been a problem in the past in some industrial environments and office buildings, has not been identified in damp home environments. The committee concludes that there is inadequate or insufficient evidence to determine whether an association exists between indoor dampness or the presence of mold or dampness-related agents and inhalation fevers in nonoccupational environments.

Respiratory Infections

Overview of the Evidence

Immune-Compromised Persons   Serious respiratory infections resulting from exposure to a variety of fungi, including *Aspergillus* spp. and *Fusarium* spp., are common in persons who undergo high-dose cancer chemotherapy, are recent recipients of a solid-organ transplant, or are otherwise immuno-compromised (DeShazo et al., 1997; Fridkin and Jarvis, 1996; Iwen et al., 2000; Patterson, 1999; Young et al., 1978). It is likely that many of these fungal infections are contracted through contact with fungi in some indoor environment because poor health leads people with severely impaired immune systems to spend most of or all their time indoors. It is less clear that the fungal exposure is related to the presence of moisture problems in the indoor spaces. Signs of water intrusion or damage are not typically noted in case reports of life-threatening fungal infections in severely immunocompromised patients, but such reports rarely indicate that there was a specific investigation of potentially problematic environmental conditions or possible mold growth. Not all forms of immune compromise are the same: the susceptibility of hosts to different fungal infections or other fungal processes depends on the nature of their deficit. Further, fungal infections experienced by immune-compromised persons do not always involve genera and species that are typically associated with damp indoor environments.

The lungs of persons with some chronic pulmonary disorders—such as cystic fibrosis, asthma, and COPD—may become colonized and potentially infected with *Aspergillus*. Clinical manifestations vary, depending on the status of the host's immune system (Marr et al., 2002). In a cystic fibrosis or asthma patient, the presence of *Aspergillus* may result in worsening of the airway disease secondary to the hypersensitivity response called allergic bronchopulmonary aspergillosis (Greenberger, 2002). This disorder occurs predominantly in atopic subjects, and its manifestations include wheeze, pulmonary infiltrates, bronchiectasis, and fibrosis (Kauffman, 2003). Atopic persons may also develop sinus disease secondary to the presence of *Aspergillus* organisms (Stevens et al., 2000). The proposed mechanism of this disorder is damage to the epithelium and underlying tissue by the exaggerated inflammatory response (Kauffman et al., 1995). It is usually not clear whether the exposure to *Aspergillus* occurred in the indoor environment or outdoors. Persons with COPD can develop respiratory failure secondary to invasive or semi-invasive pulmonary aspergillosis (Bulpa et al., 2001; Franquet et al., 2000). This outcome can also occur in persons with a variety of conditions that compromise the immune system, such as some malignancies (Stevens et al., 2000). The third disease process caused by *Aspergillus* is pulmonary aspergilloma, which is defined as a conglomeration of *Aspergillus* hyphae matted together with fibrin, mucus, and cellular debris in a pulmonary cavity or ecstatic (stretched) bronchus (Stevens et al., 2000). Persons with this problem may be without symptoms or may suffer from clinically significant hemoptysis (coughing up blood or bloody sputum). Pulmonary aspergillosis can also lead to the invasive form of the disease.

Otherwise-Healthy Persons   A few studies have examined the association of the presence of fungi or other agents in damp indoor spaces with respiratory infections or illnesses in otherwise-healthy children. Comparable studies of adults were not identified. Studies reviewed by the committee are summarized below.

A 1989 paper by Brunekreef et al. describes a study of 4,624 children 7–11 years old in homes in six U.S. cities. In analyses that controlled for other predictors of respiratory symptoms and illnesses, statistically significant associations were observed between parent-reported household mold and several respiratory illnesses (bronchitis: 1.48; 1.17–1.87; "chest illness": 1.40; 1.11–1.78; "lower respiratory illness": 1.57; 1.31–1.87). Similar results were obtained for the association with parent-reported home dampness (bronchitis: 1.32; 1.05–1.67; "chest illness": 1.52; 1.20–1.93; "lower respiratory illness": 1.68; 1.41–2.01).

Dales and Miller (1999) evaluated parent-reported "chest illness" (otherwise unspecified) during the winter months in 403 elementary-school children in Wallaceburg, Ontario. Air and dust samples from the children's bedrooms and living areas were analyzed for endotoxin; dust samples from those locations were also tested for the dust-mite allergens Der p 1 and Der f 1. After controlling for the presence of these factors and a set of child and family characteristics (child's age and sex, parental allergies, parental education, and the presence of pets or smokers in the home), an OR of 1.51 (0.76–3.02) was reported for the association between parent-reported mold or mildew in the preceding 12 months and chest illness in the subjects.

Stark and colleagues (2003) examined the incidence of four lower respiratory illnesses—croup, pneumonia, bronchitis, and bronchiolitis—in a cohort of 499 infants of asthmatic or allergic parents (that is, a birth cohort at risk for asthma). Samples were taken in the infants' bedrooms and analyzed for four airborne and 11 dustborne fungal taxa at the start of the study. The risk of lower respiratory illnesses in the year after the sampling was associated with high (>90th percentile) indoor concentrations of *Penicillium* (airborne, RR, 1.73; 1.23–2.43), *Cladosporium* (dustborne, 1.52; 1.02–2.25), *Zygomycetes* (dustborne, 1.96; 1.35–2.83), *Alternaria* (dustborne, 1.51; 1.00–2.28), or any fungus (1.86; 1.21–2.85). When the outcomes were separated by the presence or absence of wheeze, the relative risk of lower respiratory illness was greater without wheeze (3.88; 1.43–

10.52) than with wheeze (1.58; 0.95–2.64). Three more general measures of dampness or microbial agents in the home—water damage (1.30; 0.97–1.75), mold or mildew (1.33; 0.99–1.77), and either of these (1.32; 0.96–1.80)—yielded similar ORs. The authors state that "the independent effects of visible mold or mildew suggest that dampness-related factors other than exposure to the common culturable fungi are important" in the outcomes they studied.

Healthy persons exposed to damp or moldy indoor environments sometimes report that they are more prone to respiratory infections, including the common cold, sinusitis, tonsillitis, otitis, and bronchitis. Some investigators have suggested that could be due to an immunosuppressive effect (Johanning et al., 1996). Chapter 4 reviews the evidence regarding mycotoxins and immune response.

Other Evidence

$\beta(1{\rightarrow}3)$-glucans (also referred to as $(1{\rightarrow}3)$-β-D-glucans, $(1{\rightarrow}3)$-β-glucans, 1,3-beta-D-glucans, and other variants) have been implicated as a cause of the increased risk of respiratory infections in addition to serving as a marker for the presence of mold (Rylander et al., 1998). They are components of the cell walls of fungi and some bacteria. Concentrations of $\beta(1{\rightarrow}3)$-glucans can be high in dust collected from buildings that have moisture problems and housing in which residents have no complaints (Gehring et al., 2001; Rylander, 1997). Airborne $\beta(1{\rightarrow}3)$-glucans have been implicated in the presence of symptoms consistent with mucous membrane irritation syndrome, shortness of breath, other chest complaints, and lethargy and fatigue (Wan et al., 1999). There is also some evidence from animal studies that $\beta(1{\rightarrow}3)$-glucans can potentiate allergen-induced eosinophil infiltration (Wan et al., 1999). One study demonstrated a decrease in inflammatory cell numbers after experimental $\beta(1{\rightarrow}3)$-glucan exposure (Fogelmark et al., 1992). Mouse experiments of Korpi et al. (2003) suggested that—with regard to previous fungal sensitization of the animals—inhaled $\beta(1{\rightarrow}3)$-glucan may cause symptoms of respiratory tract irritation without apparent inflammation. It is not yet clear how those observations in animal studies apply to the risk of infection in humans exposed to $\beta(1{\rightarrow}3)$-glucans. Fully characterizing their effects in environmental exposures of humans has been made more difficult by technical challenges in performing the necessary measurements.

Conclusions

Exposure to fungi in indoor environments is associated with severe respiratory infections in some severely immunocompromised persons. There is sufficient evidence of an association between exposure to *Aspergillus* spp. and pulmonary aspergillosis and aspergillomas in such persons. Several less common fungi and bacteria can also induce characteristic infections; but these are uncommon disorders that have not been specifically associated with a damp indoor environment, and they are mentioned here for completeness.

Available studies of respiratory illnesses in otherwise healthy people address children. Only Brunekreef et al. (1989) specifically account for indoor dampness, but its presence may be inferred in the other studies because mold is present. Together the available studies indicate that

• There is limited or suggestive evidence of an association between exposure to a damp indoor environment and lower respiratory illness (otherwise unspecified) in otherwise-healthy children.

• There is limited or suggestive evidence of an association between the presence of "mold" (otherwise unspecified) in a damp indoor environment and lower respiratory illness in otherwise-healthy children.

• There is inadequate or insufficient evidence to determine whether an association exists between exposure to dampness or the presence of mold or other agents in damp indoor environments and respiratory illness in otherwise-healthy adults.

Pulmonary Hemorrhage or Hemosiderosis

Pulmonary hemorrhage or hemosiderosis is a pathologic condition characterized by an abnormal accumulation of hemosiderin, an iron-containing pigment, in lung tissue (Boat, 1998). It results from diffuse bleeding or hemorrhage in the alveoli, the portion of the lung where gas exchange occurs. Pulmonary hemorrhage may be reported at any age and in association with a variety of conditions. In children, those conditions include hypersensitivity to cow's milk, also known as Heiner's syndrome (Heiner et al., 1962). Autoimmune conditions—such as Goodpasture's syndrome, Wegner's granulomatosis, and celiac disease—may present with pulmonary hemorrhage (Boat, 1998). High-dose chemotherapy for cancer can cause it (Robbins et al., 1989). In some cases no etiology is identified.

Pulmonary hemorrhage or hemosiderosis is most common in newborns, infants, especially those born prematurely. It is seen in about 55% of autopsies performed on infants (van Houten et al., 1992). Pulmonary hemorrhage can also be a subclinical, chronic condition that presents as iron deficiency, and alternatively, it can be episodically symptomatic or be massive and life-threatening. Usually, the clinical picture is characterized by recurrent episodes of acute pulmonary bleeding with associated fever, tachypnea, and leukocytosis. Examination of the chest reveals bronchial breath sounds or diminished breath sounds, crackles, and wheeze. The patient may experience a cough productive of sputum and this sputum characteristically contains hemosiderin-laden macrophages. A chest x-ray picture usually demonstrates infiltrates, which are typically fleeting and eventually evolve to a reticulonodular pattern.

Treatment of pulmonary hemorrhage or hemosiderosis consists primarily of providing supportive care. Drug therapy with high-dose corticosteroids or immunosuppressive agents may be beneficial. Typically, the disease evolves into a chronic form with interstitial fibrosis, pulmonary hypertension, and right-sided heart failure. The disease can be fatal. At autopsy, hemosiderin-laden macrophages may be found in lung tissue at the site of bleeding.

Overview of the Evidence

Investigators described eight cases of acute pulmonary hemorrhage or hemosiderosis in infants presenting at a Cleveland children's hospital in 1993–1994 (CDC, 1994). Milk-protein allergies, congenital heart or vascular malformations, infectious processes, trauma, and other causes of pulmonary hemorrhage in infants were ruled out. The investigators noted that the cluster of cases was characterized by the geographic proximity of the infants' homes (all were in the eastern metropolitan area of Cleveland) and by the race or ethnicity of the cases (all were black). A later report observed that most homes in the area where the infants lived were more than 60 years old and inadequately maintained and that there was a comparatively high rate of poverty in the area: 48% of all the children in the area lived below the poverty level (Dearborn et al., 1999).

A follow-up report (CDC, 1997) and two papers by the Cleveland investigators (Etzel et al., 1998; Montaña et al., 1997) included the findings from a case-control study and an assessment of infant-death cases conducted by the county coroner. The case-control study matched 10 case infants (two additional cases were identified after the 1994 report was published)—a case was defined as an episode of acute, diffuse pulmonary hemorrhage of unknown etiology in a previously healthy infant in the first year of life and requiring hospitalization—with 30 control infants matched for age and residence in the same geographic area of Cleveland. All case infants and seven of the 30 control infants lived in homes where "major" water damage had occurred during the preceding 6 months. Later visual inspection of those homes and aggressive air sampling revealed a higher concentration of *Stachybotrys atra* (now called *S. chartarum*) in the residences of case infants than in those of control infants (1.6; 1.0–30.8)[6]

---

[6]Research later showed that some strains of *S. chartarum* can produce mycotoxins, including hemolysin, tricothecenes, and atranones (Jarvis et al., 1998; Vesper et al., 2000). In animal models, spores of these toxic strains induce pulmonary inflammation and hemorrhage (Flemming et al., 2004; Nikulin et al., 1997; Rand et al., 2002, 2003; Yike et al., 2002). Chapter 4 addresses this literature in greater detail.

(CDC, 1997). Case infants were also more likely to have been in the presence of environmental tobacco smoke before their illness (7.9; 0.9–70.6). The small numbers of cases and controls limited the statistical confidence with which any associations could be assessed.

The 1997 Centers for Disease Control and Prevention (CDC) report noted that "based on the findings of the case-control study, health authorities in Cleveland recommended prompt clean-up and disposal of all moldy materials in the water-damaged homes and have designed a prevention program focusing on water-damaged homes." An editorial note that accompanied the report concluded, however, that further efforts were needed to identify what association, if any, existed between pulmonary hemorrhage in infants and exposure to water-damaged building materials.

In 1997, CDC convened an internal scientific taskforce and a panel of outside experts was chosen to review the possible association between pulmonary hemorrhage/hemosiderosis in infants and *S. chartarum* in the indoor environment (CDC External Expert Panel, 1999; CDC Working Group, 1999). The results of those independent investigations were reviewed in a 2000 issue of *Morbidity and Mortality Weekly Report* (CDC, 2000). The publication noted that both groups of reviewers considered that the interpretation of an epidemiologic association between household water damage and the Cleveland cases was hampered by a lack of consistent criteria for defining water damage, an absence of a standardized protocol for inspecting and gathering data on the presence of fungi in individual homes, a failure to distinguish between contamination and clinically significant exposure to fungi, and a failure to obtain isolates of or serologic evidence of exposure to fungi or mycotoxin from case-infants. Reviewers were also critical of the analytic methods used in the original study. The reported OR associating *S. chartarum* concentrations for each household was judged to be "unstable" in part because of inconsistent methods for sampling and a "potential misleading" strategy for matching cases and controls. The groups concluded that *S. chartarum* was not clearly associated with acute pulmonary hemorrhage or hemosiderosis in infants, because of issues concerning environmental sampling methods and statistical methods, the fact that sampling was not performed in a blinded fashion,[7] and the small difference in the presence of culturable *S. chartarum* between water-damaged case and control homes.

The Cleveland cases (and cases in a Chicago outbreak mentioned in the

---

[7]The 2000 *MMWR* publication notes (p. 182) that "one investigator correctly inferred the identity of many case homes and wanted to be certain to identify culturable fungi in these homes if they were present. As a result, the investigator collected twice the number of air samples from case homes as were collected from control homes."

paper) appeared different from classically described idiopathic pulmonary hemosiderosis, and the term acute idiopathic pulmonary hemorrhage (in infants) (AIPH) was proposed to describe them (CDC, 2000). (CDC later proposed a case definition for "acute idiopathic pulmonary hemorrhage in infants," or AIPHI [CDC, 2001].)

The CDC working group indicated that it did not reject a possible relationship between *S. chartarum* and AIPH in the Cleveland cluster (CDC External Expert Panel, 1999). However, the editorial note that accompanied the 2000 CDC report was highly critical of the original findings in the 1997 report and advised that conclusions regarding the possible association between cases of pulmonary hemorrhage/hemosiderosis in infants in Cleveland and household water damage or exposure to *S. chartarum* are not substantiated adequately by scientific evidence produced in the CDC investigation. . . . The associations should be considered not proven; the etiology of AIPH is unresolved.

In noting those shortcomings in the collection, analysis, and reporting of study data, the editorial note indicated that a number of actions would be taken by CDC, including

• The continuing investigation of AIPH cases, particularly if clusters of such cases could be identified.
• The standardization of protocols for data collection and environmental assessment to ascertain the possible association between AIPH and environmental etiologies, such as household water damage and exposure to *S. chartarum* and other fungi.
• The implementation of surveillance of AIPH in infants, either in clusters or in individual cases, and the development of a consistent standard surveillance case definition for reporting.
• The development of enhanced sampling and laboratory analytic methods for the assessment of environmental exposures to molds and fungi.

The Cleveland researchers posted a response to the 2000 CDC report on the World Wide Web that offered rebuttals to the criticisms (Etzel et al., undated). Separately, a 2003 paper by Etzel argued that available evidence regarding *Stachybotrys* exposure and AIPHI fulfills the Bradford Hill criteria of causality (Etzel, 2003). Dearborn et al. (2002), however, maintain that although the potential role of *Stachybotrys* toxins is suggested by the data, additional evidence is needed to support causation. "The possibility remains that the presence in the infants' homes of *Stachybotrys*, a high water-requiring fungus, may simply be an indicator of water intrusion, and other unknown, related factors (in addition to ETS) may play primary or secondary causative roles."

Additional individual cases of acute pulmonary hemorrhage in infants in indoor environments where *S. chartarum* or other fungi are present have been described in Cleveland and elsewhere (CDC, 1997; Dearborn et al., 2002; Elidemir et al., 1999; Flappan et al., 1999; Knapp et al., 1999; Novotny and Dixit, 2000; Tripi et al., 2000; Weiss and Chidekel, 2002). Toxigenic fungal exposure has been suggested as a factor, but its role is yet to be determined.

Conclusions

The role of *Stachybotrys chartarum* in the Cleveland AIPH cluster and other AIPH cases remains controversial. The committee did not undertake to reanalyze the Cleveland outbreak data, and it is not in a position to second-guess either the researchers or those who reviewed their work. It does offer the following observations on the basis of published materials.

Toxicologic data (discussed in Chapter 4) indicates that exposure to at least some strains of *S. chartarum* affects the lungs of young animals, although these observations require validation from more extensive research before conclusions can be drawn. The human data are equivocal. *S. chartarum* is not uncommon, and, although theories have been put forward,[8] no compelling explanation has been presented for its causing such adverse health outcomes in the Cleveland cluster and some other isolated cases but not elsewhere. The Cleveland case infants lived in homes distinguished by a number of factors that may have contributed to poor health. The influence of some of those factors was explored in the analysis of the cluster, but the small number of cases and the retrospective nature of the data collection necessarily limited the evaluation. The possible synergistic influence of ETS mentioned by some investigators (Dearborn et al., 2002; Montaña et al., 1997; Novotny and Dixit, 2000) is especially interesting in this regard, given the known adverse effects of ETS on infants.

The committee concludes that available case-report information, taken together, constitutes inadequate or insufficient information to determine whether an association exists between AIPHI and the presence of *Stachybotrys chartarum* or agents present in damp indoor spaces in general. AIPHI is a serious health outcome, and the committee encourages the CDC to pursue surveillance and additional research on the issue to resolve outstanding questions. Epidemiologic and case studies should take a broad-based approach to gathering and evaluating information on exposures and

---

[8] Vesper and Vesper (2002) have suggested that strain differences may help to explain why pulmonary hemorrhage occurred in some homes infested with *S. chartarum* and not others. In the absence of information on the ubiquity of the strains found in the Cleveland infants' homes, it is not possible to evaluate the merit of this suggestion.

other factors that would help to elucidate the etiology of acute pulmonary hemorrhage or hemosiderosis in infants, including dampness and agents associated with damp indoor environments; ETS and other potentially adverse exposures; and social and cultural circumstances, race/ethnicity, housing conditions, and other determinants of study subjects' health.

OTHER HEALTH COMPLAINTS AND DISORDERS

Although the health problems most often associated with dampness and dampness-related agents are respiratory, concerns have also been raised about other health outcomes. This section addresses several of them. Many of the epidemiologic studies cited here are wide-ranging examinations of health complaints rather than specific assessments of particular outcomes. The committee's judgment about the association between exposure to damp indoor spaces or the presence of mold or other agents in damp indoor environments and these health outcomes is discussed at the end of the section.

Skin

It is well established that severely immunocompromised patients—for example, those treated with immunosuppressive drugs during transplantation procedures, cancer patients receiving chemotherapy, AIDS patients—can develop opportunistic cutaneous and subcutaneous fungal infections of the skin (Wald et al., 1997). Trichothecenes are established dermal irritants. (The dermal toxicity of mold, bacteria, and their constituents is addressed in Chapter 4.)

Questions have been raised about whether eczema (eczematous dermatitis) and atopic dermatitis may be related to damp indoor environments. Eczema is a characteristic inflammatory response of the skin to multiple stimuli. There is usually a primary elicitor of the response (such as an allergen), after which many factors may contribute. The unifying feature of eczema is the occurrence of raised erythematous, scaly skin lesions that are extremely pruritic (that is, they provoke itching). The initial diagnosis is based on the patient's medical history and the appearance of the skin; occasionally, skin biopsy is used. Atopic dermatitis is associated with IgE allergy and often with a family history of atopy. The striking feature of the disease is severe, spasmodic itching. The diagnosis is based on the patient's medical history and physical examination and on identification of IgE allergy with skin tests for specific antibody in the serum.

Table 5-9 summarizes studies that address dermal outcomes in the general population. The outcomes are eczema and dermatitis, which are diagnosed conditions; irritation, a symptom that may be frank; and "skin

TABLE 5-9   Selected Epidemiologic Studies—Skin Problems and
Exposure to Damp Indoor Environment or Presence of Mold or Other
Agents in Damp Indoor Environments

Reference	Subjects	Dampness or Mold Measure	Risk Estimate
Cross-sectional studies			
Engvall et al., 2002	3,241 adults	Self-reported odor and water leakage in last 5 years	2.45 (2.25–2.66) for facial skin irritation
Kilpeläinen et al., 2001	10,667 students (18–25 years old)	Self-reported visible mold	1.31 (0.96–1.79) for atopic dermatitis
		Self-reported visible mold, damp stains, or water damage	1.29 (1.06–1.56) for atopic dermatitis
Koskinen et al., 1999a	699 adults (16+ years old) in 310 households	Surveyor-assessed moisture	1.03 (0.71–1.49) for eczema 1.18 (0.76–1.83) for atopic eczema 1.06 (0.64–1.78) for allergic eczema
		Self-reported mold	1.40 (0.93–2.10) for eczema 1.18 (0.73–1.90) for atopic eczema 0.77 (0.43–1.38) for allergic eczema
Wan and Li, 1999	1,113 workers in 19 office buildings	Self-reported mold	2.97 (1.52–5.82) for skin symptoms
Li et al., 1997	612 day-care workers	Self-reported mold	0.58 (0.21–1.60) for allergic eczema 1.95 (1.30–2.91) for skin symptoms
		Self-reported water damage	1.55 (0.58–4.14) for allergic eczema 2.04 (1.35–3.08) for skin symptoms
Pirhonen et al., 1996	1,460 adults (25–64 years old)	Self-reported damp or mold problem	1.21 (0.86–1.71) for allergic eczema 1.31 (0.97–1.76) for atopic eczema
Waegemaekers et al., 1989	190 children	Self-reported damp homes	1.65 (1.11–2.46) for skin irritation

symptoms," a nonspecific complaint. It should be noted that *eczema* and *atopic dermatitis* are sometimes (incorrectly) used interchangeably.

Studies that examine eczema and dermatitis report ORs of 0.58–1.55; the confidence intervals of all but one include unity. The four papers that address less specific outcomes—irritation or "skin symptoms"—note statistically significant associations with measures of dampness or mold, but it is difficult to assess the implications of these results in the absence of details on the nature of the skin problem or controls for the possibility that the symptom is a characteristic of some other health outcome. The committee notes that future research examining the possible association of microbial exposure with skin symptoms should include stratification by allergic status, or by preexisting eczema or allergic disease when possible.

Gastrointestinal Tract

A review article by Peraica et al. (1999) notes that ingestion of mycotoxins in moldy foodstuffs can lead to gastrointestinal symptoms in humans. Nausea, vomiting, and diarrhea are sometimes reported by persons exposed to dampness and molds in indoor spaces (Dales et al., 1991a), but relatively few epidemiologic studies have evaluated the association between these symptoms and damp indoor conditions. Table 5-10 summarizes the results of some of the studies that have examined nausea as an outcome.

Most of the studies report nonsignificant increases in incidence in self-reported symptoms. An early study by Waegemaekers et al. (1989) yielded a statistically significant increase in parent-reported nausea and vomiting in their children, but the authors note that the sample was self-selected among people who expressed health concerns and that respondent bias might be influencing the results. In a controlled human-exposure study, Thorn and Rylander (1998b) examined the responses of 21 healthy adult volunteers who inhaled 40 µg of lipopolysaccharide (endotoxin) under controlled conditions. Four reported nausea 24 hrs later; two, diarrhea.

Gastrointestinal complaints have a large differential diagnosis, which complicates analyses. Other causes of the symptoms need to be ruled out before it is concluded that these outcomes are secondary to a dampness-related agent. The independent influence of mold odors, which was not controlled for in many of the cited studies, also needs to be considered.

Fatigue

Fatigue is a common nonspecific complaint that can be caused by a long list of physical and psychiatric disorders. Lifestyle choices, such as long work hours and irregular sleep habits, can contribute to fatigue. Clinicians approach this problem by ruling out a variety of disorders.

TABLE 5-10   Selected Epidemiologic Studies—Nausea and Related
Symptoms and Exposure to Damp Indoor Environment or Presence of
Mold or Other Agents in Damp Indoor Environments

Reference	Subjects	Dampness or Mold Measure	Risk Estimate
Cross-sectional studies			
Koskinen et al., 1999a	699 adults (16+ years old) in 310 households	Surveyor-assessed moisture	1.23 (0.85–1.80) for nausea
		Self-reported mold	1.36 (0.90–2.06) for nausea
Koskinen et al., 1999b	204 children (≤ 15 years old) in 310 households	Surveyor-assessed moisture	1.62 (0.78–3.37) for nausea, 7–15 years old
		Self-reported mold	2.11 (1.01–4.42) for nausea, 7–15 years old
Pirhonen et al., 1996	1,460 adults (25–64 years old)	Self-reported damp or mold problem	1.80 (0.87–3.71) for nausea 1.65 (1.24–2.20) for recurrent stomach ache
Waegemaekers et al., 1989	190 children	Self-reported damp homes	3.13 (1.29–5.26) for nausea, vomiting

Table 5-11 summaries the results of studies of people in damp and
moldy environments in which fatigue was examined. As evidenced by rather
broad confidence intervals, data on relatively small numbers of people were
analyzed in some of the studies.

There is evidence that exposure to molds causes release of proinflammatory cytokines, including TNFα and IL-6. (Hirvonen et al., 1999). Thorn
and Rylander (1998b) reported that 12 of 21 subjects (p < 0.001) who
voluntarily inhaled 40 µg of lipopolysaccharide (endotoxin) reported "unusual tiredness" 24 hrs later. The researchers speculated that the outcome
might have been mediated by TNFα activity. A variety of disorders characterized by release of cytokines into peripheral blood are associated with fatigue,
but it is not known whether the cytokines are the cause of that symptom.
Bhattacharyya (2003) reports a coincidence between fatigue and chronic
rhinosinusitis, noting that 32% of 322 chronic-rhinosinusitis patients surveyed report the symptom (the presence of dampness or mold was not
addressed); it is not clear whether this suggests a confounding factor in some
reports of fatigue or whether the outcomes might have a common cause.

Neuropsychiatric Symptoms

Self-reported cognitive defects and difficulties in concentrating are sometimes noted by people who occupy damp or moldy buildings (Johanning et al., 1996; Koskinen et al., 1999a,b; Sudakin, 1998). Such complaints, however, are notoriously difficult to quantify.

The committee did not identify any well-defined epidemiologic studies that conducted neuropsychiatric testing of individuals and accounted explicitly for dampness or the presence of mold. It did, however, find three small-sample case studies in the peer-reviewed, medical literature, and they are mentioned here for completeness. (Chapter 4 addresses the literature on the neurologic effects of mycotoxin exposure in animals.)

Hodgson et al. (1998) investigated a set of three buildings in which occupant health complaints had been reported. *Stachybotrys chartarum* and *Aspergillus* spp. were identified on moisture-damaged interior surfaces and in air samples taken in these structures. Three neuropsychologic tests (in addition to medical testing and a questionnaire) were administered to 14 occupants selected by the building's insurance carrier and to 47 volunteer occupants. The subjects' results on trial I of the California Verbal Learning Test (CVLT), the dominant-hand time for the Grooved Pegboard Test, and the CVLT long-delay cued-recall test were better than or similar to results of two sets of undocumented controls.

More recently, Baldo et al. (2002) administered the San Diego Neuropsychological Test Battery to 10 adults who had been involved in litigation regarding their exposure to mold. The type of molds they had been exposed to and the duration and setting of the exposure varied among the subjects. The authors found no consistent pattern of test results among the subjects but noted that deficits (relative to normative data) in visuospatial learning, visuospatial memory, verbal learning, and psychomotor speed were observed more frequently than deficits in other assessed functions. There was also a significant correlation (Pearson product moment correlation, one-tailed, 0.47; $p < 0.05$) between a measure of depression (the Beck Depression Inventory, second edition) and the number of cognitive impairments identified in testing of a subset of seven subjects.

Anyanwu et al. (2002) identify a unique disorder that they term acoustic mycotic neuroma and that they state is frequently observed in their clinic. The authors report that four adolescent boys with a variety of physical and mental symptoms and mold exposure were evaluated with brainstem auditory evoked response testing. Waveform abnormalities were reported in the subjects; specifics varied except for one to three interpeak latencies (a measure of acoustic-nerve dysfunction), which were outside the expected range of results in all subjects.

It has been shown that inhalation of odorants can activate the temporal

TABLE 5-11 Selected Epidemiologic Studies—Fatigue and Related Symptoms and Exposure to Damp Indoor Environment or Presence of Mold or Other Agents in Damp Indoor Environments

Reference	Subjects	Dampness or Mold Measure	Risk Estimate
Cross-sectional studies			
Engvall et al., 2002	3,241 adults	Self-reported odor and water leakage in last 5 years	2.54 (2.38–2.71) for tiredness
Lander et al., 2001	86 workers in moist buildings with mold growth	Histamine-release test (HRT) positive for 1 or more molds	1.81 (0.70–4.65) for fatigue
		Test positive for *Penicillium chrysogenum*	3.60 (1.09–11.86) for fatigue
Koskinen et al., 1999a	699 adults (16+ years old) in 310 households	Surveyor-assessed moisture	1.50 (0.88–2.56) for fatigue
		Self-reported mold	3.97 (1.67–9.46) for fatigue
Koskinen et al., 1999b	204 children (≤ 15 years old) in 310 households	Surveyor-assessed moisture	0.93 (0.37–2.33) for fatigue 7–15 years old
		Self-reported mold	1.48 (0.55–4.01) for fatigue 7–15 years old

Wan and Li, 1999	1,113 workers in 19 office buildings	Self-reported mold	1.32 (0.81–2.17) for lethargy/fatigue
		Self-reported water damage	0.96 (0.56–1.63) for lethargy/fatigue
		Self-reported flooding	1.12 (0.60–2.03) for lethargy/fatigue
Thorn and Rylander, 1998a	129 adults (18–83 years)	Measured airborne glucan (>2–4 ng/m^3)	1.11 (0.79–1.55) for unusual tiredness
		Measured airborne glucan (>4 ng/m^3)	0.95 (0.68–1.33) for unusual tiredness
Li et al., 1997	612 day-care workers	Self-reported mold	1.89 (1.25–2.87) for fatigue
		Self-reported water damage	1.99 (1.35–2.92) for fatigue
Pirhonen et al., 1996	1,460 adults (25–64 years)	Self-reported damp or mold problem	1.81 (1.37–2.39) for fatigue or tiredness
Waegemaekers et al., 1989	190 children	Self-reported damp homes	2.40 (1.68–3.44) for tiredness

lobe of the cerebral cortex (Kettenmann et al., 1996), and exposure to odorous emissions from animal operations, wastewater treatment, and recycling of biosolids has been associated with stress and alterations of mood (Schiffman et al., 2000). Damp and mold-contaminated buildings commonly have a musty odor that people find unpleasant, but the neurophysiologic effect of such odors has not been studied. Odor perception, however, does offer a possible explanation of some reported problems.

Sick Building Syndrome

Sick building syndrome (SBS) is a term used to describe a combination of nonspecific symptoms related to residence or work in a particular building. Thörn (1999) identifies the core symptoms associated with SBS as

- Irritation of the eyes, nose and throat, cough.
- Experience of dry skin, rash, pruritus.
- Fatigue, headache, lack of concentration.
- High frequency of respiratory tract infections.
- Hoarseness, wheezing, shortness of breath.
- Nausea, dizziness.
- Enhanced or abnormal odor perception.

Researchers studying SBS incidence and etiology vary in their requirements for which or how many symptoms must be present over what period of time in order for a subject to be defined as a case. Chemical contaminants, biologic contaminants (including molds, bacteria, and viruses), inadequate ventilation, odor perception, thermal comfort, and psychological factors have all been suggested as putative causes (Ebbehøj et al., 2002; Hodgson, 2002; Seppänen and Fisk, 2002). Because of the lack of consistent diagnostic criteria, the committee chose not to address SBS as a separate clinical outcome. The report does address individual symptoms that have been associated with SBS in the discussions above.

Cancers

Several common fungi can produce mycotoxins that are carcinogenic in at least some animal species (Rao, 2000). As discussed in Chapter 4, ingestion of some molds—typically as contaminants in grain—has been associated with cancers in humans. The committee did not identify any peer-reviewed scientific literature that addressed microbial agents found indoors and the inhalation route of exposure. A 2002 study showed that human lung cells can activate cytotoxic and DNA-reactive intermediates from afla-toxin (Van Vleet et al., 2002). However, a laboratory study of microbial volatile organic compounds found that DNA damage occurred only in circumstances where cytotoxic effects were also seen; neither clastogenic nor mutagenic effects were observed (Kreja and Seidel, 2002).

Reproductive Effects

Measuring the effects of exposures on reproductive outcomes is chal-lenging because of the large number of factors that contribute to these outcomes. The committee did not identify any peer-reviewed published data on agents or exposure routes relevant to damp conditions indoors.

Rheumatologic and Other Immune Diseases

Rheumatologic diseases are characterized by inflammation and stiff-ness or pain in muscles, joints, or fibrous tissue and other symptoms. People suffering from such diseases often note that they are exacerbated by envi-ronmental conditions, including dampness, and changes in weather. Atten-tion has been paid to the possibility that exposure to damp indoor spaces might cause rheumatologic conditions because fungi and their constituents can trigger immune responses that result in inflammation.

There is some evidence that cold, damp indoor spaces causes erythema-tous skin lesions known as chilblain lupus erythematosus of Hutchinson, but there is little evidence that this problem is associated with systemic lupus erythematosus (Franceschini et al., 1999). In patients with existing rheumatologic arthritis, damp weather increases complaints of stiffness (Rasker et al., 1986). A 2002 study described a cluster of persons with several different rheumatologic diseases (rheumatoid arthritis, ankylosing spondylitis, Sjogren's syndrome, and psoriatric arthritis) who worked in a water-damaged building contaminated by microbial growth, including molds (Myllykangas-Luosujärvi et al., 2002). It could not be discerned whether particular agents had accounted for this cluster. A broader study of workers in the building measured a series of potential biomarkers in nasal lavage fluids, induced sputum, and serum of occupants who had rheumatic or respiratory disorders and of controls in the same workplace (Roponen et al., 2001b). It was concluded that IL-4 was significantly higher in cases than in control subjects and in all workers when at work vs away on vacation. That implied that IL-4—which down-regulates Th-1-mediated inflammatory responses and up-regulates Th-2—might play a role in both the respiratory and rheumatologic disorders that were observed. It was also observed that higher exhaled NO concentrations were associated with respiratory symptoms. A follow-up study (Ropenen et al., 2003) monitored concentrations of inflammatory mediators in nasal lavage fluids from 12 healthy volunteers over a 1-year period. Substantial individual and sex variability suggested that the measure was most appropriate when subjects could be used as their own controls. Additional studies are needed to explore the relationship between IL-4, exhaled NO, respiratory and rheumatologic disease, and microorganism-contaminated indoor spaces.

More generally, studies by Hirvonen et al. (2001) and Roponen et al. (2001a) indicate experimentally that the ability of spores of a bacterium found in some damp environments (*Streptomyces anulatus*) to produce an inflammatory response depended on the building material it grew on. Beijer et al. (2003) examined inflammatory markers in the blood of non-smoking persons in homes with either "high" (>4.0 ng/m^3) or "low" (<2.0 ng/m^3) airborne concentrations of $\beta(1\rightarrow3)$-D-glucan, which was used as a surrogate of mold exposure. Among nonatopic subjects, the ratio between interferon gamma and IL-4 was significantly higher in the high-airborne-concentration group than in the low-airborne-concentration group. The authors suggest that this effect on inflammatory markers indicates that mold exposure may stimulate some parts of the inflammatory-immuno-logic system.

Conclusions

Few epidemiologic studies have examined damp or "moldy" environments and skin, gastrointestinal tract, fatigue, neuropsychiatric, cancers, reproductive, and rheumatologic outcomes. Studies that are available have tended to address the outcomes only as a part of a survey of signs, symptoms, and diseases. In some cases, the outcomes have been anecdotally related to the environments; in others, effects have been noted in high-level (often ingestion) exposure scenarios in humans or animals. Generally, relatively little attention has been paid to evaluating or controlling for other potential explanations of the reported problem.

The association between fungal exposures and opportunistic fungal infections of the skin of severely immunocompromised persons is well established. For all the other listed outcomes, the committee concludes that there is inadequate or insufficient information to determine whether an association exists between them and exposure to a damp indoor environment or the presence of mold or other agents associated with damp indoor environments. A small number of case studies have associated those adverse health outcomes with damp or moldy environments but only in persons with highly compromised immune systems or when the circumstances, such as ingestion of contaminated foodstuffs, are not relevant to this report.

FINDINGS, RECOMMENDATIONS, AND RESEARCH NEEDS

Research on health outcomes that may be associated with damp indoor environments is challenging: formidable study-design issues and prohibitive costs are associated with comprehensive investigations. Chapter 3 delineates the committee's finding that the limitations of knowledge regarding indoor microbial agents and related health problems are due primarily to a lack of valid quantitative methods for assessing exposure. It also offers research recommendations to address that problem.

On the basis of its review of the papers, reports, and other information presented in this chapter, the committee has reached the following findings and recommendations and has identified the following research needs regarding the human health effects of damp indoor environments.

Findings

Tables 5-12 and 5-13 summarize the committee's findings on the state of the scientific evidence regarding the association between various health outcomes and exposure to damp indoor environments or the presence of

TABLE 5-12   Summary of Findings Regarding the Association Between Health Outcomes and Exposure to Damp Indoor Environments[a]

Sufficient Evidence of a Causal Relationship
(no outcomes met this definition)

Sufficient Evidence of an Association

Upper respiratory (nasal and throat) tract symptoms	Wheeze
Cough	Asthma symptoms in sensitized asthmatic persons

Limited or Suggestive Evidence of an Association

Dyspnea (shortness of breath)	Asthma development
Lower respiratory illness in otherwise-healthy children	

Inadequate or Insufficient Evidence to Determine Whether an Association Exists

Airflow obstruction (in otherwise-healthy persons)	Skin symptoms
Mucous membrane irritation syndrome	Gastrointestinal tract problems
Chronic obstructive pulmonary disease	Fatigue
Inhalation fevers (nonoccupational exposures)	Neuropsychiatric symptoms
Lower respiratory illness in otherwise-healthy adults	Cancer
Acute idiopathic pulmonary hemorrhage in infants	Reproductive effects
	Rheumatologic and other immune diseases

[a]These conclusions are not applicable to immunocompromised persons, who are at increased risk for fungal colonization or opportunistic infections.

TABLE 5-13 Summary of Findings Regarding the Association Between Health Outcomes and the Presence of Mold or Other Agents in Damp Indoor Environments[a]

Sufficient Evidence of a Causal Relationship (no outcomes met this definition)	
Sufficient Evidence of an Association	
Upper respiratory (nasal and throat) tract symptoms	Wheeze
Asthma symptoms in sensitized asthmatic persons	Cough
Hypersensitivity pneumonitis in susceptible persons[b]	
Limited or Suggestive Evidence of an Association	
Lower respiratory illness in otherwise-healthy children	
Inadequate or Insufficient Evidence to Determine Whether an Association Exists	
Dyspnea (shortness of breath)	Skin symptoms
Airflow obstruction (in otherwise-healthy persons)	Asthma development
Mucous membrane irritation syndrome	Gastrointestinal tract problems
Chronic obstructive pulmonary disease	Fatigue
Inhalation fevers (nonoccupational exposures)	Neuropsychiatric symptoms
Lower respiratory illness in otherwise-healthy adults	Cancer
Rheumatologic and other immune diseases	Reproductive effects
Acute idiopathic pulmonary hemorrhage in infants	

[a]These conclusions are not applicable to immunocompromised persons, who are at increased risk for fungal colonization or opportunistic infections.
[b]For mold or bacteria in damp indoor environments.

mold or other agents in damp indoor environments. As already noted, fungi and bacteria are omnipresent in the environment, and the committee restricted its evaluation to circumstances that could be reasonably associated with damp indoor environments. Studies regarding homes, schools, and office buildings were considered; such other indoor environments as barns, silos, and factories—which may subject people to high occupational exposures to organic dusts and other microbial contaminants—were not. The tables' conclusions are not applicable to persons with compromised immune systems, who are at risk for fungal colonization and opportunistic infections. The committee considered whether any of the health outcomes addressed in this chapter met the definitions for the categories "sufficient evidence of a causal relationship" and "limited or suggestive evidence of no association" defined in Chapter 1, and concluded that none did.

Recommendations and Research Needs

• Indoor environments subject occupants to multiple exposures that may interact physically or chemically with one another and with the other characteristics of the environment, such as humidity, temperature, and ventilation. Few studies to date have considered whether there are additive or synergistic interactions among these factors. The committee encourages researchers to collect and analyze data on a broad range of exposures and factors characterizing indoor environments in order to inform these questions and possibly point the way toward more effective and efficient intervention strategies.

• The committee encourages the CDC to pursue surveillance and additional research on acute pulmonary hemorrhage or hemosiderosis in infants to resolve questions regarding this serious health outcome. Epidemiologic and case studies should take a broad-based approach to gathering and evaluating information on exposures and other factors that would help to elucidate the etiology of acute pulmonary hemorrhage or hemosiderosis in infants, including dampness and agents associated with damp indoor environments; ETS and other potentially adverse exposures; and social and cultural circumstances, race/ethnicity, housing conditions, and other determinants of study subjects' health.

• Concentrations of organic dust consistent with the development of organic dust toxic syndrome are very unlikely to be found in homes or public buildings. However, clinicians should consider the syndrome as a possible explanation of symptoms experienced by some occupants of highly contaminated indoor environments.

• Greater research attention to the possible role of damp indoor environments and the agents associated with them in less well understood disease entities is needed to address gaps in scientific knowledge and concerns among the public.

REFERENCES

Ando M, Suga M, Nishiura Y, Miyajima M. 1995. Summer-type hypersensitivity pneumonitis. International Medicine 34(8):707–712.

Andriessen J, Brunekreef B, Roemer W. 1998. Home dampness and respiratory health status in European children. Clinical and Experimental Allergy 28:1191–1200.

Anyanwu E, Campbell AW, High W. 2002. Brainstem auditory evoked response in adolescents with acoustic mycotic neuroma due to environmental exposure to toxic molds. International Journal of Adolescent Medicine and Health 14(1):67–76.

Apostolakos MJ, Rossmoore H, Beckett WS. 2001. Hypersensitivity pneumonitis from ordinary residential exposures. Environmental Health Perspectives 109(9):979–981.

Arundel AV, Sterling EM, Biggin JH, Sterling TD. 1986. Indirect health effects of relative humidity in indoor environments. Environmental Health Perspectives 65:351–361.

Asero R, Bottazzi G. 2000. Hypersensitivity to molds in patients with nasal polyposis: a clinical study. Journal of Allergy and Clinical Immunology 105(1 Pt 1):186–188.

Austin JB, Russell G. 1997. Wheeze, cough, atopy, and indoor environment in the Scottish Highlands. Archives of Disease in Childhood 76(1):22–26.

Baldo JV, Ahmad L, Ruff R. 2002. Neuropsychological performance of patients following mold exposure. Applied Neuropsychology 9(4):193–202.

Balmes JR. 2002. Occupational airways diseases from chronic low-level exposures to irritants. Clinics in Chest Medicine 23(4):727–735.

Baur X, Behr J, Dewair M, Ehret W, Fruhmann G, Vogelmeier C, Weiss W, Zinkernagel V. 1988. Humidifier lung and humidifier fever. Lung 166(2):113–124.

Beijer L, Thorn J, Rylander R. 2003. Mould exposure at home relates to inflammatory markers in blood. European Respiratory Journal 21(2):317–322.

Belanger K, Beckett W, Triche E, Bracken MB, Holford T, Ren P, McSharry JE, Gold DR, Platts-Mills TA, Leaderer BP. 2003. Symptoms of wheeze and persistent cough in the first year of life: associations with indoor allergens, air contaminants, and maternal history of asthma. American Journal of Epidemiology 158(3):195–202.

Berglund LG. 1998. Comfort and humidity. ASHRAE Journal 40(8):35–41.

Bhattacharyya N. 2003. The economic burden and symptom manifestations of chronic rhinosinusitis. American Journal of Rhinology 17(1):27–32.

Black P, Udy A, Brodie S. 2000. Sensitivity to fungal allergens is a risk factor for life-threatening asthma. Allergy 55(5):501–504.

Blanc PD, Burney P, Janson C, Toren K. 2003. The prevalence and predictors of respiratory-related work limitation and occupational disability in an international study. Chest 124(3):1153–1159.

Blanc PM. 1997. Lesson 1, Volume 12—Inhalation Fever. in Pulmonary & Critical Care Update, American College of Chest Physicians. http://www.chestnet.org/education/online/pccu/vol12/lessons1_2/lesson01.php.

Blumenthal MM. 2002. What we know about the genetics of asthma at the beginning of the 21st century. Clinical Reviews in Allergy & Immunology 22(1):11–31.

Boat TF. 1998. Pulmonary Hemorrhage and Hemoptysis. Philadelphia, PA: W.B. Saunders Company.

Bornehag CG, Blomquist G, Gyntelberg F, Järvholm B, Malmberg P, Nordvall L, Nielsen A, Pershagen G, Sundell J. 2001. Dampness in buildings and health. Nordic interdisciplinary review of the scientific evidence on associations between exposure to "dampness" in buildings and health effects (NORDDAMP). Indoor Air 11(2):72–86.

Böttcher MF, Björkstén B, Gustafson S, Voor T, Jenmalm MC. 2003. Endotoxin levels in Estonian and Swedish house dust and atopy in infancy. Clinical & Experimental Allergy 33(3):273–276.

Braun H, Buzina W, Freudenschuss K, Beham A, Stammberger H. 2003. 'Eosinophilic fungal rhinosinusitis': a common disorder in Europe? The Laryngoscope 113(2):264–269.

Braun-Fahrländer C, Riedler J, Herz U, Eder W, Waser M, Grize L, Maisch S, Carr D, Gerlach F, Bufe A, Lauener RP, Schierl R, Renz H, Nowak D, von Mutius E; Allergy and Endotoxin Study Team. 2002. Environmental exposure to endotoxin and its relation to asthma in school-age children. New England Journal of Medicine 347(12):869–877.

Braunwald E, Fauci AS, Kasper DL, Hauser SL, Longo DL, Jameson JL, eds. 2001. Harrison's Principles of Internal Medicine, 15th Edition. New York: McGraw-Hill.

Brunekreef B. 1992. Damp housing and adult respiratory symptoms. Allergy 47(5):498–502.

Brunekreef B, Dockery DW, Speizer FE, Ware JH, Spengler JD, Ferris BG. 1989. Home dampness and respiratory morbidity in children. American Review of Respiratory Disease 140(5):1363–1367.

Bulpa PA, Dive AM, Garrino MG, Delos MA, Gonzalez MR, Evrard PA, Glupczynski Y, Installé EJ. 2001. Chronic obstructive pulmonary disease patients with invasive pulmonary aspergillosis: benefits of intensive care? Intensive Care Medicine 27(1):59–67.

Burge, H, Solomon WR, Boise JR. 1980. Microbial prevalence in domestic humidifiers. Applied and Environmental Microbiology 39(4):840–844.

Burke W, Fesinmeyer M, Reed K, Hampson L, Carlsten C. 2003. Family history as a predictor of asthma risk. American Journal of Preventive Medicine 24(2):160–169.

Carvalheiro MF, Peterson Y, Rubenowitz E, Rylander R. 1995. Bronchial reactivity and work-related symptoms in farmers. American Journal of Industrial Medicine 27(1):65–74.

CDC (Centers for Disease Control and Prevention). 1994. Acute pulmonary hemorrhage/hemosiderosis among infants—Cleveland, January 1993–November 1994. Morbidity and Mortality Weekly Report 43(48):881–883.

CDC. 1997. Update: pulmonary hemorrhage/hemosiderosis among infants—Cleveland, Ohio, 1993–1996. Morbidity and Mortality Weekly Report 46(2):33–35.

CDC. 2000. Update: pulmonary hemorrhage/hemosiderosis among infants—Cleveland, Ohio, 1993–1996. Morbidity and Mortality Weekly Report 49(9):180–185. [Errata in 49(10):213.]

CDC. 2001. Availability of Case Definition for Acute Idiopathic Pulmonary Hemorrhage in Infants. Morbidity and Mortality Weekly Report 50(23):494–495.

CDC External Expert Panel (CDC External Expert Panel on Acute Idiopathic Pulmonary Hemorrhage in Infants). 1999. Reports of Members of the CDC External Expert Panel on Acute Idiopathic Pulmonary Hemorrhage in Infants: A Synthesis. December 1999. http://www.cdc.gov/od/ads/ref30.pdf. accessed July 17, 2003.

CDC Working Group (CDC Working Group on Pulmonary Hemosiderosis). 1999. Report of the CDC Working Group on Pulmonary Hemorrhage/Hemosiderosis. June 17, 1999. http://www.cdc.gov/od/ads/ref29.pdf. accessed July 17, 2003.

Chinn S, Jarvis D, Luczynska C, Burney P. 1998. Individual allergens as risk factors for bronchial responsiveness in young adults. Thorax 53(8):662–667.

Cookson W. 2002. Genetics and genomics of asthma and allergic diseases. Immunological Reviews 190(1):195–206.

Cormier Y, Bélanger J. 1989. The fluctuant nature of precipitating antibodies in dairy farmers. Thorax 44:469–473.

Dales RE, Miller D. 1999. Residential fungal contamination and health: microbial cohabitants as covariates. Environmental Health Perspectives 107(Supplement 3):481–483.

Dales R, Zwanenburg H, Burnett R, Franklin C. 1991a. Respiratory health effects of home dampness and molds among Canadian children. American Journal of Epidemiology 134(2):196–203.

Dales RE, Burnett R, Zwanenburg H. 1991b. Adverse health effects among adults exposed to home dampness and molds. American Review of Respiratory Disease 143(3):505–509.

Dales RE, Cakmak S, Judek S, Dann T, Coates F, Brook JR, Burnett RT. 2003. The role of fungal spores in thunderstorm asthma. Chest 123(3):745–750.

Dales RE, Cakmak S, Judek S, Dann T, Coates F, Brook JR, Burnett RT. 2004. Influence of outdoor aeroallergens on hospitalization for asthma in Canada. Journal of Allergy and Clinical Immunology 113(2):303–306.

Dearborn D, Yike I, Sorenson W, Miller M, Etzel R. 1999. Overview of investigations into pulmonary hemorrhage among infants in Cleveland, Ohio. Environmental Health Perspectives 107(Supplement 3):495–499.

Dearborn DG, Smith PG, Dahms BB, Allan TM, Sorenson WG, Montaña E, Etzel RA. 2002. Clinical profile of 30 infants with acute pulmonary hemorrhage in Cleveland. Pediatrics 110(3):627–637.

Dekker C, Dales R, Bartlett S, Brunekreff B, Zwanenburg H. 1991. Childhood asthma and the indoor environment. Chest 100:922–926.

Delfino RJ, Zeiger RS, Seltzer JM, Street DH, Matteucci RM, Anderson PR, Koutrakis P. 1997. The effect of outdoor fungal spore concentrations on daily asthma severity. Environmental Health Perspectives 105(6):622–635.

DeShazo RD, Chapin K, Swain RE. 1997. Fungal sinusitis. New England Journal of Medicine 337(4):254–259.

Dharmage S, Bailey M, Raven J, Mitakakis T, Cheng A, Guest D, Rolland J, Forbes A, Thien F, Abramson M, Walters EH. 2001. Current indoor allergen levels of fungi and cats, but not house dust mites, influence allergy and asthma in adults with high dust mite exposure. American Journal of Respiratory and Critical Care Medicine 164(1):65–71.

Dijkstra L, Houthuijs D, Brunekreff B, Akkerman I, Boleij J. 1990. Respiratory health effects of the indoor environment in a population of Dutch children. American Review of Respiratory Disease 142:1172–1178.

Douwes J, Pearce N. 2003. Invited commentary: is indoor mold exposure a risk factor for asthma? American Journal of Epidemiology 158(3):203–206.

Douwes J, Gibson P, Pekkanen J, Pearce N. 2002. Non-eosinophilic asthma: importance and possible mechanisms. Thorax 57(7):643–648.

Ebbehøj NE, Hansen MØ, Sigsgaard T, Larsen L. 2002. Building-related symptoms and molds: a two-step intervention study. Indoor Air 12(4):273–277.

Elidemir O, Colasurdo GN, Rossmann SN, Fan LL. 1999. Isolation of *Stachybotrys* from the lung of a child with pulmonary hemosiderosis. Pediatrics 104(4 Pt 1):964–966.

Emanuel DA, Wenzel FJ, Lawton BR. 1975. Pulmonary mycotoxicosis. Chest 67(3):293–297.

Engvall K, Norrby C, Norbäck D. 2001. Asthma symptoms in relation to building dampness and odour in older multifamily houses in Stockholm. International Journal of Tuberculosis and Lung Disease 5(5):468–477.

Engvall K, Norrby C, Norbäck D. 2002. Ocular, airway, and dermal symptoms related to building dampness and odors in dwellings. Archives of Environmental Health 57(4):304–310.

Engvall K, Norrby C, Norbäck D. 2003. Ocular, nasal, dermal and respiratory symptoms in relation to heating, ventilation, energy conservation, and reconstruction of older multifamily houses. Indoor Air 13(3):206–211.

Erel F, Karaayvaz M, Caliskaner Z, Ozanguc N. 1998. The allergen spectrum in Turkey and the relationships between allergens and age, sex, birth month, birthplace, blood groups and family history of atopy. Journal of Investigational Allergology and Clinical Immunology 8(4):226–233.

Etzel RA. 2003. *Stachybotrys*. Current Opinion in Pediatrics 15(1):103–106.

Etzel R, Montaña E, Sorenson W, Kullman G, Allan T, Dearborn D. 1998. Acute pulmonary hemorrhage in infants associated with exposure to *Stachybotrys atra* and other fungi. Archives of Pediatric and Adolescence Medicine 152(8):757–762. [Erratum: 152(11): 1055; Comment: 153(2):205–206.]

Etzel RA, Dearborn DG, Allan TM, Horgan TE, Sorenson WG, Jarvis BB, Miller JD. undated (circa 2000). Investigator Team's Response to MMWR Report. http://gcrc.meds.cwru.edu/stachy/InvestTeamResponse.html.

Flappan SM, Portnoy J, Jones P, Barnes C. 1999. Infant pulmonary hemorrhage in a suburban home with water damage and mold (*Stachybotrys atra*). Environmental Health Perspectives 107(11):927–930.

Flemming J, Hudson B, Rand TG. 2004. Comparison of inflammatory and cytotoxic lung responses in mice after intratracheal exposure to spores of two different *Stachybotrys chartarum* strains. Toxicologic Sciences, January 12 (Epub ahead of print).

Fogelmark B, Goto H, Yuasa K, Marchat B, Rylander R. 1992. Acute pulmonary toxicity of inhaled beta-1,3-glucan and endotoxin. Agents Actions 35(1-2):50–56.

Forsgren A, Persson K, Ursing J, Walder M, Borg I. 1984. Immunological aspects of humidifier fever. European Journal of Clinical Microbiology 3(5):411–418.

Franceschini F, Calzavara-Pinton P, Valsecchi L, Quinzanini M, Zane C, Facchetti F, Airo P, Cattaneo R. 1999. Chilblain lupus erythematosus is associated with antibodies to SSA/Ro. Advances in Experimental Medicine and Biology 455:167–171.

Franquet T, Muller NL, Gimenez A, Domingo P, Plaza V, Bordes R. 2000. Semiinvasive pulmonary aspergillosis in chronic obstructive pulmonary disease: radiologic and pathologic findings in nine patients. American Journal of Roentgenology 174(1):51–56.

Fraser RS, Muller NL, Colman N, Pare PD, Bralow L, eds. 1999. Fraser and Pare's Inhalation of Organic Dust. In: Diagnosis of Diseases of the Chest. Philadelphia, PA: W.B. Saunders Co. pp. 2361–2385.

Fridkin SK, Jarvis WR. 1996. Epidemiology of nosocomial fungal infections. Clinical Microbiology Review 9(4):499–511.

Fung F, Hughson WG. 2003. Health effects of indoor fungal bioaerosol exposure. Applied Occupational and Environmental Hygiene 18(7):535–544.

Galant S, Berger W, Gillman S, Goldsobel A, Incaudo G, Kanter L, Machtinger S, McLean A, Prenner B, Sokol W, Spector S, Welch M, Ziering W. 1998. Prevalence of sensitization to aeroallergens in California patients with respiratory allergy. Allergy Skin Test Project Team. Annals of Allergy, Asthma, and Immunology 81(3):203–210.

Gehring U, Douwes J, Doekes G, Koch A, Bischof W, Fahlbusch B, Richter K, Wichmann HE, Heinrich J. 2001. β(1→3)-glucan in house dust of German homes: housing characteristics, occupant behavior, and relations with endotoxins, allergens, and molds. Environmental Health Perspectives 109(2):139–144.

Gehring U, Bischof W, Fahlbusch B, Wichmann HE, Heinrich J. 2002. House dust endotoxin and allergic sensitization in children. American Journal of Respiratory and Critical Care Medicine 166(7):939–944.

Gent JF, Ren P, Belanger K, Triche E, Bracken MB, Holford TR, Leaderer BP. 2002. Levels of household mold associated with respiratory symptoms in the first year of life in a cohort at risk for asthma. Environmental Health Perspectives 110(12):A781–A786.

Gern JE, Lemanske RF Jr, Busse WW. 1999. Early life origins of asthma. The Journal of Clinical Investigation 104(7):837–843.

Greenberger PA. 2002. Allergic bronchopulmonary aspergillosis. Journal of Allergy and Clinical Immunology 110(5):685–692.

Guernsey JR, Morgan DP, Marx JJ, Horvath EP, Pierce WE, Merchant JA. 1989. Respiratory disease risk relative to farmer's lung disease antibody status. In: Dosman JA Cockroft DW, eds. Principles of Health and Safety in Agriculture. Boca Raton, FL: CRC Press. pp. 81–84.

Gunnbjörnsdottir MI, Norbäck D, Plaschke P, Norrman E, Björnsson E, Janson C. 2003. The relationship between indicators of building dampness and respiratory health in young Swedish adults. Respiratory Medicine 97(4):302–307.

Halonen M, Stern DA, Wright AL, Taussig LM, Martinez FD. 1997. Alternaria as a major allergen for asthma in children raised in a desert environment. American Journal of Respiratory and Critical Care Medicine 155:1356–1361.

Heiner DC, Sears JW, Knicker WT. 1962. Multiple precipitins to cows milk in chronic pulmonary hemosiderosis. American Journal of Diseases of Children 103:635–640.

Hirvonen MR, Ruotsolainen M, Savolainen K, Nevalainen A. 1997. Effect of viability of actinomycete spores on their ability to stimulate production of nitric oxide and reactive oxygen species in RAW 264.7 macrophages. Toxicology 124:105–114.

Hirvonen M-R, Ruotsalainen M, Roponen M, Hyvärinen A, Husman T, Kosma V-M, Komulainen H, Savolainen K, Nevalainen A. 1999. Nitric oxide and proinflammatory cytokines in nasal lavage fluid associated with symptoms and exposure to moldy building microbes. American Journal of Respiratory and Critical Care Medicine 160(6):1943–1946.

Hirvonen MR, Suutari M, Ruotsalainen M, Lignell U, Nevalainen A. 2001. Effect of growth medium on potential of Streptomyces anulatus spores to induce inflammatory responses and cytotoxicity in RAW264.7 macrophages. Inhalation Toxicology 13(1):55–68.

Hodgson M. 2002. Indoor environmental exposures and symptoms. Environmental Health Perspectives 110(Supplement 4):663–667.

Hodgson MJ, Morey P, Leung WY, Morrow L, Miller D, Jarvis BB, Robbins H, Halsey JF, Storey E. 1998. Building-associated pulmonary disease from exposure to *Stachybotrys chartarum* and *Aspergillus versicolor*. Journal of Occupational and Environmental Medicine 40(3):241–249.

Hogan MB, Patterson R, Pore RS, Corder WT, Wilson NW. 1996. Basement shower hypersensitivity pneumonitis secondary to *Epicoccum nigrum*. Chest 110(3):854–856.

Hu F, Persky V, Flay B, Phil D, Richardson PH. 1997. An epidemiological study of asthma prevalence and related factors amoung young adults. Journal of Asthma 34(1):67–76.

Hyndman SJ. 1990. Housing dampness and health amongst British Bengalis in east London. Social Science & Medicine 30(1):131–141.

Infante-Rivard C. 1993. Childhood asthma and indoor environmental risk factors. American Journal of Epidemiology 137(8):834–844.

IOM (Institute of Medicine). 1993. Indoor Allergens: Assessing and Controlling Adverse Health Effects. Washington, DC: National Academy Press.

IOM. 2000. Clearing the Air: Asthma and Indoor Air Exposures. Washington, DC: National Academy Press.

Iwen PC, Sigler L, Tarantolo S, Sutton DA, Rinaldi MG, Lackner RP, McCarthy DI, Hinrichs SH. 2000. Pulmonary infection caused by *Gymnascella hyalinospora* in a patient with acute myelogenous leukemia. Journal of Clinical Microbiology 38(1):375–381.

Jaakkola JJ. 1998. The office environment model: a conceptual analysis of the sick building syndrome. Indoor Air 8(Supplement 4):7–16.

Jaakkola JJ, Jaakkola N, Ruotsalainen R. 1993. Home dampness and molds as determinants of respiratory symptoms and asthma in pre-school children. Journal of Exposure Analysis and Environmental Epidemiology 3(Supplement 1):129–142.

Jaakkola MS, Nordman H, Piipari R, Uitti J, Laitinen J, Karjalainen A, Hahtola P, Jaakkola JJ. 2002a. Indoor dampness and molds and development of adult-onset asthma: a population-based incident case-control study. Environmental Health Perspectives 110(5):543–547.

Jaakkola MS, Laitinen S, Piipari R, Utti J, Nordman H, Haapala AM, Jaakkola JJ. 2002b. Immunoglobulin G antibodies against indoor dampness-related microbes and adult-onset asthma: a population-based incident case-control study. Clinical and Experimental Immunology 129:107–112.

Jagielo PJ, Thorne PS, Watt JL, Frees KL, Quinn TJ, Schwartz DA. 1996. Grain dust and endotoxin inhalation challenges produce similar inflammatory responses in normal subjects. Chest 110(1):263–270.

Jarvis BB, Sorenson WG, Hintikka EL, Nikulin M, Zhou Y, Jiang J, Wang S, Hinkley S, Etzel RA, Dearborn D. 1998. Study of toxin production by isolates of *Stachybotrys chartarum* and *Memnoniella exhinata* isolated during a study of pulmonary hemosiderosis in infants. Applied and Environmental Microbiology 64(10):3620–3625.

Jędrychowski W, Flak E. 1998. Separate and combined effects of the outdoor and indoor air quality on chronic respiratory symptoms adjusted for allergy among preadolescent children. International Journal of Occupational Medicine and Environmental Health 11(1): 19–35.

Johanning E, Biagini R, Hull D, Morey P, Jarvis B, Landsbergis P. 1996. Health and immunology study following exposure to toxigenic fungi (*Stachybotrys chartarum*) in a water-damaged office environment. International Archives of Occupational and Environmental Health 68(4):207–218.

Jussila J, Komulainen H, Huttunen K, Roponen M, Hälinen A, Hyvärinen A, Kosma V-M, Pelkonen J, Hirvonen M-R. 2001. Inflammatory responses in mice after intratracheal instillation of spores of *Streptomyces californicus* isolated from indoor air of a moldy building. Toxicology and Applied Pharmacology 171(1):61–69.

Jussila J, Komulainen H, Kosma V-M, Nevalainen A, Pelkonen J, Hirvonen M-R. 2002a. Spores of *Aspergillus versicolor* isolated from indoor air of a moisture-damaged building provoke acute inflammation in mouse lungs. Inhalation Toxicology 14(12):1261–1277.

Jussila J, Komulainen H, Kosma V-M, Pelkonen J, Hirvonen M-R. 2002b. Inflammatory potential of the spores of *Penicillium spinulosum* isolated from indoor air of a moisture-damaged building in mouse lungs. Environmental Toxicology and Pharmacology 12(3): 137–145.

Jussila J, Komulainen H, Huttunen K, Roponen M, Iivanainen E, Torkko P, Kosma V-M, Pelkonen J, Hirvonen M-R. 2002c. *Mycobacterium terrae* isolated from indoor air of a moisture-damaged building induces sustained biphasic inflammatory response in mouse lungs. Environmental Health Perspectives 110(11):1119–1125.

Kane GC, Marx JJ, Prince DS. 1993. Hypersensitivity pneumonitis secondary to *Klebsiella oxytoca*. A new cause of humidifier lung. Chest 104(2):627–629.

Kauffman HF, Tomee JF, van der Werf TS, de Monchy JG, Koeter GK. 1995. American Journal of Respiratory and Critical Care Medicine 151(6):2109–2115.

Kauffman HK. 2003. Immunopathogenesis of allergic bronchopulmonary aspergillosis and airway remodeling. Frontiers in Bioscience 8:e190–e196.

Kettenmann B, Jousmaki V, Portin K, Salmelin R, Kobal G, Hari R. 1996. Odorants activate the human superior temporal sulcus. Neuroscience Letters 203(2):143–145.

Kilpeläinen M, Terho EO, Helenius H, Koskenvuo M. 2001. Home dampness, current allergic diseases, and respiratory infections among young adults. Thorax 56(6):462–467.

Kistemann T, Huneburg H, Exner M, Vacata V, Engelhart S. 2002. Role of increased environmental *Aspergillus* exposure for patients with chronic obstructive pulmonary disease (COPD) treated with corticosteroids in an intensive care unit. International Journal of Hygiene and Environmental Health 204(5-6):347–351.

Knapp JF, Michael JG, Hegenbarth MA, Jones PE, Black PG. 1999. Case records of the Children's Mercy Hospital, Case 02-1999: a 1-month-old infant with respiratory distress and shock. Pediatric Emergency Care 15(4):288–293.

Kolmodin-Hedman B, Blomquist G, Sikström E. 1986. Mould exposure in museum personnel. International Archives of Occupational and Environmental Health 57(4):321–323.

Kolstad HA, Brauer C, Iversen M, Sigsgaard T, Mikkelsen S. 2002. Do indoor molds in nonindustrial environments threaten workers' health? A review of the epidemiologic evidence. Epidemiologic Reviews 24(2):203–217.

Korpi A, Kasanen JP, Kosma VM, Rylander R, Pasanen AL. 2003. Slight respiratory irritation but not inflammation in mice exposed to $(1\rightarrow3)$-β-D-glucan aerosols. Mediators of Inflammation 12(3):139–146.

Koskinen OM. 1999. Moisture, mold and health. Doctoral Dissertation. National Public Health Institute, Kuopio, Finland.

Koskinen OM, Husman TM, Meklin TM, Nevalainen AI. 1999a. The relationship between moisture or mould observations in houses and the state of health of their occupants. European Respiratory Journal 14(6):1363–1367.

Koskinen OM, Husman TM, Meklin TM, Nevalainen AI. 1999b. Adverse health effects in children associated with moisture and mold observations in houses. International Journal of Environmental Health Research 9(2):143–156.

Kreja L, Seidel HJ. 2002. Evaluation of the genotoxic potential of some microbial volatile organic compounds (MVOC) with the comet assay, the micronucleus assay and the HPRT gene mutation assay. Mutation Research/Genetic Toxicology and Environmental Mutagenesis 513(1-2):143–150.

Kuhn D, Ghannoum M. 2003. Indoor mold, toxigenic fungi, and *Stachybotrys chartarum*: infectious disease perspective. Clinical Microbiology Reviews 16(1):144–172.

Lacasse Y, Selman M, Costabel U, Dalphin JC, Ando M, Morell F, Erkinjuntti-Pekkanen R, Müller N, Colby TV, Schuyler M, Cormier Y; HP Study Group. 2003. Clinical diagnosis of hypersensitivity pneumonitis. American Journal of Respiratory and Critical Care Medicine 168(8):952–958.

Lander F, Meyer HW, Norn S. 2001. Serum IgE specific to indoor moulds, measured by basophil histamine release, is associated with building-related symptoms in damp buildings. 2001. Inflammation Research 50(4):227–231.

Lee SK, Kim SS, Nahm DH, Park HS, Oh YJ, Park KJ, Kim SO, Kim SJ. 2000. Hypersensitivity pneumonitis caused by *Fusarium napiforme* in a home environment. Allergy 55(12): 1190–1193.

Lewis RG. 2000. Pesticides. In Indoor Air Quality Handbook. Spengler JD, McCarthy JF, Samet JM, eds. New York: McGraw-Hill.

Li CS, Hsu CW, Lu CH. 1997. Dampness and respiratory symptoms among workers in daycare centers in a subtropical climate. Archives of Environmental Health 52(1):68–71.

Lierl MB, Hornung RW. 2003. Relationship of outdoor air quality to pediatric asthma exacerbations. Annals of Allergy, Asthma and Immunology 90(1):28–33.

Maier WC, Arrighi HM, Morray B, Llewellyn C, Redding GJ. 1997. Indoor risk factors for asthma and wheezing among Seattle school children. Environmental Health Perspectives 105(2):208–214.

Malani PN, Kauffman CA. 2002. Invasive and allergic fungal sinusitis. Current Infectious Disease Reports 4(3):225–232.

Mari A, Schneider P, Wally V, Breitenbach M, Simon-Nobbe B. 2003. Sensitization to fungi: epidemiology, comparative skin tests, and IgE reactivity of fungal extracts. Clinical & Experimental Allergy 33(10):1429–1438.

Marple BF. 2002. Allergic fungal rhinosinusitis: current theories and management strategies. The Laryngoscope 111(6):1006–1019.

Marr KA, Patterson T, Denning D. 2002. Aspergillosis. Pathogenesis, clinical manifestations, and therapy. Infectious Disease Clinics of North America 16(4):875–894.

Marx JJ Jr, Arden-Jones MP, Treuhaft MW, Gray RL, Motszko CS, Hahn FF. 1981. The pathogenetic role of inhaled microbial material in pulmonary mycotoxicosis as demonstrated in an animal model. Chest 80(1 Supplement):76–78.

Marx JJ, Guernsey J, Emanuel DA, Merchant JA, Morgan DP, Kryda M. 1990. Cohort studies of immunologic lung disease among Wisconsin dairy farmers. American Journal of Industrial Medicine 18(3):263–268.

May JJ, Stallones L, Darrow D, Pratt DS. 1986. Organic dust toxicity (pulmonary mycotoxicosis) associated with silo unloading. Thorax 41(12):919–923.

Maziak W, Perzanowski MS, Platts-Mills TAE, Speiser DE, Zippelius A, Braun-Fahrländer C, Lauener RP, von Mutius E, Weiss ST. 2003. Endotoxin and asthma (correspondence). New England Journal of Medicine 348(2):171–174.

McConnell R, Berhane K, Gilliland F, Islam T, Gauderman WJ, London SJ, Avol E, Rappaport EB, Margolis HG, Peters JM. 2002. Indoor risk factors for asthma in a prospective study of adolescents. Epidemiology 13(3):288–295.

Michel O, Kips J, Duchateau J, Vertongen F, Robert L, Collet H, Pauwels R, Sergysels R. 1996. Severity of asthma is related to endotoxin in house dust. American Journal of Respiratory and Critical Care Medicine 154(6 Pt 1):1641–1646.

Mohamed N, Ng'ang'a L, Odhiambo J, Nyamwaya J, Menzies R. 1995. Home environment and asthma in Kenyan schoolchildren: a case-control study. Thorax 50(1):74–78.

Montaña E, Etzel R, Allan T, Horgan T, Dearborn D. 1997. Environmental risk factors associated with pediatric idiopathic pulmonary hemorrhage and hemosiderosis in a Cleveland community. Pediatrics 99(1):e5. http://pediatrics.aappublications.org/cgi/content/full/99/1/e5. accessed 15 July, 2002.

Murtoniemi T, Nevalainen A, Suutari M, Toivola M, Komulainen H, Hirvonen M. 2001. Induction of cytotoxicity and production of inflammatory mediators in RAW264.7 macrophages by spores grown on six different plasterboards. Inhalation Toxicology 13:233–247.

Mutli GM, Garey KW, Robbins RA, Danziger LH, Rubinstein I. 2001. Collection and analysis of exhaled breath condensate in humans. American Journal of Respiratory and Critical Care Medicine 164(5):731–737.

Myllykangas-Luosujärvi R, Seuri M, Husman T, Korhonen R, Pakkala K, Aho K. 2002. A cluster of inflammatory rheumatic diseases in a moisture-damaged office. Clinical and Experimental Rheumatology 20(6):833–866.

Nafstad P, Øie L, Mehl R, Gaarder P, Lodrup-Carlsen K, Botten G, Magnus P, Jaakkola J. 1998. Residential dampness problems and symptoms and signs of bronchial obstruction in young Norwegian children. American Journal of Respiratory and Critical Care Medicine 157:410–414.

Nagda NL, Hodgson M. 2001. Low relative humidity and aircraft cabin air quality. Indoor Air 11(3):200–214.

Neas LM, Dockery DW, Ware JH, Spengler JD, Speizer FE, Ferris BG Jr. 1991. Association of indoor nitrogen dioxide with respiratory symptoms and pulmonary function in children. American Journal of Epidemiology 134(2):204–219.

Nelson HS, Szefler SJ, Jacobs J, Huss K, Shapiro G, Sternberg AL. 1999. The relationships among environmental allergen sensitization, allergen exposure, pulmonary function, and bronchial hyperresponsiveness in the Childhood Asthma Management Program. Journal of Allergy and Clinical Immunology 104(4 Pt 1):775–785.

Nevalainen A, Vahteristo M, Koivisto J, Meklin T, Hyvärinen A, Keski-karhu J, Husman T. 2001. Moisture, mold and health in apartment homes. In: Bioaerosols, Fungi and Mycotoxins: Health Effects, Assessment, Prevention and Control. Johanning E, ed. Albany, NY: Boyd Printing.

Nicolai T, Illi S, von Mutius E. 1998. Effect of dampness at home in childhood on bronchial hyperreactivity in adolescence. Thorax 53(12):1035–1040.

Nikulin M, Reijula K, Jarvis BB, Veijalainen P, Hintikka EL. 1997. Effects of intranasal exposure to spores of Stachybotrys atra in mice. Fundamental and Applied Toxicology 35(2):182–188.

Nolles G, Hoekstra MO, Schouten JP, Gerritsen J, Kauffman HF. 2001. Prevalence of immunoglobulin E for fungi in atopic children. Clinical and Experimental Allergy 31(10): 1564–1570.

Norbäck D, Edling C. 1991. Environmental, occupational, and personal factors related to the prevalence of sick building syndrome in the general population. British Journal of Industrial Medicine 48(7):451–462.

Norbäck D, Björnsson E, Janson C, Widstrom J, Boman G. 1995. Asthmatic symptoms and volatile organic compounds, formaldehyde, and carbon dioxide in dwellings. Occupational and Environmental Medicine 52(6):388–395.

Norbäck D, Björnsson E, Janson C, Palmgren U, Boman G. 1999. Current asthma and biochemical signs of inflammation in relation to building dampness in dwellings. International Journal of Tuberculosis and Lung Disease 3(5):368–376.

Norbäck D, Wieslander G, Nordstrom K, Wålinder R. 2000. Asthma symptoms in relation to measured building dampness in upper concrete floor construction, and 2-ethyl-1-hexanol in indoor air. International Journal of Tuberculosis and Lung Disease 4(11): 1016–1025.

Novak N, Bieber T. 2003. Allergic and nonallergic forms of atopic diseases. Journal of Allergy and Clinical Immunology 112(2):252–262.

Novotny W, Dixit A. 2000. Pulmonary hemorrhage in an infant following 2 weeks of fungal exposure. Archives of Pediatric and Adolescent Medicine 154:271–275.

NYCDOH (New York City Department of Health). 2000. Guidelines on Assessment and Remediation of Fungi in Indoor Environments. http://www.ci.nyc.ny.us/html/doh/html/epi/moldrpt1.html.

O'Byrne PM, Inman MD. 2000. New considerations about measuring airway hyperresponsiveness. Journal of Asthma 37(4):293–302.

Oh SW, Pae CI, Lee DK, Jones F, Chiang GK, Kim HO, Moon SH, Cao B, Ogbu C, Jeong KW, Kozu G, Nakanishi H, Kahn M, Chi EY, Henderson WR. 2002. Tryptase inhibition blocks airway inflammation in a mouse asthma model. The Journal of Immunology 168(4):1992–2000.

Øie L, Nafstad P, Botten G, Magnus P, Jaakkola J. 1999. Ventilation in homes and bronchial obstruction in young children. Epidemiology 10(3):294–299.

Park JH, Gold DR, Spiegelman DL, Burge HA, Milton DK. 2001. House dust endotoxin and wheeze in the first year of life. American Journal of Respiratory and Critical Care Medicine 163(2):322–328.

Patterson JE. 1999. Epidemiology of fungal infections in solid organ transplant patients. Transplant Infectious Disease 1(4):229–236.

Pauwels RA, Buist AS, Calverley PM, Jenkins CR, Hurd SS, GOLD Scientific Committee. 2001. Global strategy for the diagnosis, management, and prevention of chronic obstructive pulmonary disease. NHLBI/WHO Global Initiative for Chronic Obstructive Lung Disease (GOLD) Workshop summary. American Journal of Respiratory and Critical Care Medicine 163(5):1256–1276. [Comment in: Am J Respir Crit Care Med. (2001) 163(5):1047–1048.]

Peat JK, Dickerson J, Li J. 1998. Effects of damp and mould in the home on respiratory health: a review of the literature. Allergy 53(2):120–128.

Peraica M, Radić B, Lucić A, Pavlović M. 1999. Toxic effects of mycotoxins in humans. Bulletin of the World Health Organization 77(9):754–766.

Pieckova E, Jesenska Z. 1999. Microscopic fungi in dwellings and their health implications in humans. Annals of Agriculture and Environmental Medicine 6(1):1–11.

Pirhonen I, Nevalainen A, Husman T, Pekkanen J. 1996. Home dampness, moulds and their influence on respiratory infections and symptoms in adults in Finland. European Respiratory Journal 9(12):2618–2622.

Purokivi MK, Hirvonen MR, Randell JT, Roponen MH, Meklin TM, Nevalainen A, Husman TM, Tukiainen HO. 2001. Changes in pro-inflammatory cytokines in association with exposure to moisture-damaged building microbes. European Respiratory Journal 18(6): 951–958.

Rand TG, Mahoney M, White K, Oulton M. 2002. Microanatomical changes in alveolar type II cells in juvenile mice intratracheally exposed to Stachybotrys chartarum spores and toxin. Toxicological Sciences 65(2):239–245.

Rand TG, White K, Logan A, Gregory L. 2003. Histological, immunohistochemical and morphometric changes in lung tissue in juvenile mice experimentally exposed to Stachybotrys chartarum spores. Mycopathologia 156(2):119–131.

Rao CY. 2000. Toxigenic fungi in the indoor environment. In: The IAQ Handbook. Spengler JD, Samet JM, McCarthy JF, eds. New York: McGraw-Hill.

Rasker JJ, Peters HJG, Boon KL. 1986. Influence of weather on stiffness and force in patients with rheumatoid arthritis. Scandinavian Journal of Rheumatology 15:27–36.

Reed CE, Milton DK. 2001. Endotoxin-stimulated innate immunity: A contributing factor for asthma. Journal of Allergy and Clinical Immunology 108(2):157–166.

Reinikainen LM, Jaakkola JJ. 2003. Significance of humidity and temperature on skin and upper airway symptoms. Indoor Air 13(4):344–352.

Richerson HB. 1990. Unifying concepts underlying the effects of organic dust exposures. American Journal of Industrial Medicine 17(1):139–142.

Robbins C, Swenson L, Nealley M, Gots R, Kelman B. 2000. Health effects of mycotoxins in indoor air: a critical review. Applied Occupational and Environmental Hygiene 15(10): 773–784.

Robbins RA, Linder J, Stahl MG, Thompson AB III, Haire W, Kessinger A, Armitage JO, Arneson M, Woods G, Vaughan WP, et al. 1989. Diffuse alveolar hemorrhage in autologous bone marrow transplant recipients. American Journal of Medicine 87(5):511–518.

Robbins RA, Floreani AA, Von Essen SG, Sisson JH, Hill GE, Rubinstein I, Townley RG. 1996. Measurement of exhaled nitric oxide by three different techniques. American Journal of Respiratory and Critical Care Medicine 153(5):1631–1635.

Robertson AS, Burge PS, Hedge A, Sims J, Gill FS, Finnegan M, Pickering CA, Dalton G. 1985. Comparison of health problems related to work and environmental measurements in two office buildings with different ventilation systems. British Medical Journal (Clinical Research Edition) 291(6492):373–376.

Rönmark E, Jönsson E, Platts-Mills T, Lundbäck B. 1999. Different pattern of risk factors for atopic and nonatopic asthma among children—report from the Obstructive Lung Disease in Northern Sweden Study. Allergy 54(9):926–935.

Roponen M, Toivola M, Meklin T, Ruotsalainen M, Komulainen H, Nevalainen A, Hirvonen MR. 2001a. Differences in inflammatory responses and cytotoxicity in RAW264.7 macrophages induced by *Streptomyces Anulatus* grown on different building materials. Indoor Air 11(3):179–184.

Roponen M, Kiviranta J, Seuri M, Tukiainen H, Myllykangas-Luosujärvi R, Hirvonen MR. 2001b. Inflammatory mediators in nasal lavage, induced sputum and serum of employees with rheumatic and respiratory disorders. European Respiratory Journal 18(3):542–548.

Roponen M, Seuri M, Nevalainen A, Hirvonen MR. 2002. Fungal spores as such do not cause nasal inflammation in mold exposure. Inhalation Toxicology 14(5):541–549.

Roponen M, Seuri M, Nevalainen A, Randell J, Hirvonen MR. 2003. Nasal lavage method in the monitoring of upper airway inflammation: seasonal and individual variation. Inhalation Toxicology 15(7):649–661.

Ross A, Collins M, Sanders C. 1990. Upper respiratory tract infection in children, domestic temperatures, and humidity. Journal of Epidemiology and Community Health 44(2):142–146.

Rudblad S, Andersson K, Stridh G, Bodin L, Juto JE. 2001. Nasal hyperreactivity among teachers in a school with a long history of moisture problems. American Journal of Rhinology 15(2):135–141.

Rudblad S, Andersson K, Stridh G, Bodin L, Juto JE. 2002. Slowly decreasing mucosal hyperreactivity years after working in a school with moisture problems. Indoor Air 12(2):138–144.

Ruotsalainen M, Hyvärinen A, Nevalainen A, Savolainen, KM. 1995. Production of reactive oxygen metabolites by opsonized fungi and bacteria isolated from indoor air, and their interactions with soluble stimuli, fMLP or PMA. Environmental Research 69:122–131.

Rylander R. 1997. Airborne $(1{\rightarrow}3)$-β-D-glucan and airway disease in a day-care center before and after renovation. Archives of Environmental Health 52(4):281–285.

Rylander R, Mégevand Y. 2000. Environmental risk factors for respiratory infections. Archives of Environmental Health 55(5):300–303.

Rylander R, Norrhall M, Engdahl E, Tunsäter A, Holt P. 1998. Airways inflammation, atopy, and $(1{\rightarrow}3)$-β-D-glucan exposure in two schools. American Journal of Respiratory and Critical Care Medicine 158:1685–1687.

Savilahti R, Uitti J, Roto P, Laippala P, Husman T. 2001. Increased prevalence of atopy among children exposed to mold in a school building. Allergy 56:175–179.

Savilahti R, Uitti J, Laippala P, Husman T, Reiman M. 2002. Immunoglobulin G antibodies of children exposed to microorganisms in a water-damaged school. Pediatric Allergy and Immunology 13(6):438–442.

Schiffman SS, Walker JM, Dalton P, Lorig TS, Raymer JH, Shusterman D, Williams CM. 2000. Potential health effects of odor from animal operations, wastewater treatment, and recycling of byproducts. Journal of Agromedicine 1(7):7–81.

Schubert MS. 2000. Medical treatment of allergic fungal sinusitis. Annals of Allergy, Asthma & Immunology 85(2):90–97.

Schuyler M. 2001. Are polymorphisms the answer in hypersensitivity pneumonitis? American Journal of Respiratory and Critical Care Medicine 163(7):1528–1533.

Scientific Committee on Tobacco and Health. 1998. Report of the Scientific Committee on Tobacco and Health. Her Majesty's Stationery Office (United Kingdom). http://www.official-documents.co.uk/document/doh/tobacco/contents.htm.

Seifert SA, Von Essen S, Jacobitz K, Crouch R, Lintner CP. 2003a. Organic dust toxic syndrome: a review. Journal of Toxicology—Clinical Toxicology 41(2):185–193.

Seifert SA, Von Essen S, Jacobitz K, Crouch R, Lintner CP. 2003b. Do poison centers diagnose organic dust toxic syndrome? Journal of Toxicology—Clinical Toxicology 41(2):115–117.

Seppänen O, Fisk WJ. 2002. Association of ventilation system type with SBS symptoms in office workers. Indoor Air 12(2):98–112.

Shahan TA, Sorenson WG, Paulauskis JD, Morey R, Lewis DM. 1998. Concentration- and time-dependent upregulation and release of the cytokines MIP-2, KC, TNF, and MIP-1alpha in rat alveolar macrophages by fungal spores implicated in airway inflammation. American Journal of Respiratory Cell and Molecular Biology 18(3):435–440.

Slezak JA, Persky VW, Kviz FJ, Ramakrishnan V, Byers C. 1998. Asthma prevalence and risk factors in selected Head Start sites in Chicago. Journal of Asthma 35(2):203–212.

Stark PC, Burge HA, Ryan LM, Milton DK, Gold DR. 2003. Fungal levels in the home and lower respiratory tract illnesses in the first year of life. American Journal of Respiratory and Critical Care Medicine 168(2):232–237.

Stevens DA, Kan VL, Judson MA, Morrison VA, Dummer S, Denning DW, Bennett JE, Walsh TJ, Patterson TF, Pankey GA. 2000. Practice guidelines for diseases caused by *Aspergillus*. Infectious Diseases Society of America. Clinical Infectious Diseases 30(4):696–709.

Strachan DP, Flannigan B, McCabe EM, McGarry F. 1990. Quantification of airborne moulds in the homes of children with and without wheeze. Thorax 45(5):382–387.

Su HJ, Burge H, Spengler JD. 1989. Microbiological contamination in the residential environment. Joint Canadian and Pan American Symposium on Aerobiology and Health. Ottawa, Ontario, Canada, June 7–9, 1989.

Su HJ, Burge H, Spengler JD. 1990. Indoor saprophytic aerosols and respiratory health. Journal of Allergy and Clinical Immunology 85(1, Pt 2):248.

Suda H, Sato A, Ida M, Gemma H, Hayakawa H, Chida K. 1995. Hypersensitivity pneumonitis associated with home ultrasonic humidifiers. Chest 107(3):711–717.

Sudakin DL. 1998. Toxigenic fungi in a water-damaged building: an intervention study. American Journal of Industrial Medicine 34:183–190.

Sunyer J, Soriano J, Antó, JM, Burgos F, Pereira A, Payo F, Martínez-Moratalla J, Ramos J. 2000. Sensitization to individual allergens as risk factors for lower FEV_1 in young adults. European Community Respiratory Health Survey. International Journal of Epidemiology 29(1):125–130.

Taskinen T, Meklin T, Nousiainen M, Husman T, Nevalainen A, Korppi M. 1997. Moisture and mould problems in schools and respiratory manifestations in school children: clinical and skin test findings. Acta Paediatrica 86:1181–1187.

Taskinen T, Hyvärinen A, Meklin T, Husman T, Nevalainen A, Korppi M. 1999. Asthma and respiratory infections in school children with special reference to moisture and mold problems in the school. Acta Paediatrica 88(12):1373–1379.

Taskinen T, Laitinen S, Nevalainen A, Vepsäläinen A, Meklin T, Reiman M, Korppi M, Husman T. 2002. Immunoglobulin G antibodies to moulds in school children from moisture problem schools. Allergy 57:9–16.

Terho EO. 1982. Extrinsic allergic alveolitis—the state of the art. European Journal of Respiratory Disease 63(Supplement 124):10–26.

Thörn Å. 1999. The emergence and preservation of sick building syndrome: research challenges of a modern age disease. PhD Thesis, Stockholm: Karolinska Institutet, Department of Public Health Sciences, Division of Social Medicine. http://diss.kib.ki.se/1999/91-628-3555-6/thesis.pdf

Thorn J, Rylander R. 1998a. Airways inflammation and glucan in a rowhouse area. American Journal of Respiratory and Critical Care Medicine 157(6 Pt 1):1798–1803.

Thorn J, Rylander R. 1998b. Inflammatory response after inhalation of bacterial endotoxin assessed by the induced sputum technique. Thorax 53:1047–1052.

Thorn J, Brisman J, Torén K. 2001. Adult-onset asthma is associated with self-reported mold or environmental tobacco smoke exposures in the home. Allergy 56(4):287–292.

Tilles SA, Bardana EJ. 1997. Seasonal variation in bronchial hyperreactivity (BHR) in allergic patients. Clinical Reviews in Allergy & Immunology 15(2):169–185.

Tripi PA, Modlin S, Sorensen WG, Dearborn DG. 2000. Acute pulmonary haemorrhage in an infant during induction of general anesthesia. Paediatric Anaesthia 10(1):92–94.

Trout D, Bernstein J, Martinez K, Biagini R, Wallingford K. 2001. Bioaerosol lung damage in a worker with repeated exposure to fungi in a water-damaged building. Environmental Health Perspectives 109(6):641–644.

Tyndall RL, Lehman E, Bowman EK, Milton DK, Barbaree J. 1995. Home humidifiers as a potential source of exposure to microbial pathogens, endotoxins and allergens. Indoor Air 5:171–178.

U.S. EPA (U.S. Environmental Protection Agency). 1989. Report to Congress on Indoor Air Quality, Volume II, Assessment and Control of Indoor Air Pollution.

U.S. EPA. 1992. Respiratory Health Effects of Passive Smoking: Lung Cancer and Other Disorders. EPA/600/6-90/006F. Washington, DC.

Van Emon JM, Reed AW, Yike I, Vesper SJ. 2003. ELISA measurement of Stachylysin™ in serum to quantify human exposures to the indoor mold *Stachybotrys chartarum*. Journal of Occupational and Environmental Medicine 45(6):582–591.

van Houten J, Long W, Mullett M, Finer N, Derleth D, McMurray B, Peliowski A, Walker D, Wold D, Sankaran K. 1992. Pulmonary hemorrhage in premature infants after treatment with synthetic surfactant: an autopsy evaluation. The American Exosurf Neonatal Study Group I, and the Canadian Exosurf Neonatal Study Group. Journal of Pediatrics 120(2 Pt 2):S40–44. [Erratum:120(5):762.]

Van Vleet TR, Klein PJ, Coulombe RA Jr. 2002. Metabolism and cytotoxicity of aflatoxin B_1 in cytochrome P-450-expressing human lung cells. Journal of Toxicology and Environmental Health A 65(12):853–867.

Vercelli D. 2003. Learning from discrepancies: CD14 polymorphisms, atopy and the endotoxin switch. Clinical & Experimental Allergy 33(2):153–155.

Verhoeff AP, van Strien RT, van Wijnen JH, Brunekreef B. 1995. Damp housing and childhood respiratory symptoms: the role of sensitization to dust mites and molds. American Journal of Epidemiology 141(2):103–110.

Vernooy JH, Dentener MA, van Suylen RJ, Buurman WA, Wouters EF. 2002. Long-term intratracheal lipopolysaccharide exposure in mice results in chronic lung inflammation and persistent pathology. American Journal of Respiratory Cell and Molecular Biology 26(1):152–159.

Vesper SJ, Vesper MJ. 2002. Stachylysin may be a cause of hemorrhaging in humans exposed to *Stachybotrys chartarum*. Infection and Immunity 70(4):2065–2069.

Vesper SJ, Dearborn DG, Elidemir O, Haugland RA. 2000. Quantification of siderophore and hemolysin from *Stachybotrys chartarum* strains, including a strain isolated from the lung of a child with pulmonary hemorrhage and hemosiderosis. Applied Environmental Microbiology 66(6):2678–2681.

Von Essen S, Romberger D. 2003. The respiratory inflammatory response to the swine confinement building environment: the adaptation to respiratory exposures in the chronically exposed worker. Journal of Agricultural Safety and Health 9(3):185–196.

Von Essen S, Robbins RA, Thompson AB, Rennard SI. 1990. Organic dust toxic syndrome: an acute febrile reaction to organic dust exposure distinct from hypersensitivity pneumonitis. Journal of Toxicology—Clinical Toxicology 28(4):389–420.

Waegemaekers M, van Wageningen N, Brunekreef B, Boleij JS. 1989. Respiratory symptoms in damp homes. A pilot study. Allergy 44(3):192–198.

Wald A, Leisenring W, van Burik JA, Bowden RA. 1997. Epidemiology of *Aspergillus* infection in a large cohort of patients undergoing bone marrow transplantation. The Journal of Infectious Diseases 175(6):1459–1466.

Wålinder R, Norbäck D, Wessen B, Venge P. 2001. Nasal lavage biomarkers: effect of water damage and microbial growth in an office building. Archives of Environmental Health 56(1):30–36.

Wan G, Li C. 1999. Dampness and airway inflammation and systemic symptoms in office building workers. Archives of Environmental Health 54(1):58–63.

Wan GH, Li CS, Guo SP, Rylander R, Lin RH. 1999. An airborne mold-derived product, β-1,3-D-glucan, potentiates airway allergic responses. European Journal of Immunology 29(8):2491–2497.

Weiss A, Chidekel AS. 2002. Acute pulmonary hemorrhage in a Delaware infant after exposure to *Stachybotrys atra*. Delaware Medical Journal 74(9):363–368.

Weiss ST. 2002. Eat dirt—the hygiene hypothesis and allergic diseases. New England Journal of Medicine 347(12):930–931.

Wever-Hess J, Kouwenberg JM, Duiverman EJ, Hermans J, Wever AM. 2000. Risk factors for exacerbations and hospital admissions in asthma of early childhood. Pediatric Pulmonology 29(4):250–256.

Wieslander G, Norbäck D, Nordstrom K, Wålinder R, Venge P. 1999. Nasal and ocular symptoms, tear film stability, and biomarkers in nasal lavage, in relation to building-dampness and building design in hospitals. International Archives of Occupational and Environmental Health 72(7):451–461.

Wild LG, Lopez M. 2001. Hypersensitivity pneumonitis: a comprehensive review. Journal of Investigational Allergology and Clinical Immunology 11(1):3–15.

Wilk JB, DeStefano AL, Arnett DK, Rich SS, Djousse L, Crapo RO, Leppert MF, Province MA, Cupples LA, Gottlieb DJ, Myers RH. 2003. A genome-wide scan of pulmonary function measures in the National Heart, Lung, and Blood Institute Family Heart Study. American Journal of Respiratory and Critical Care Medicine 167(11):1528–1533.

Williamson IJ, Martin C, McGill G, Monie RDH, Fennery AG. 1997. Damp housing and asthma: a case-control study. Thorax 52(3):229–234.

Wills-Karp M, Chiaramonte M. 2003. Interleukin-13 in asthma. Current Opinion in Pulmonary Medicine 9(1):21–27.

Witteman AM, Stapel SO, Sjamsoedin DH, Jansen HM, Aalberse RC, van der Zee JS. 1996. *Fel d* I-specific IgG antibodies induced by natural exposure have blocking activity in skin tests. International Archives of Allergy and Immunology 109(4):369–375.

Yang CY, Chiu JF, Chiu HF, Kao WY. 1997. Damp house conditions and respiratory symptoms in primary school children. Pediatric Pulmonology 24(2):73–77.

Yang CY, Tien YC, Hsieh HJ, Kao WY, Lin MC. 1998. Indoor environmental risk factors and childhood asthma: a case-control study in a subtropical area. Pediatric Pulmonology 26(2):120–124.

Yazicioglu M, Saltik A, Öne Ü, Sam A, Ekerbiçer HÇ, Kirçuval O. 1998. Home environment and asthma in school children from the Edire region in Turkey (Factores ambientales y asma en escolares de la región de Edirne, Turquía). Allergy and Immunopathology (Allergologia et Immunopathologia) 26(1):5–8.

Yi ES. 2002. Hypersensitivity pneumonitis. Critical Reviews in Clinical Laboratory Sciences 39(6):581–629.

Yike I, Miller MJ, Sorenson WG, Walenga R, Tomashefski JF Jr, Dearborn DG. 2002. Infant animal model of pulmonary mycotoxicosis induced by *Stachybotrys chartarum*. Mycopathologia 154(3):139–152.

Young NA, Kwon-Chung KJ, Kubota TT, Jennings AE, Fisher RI. 1978. Disseminated infection by *Fusarium moniliforme* during treatment for malignant lymphoma. Journal of Clinical Microbiology 7(6):589–594.

Zacharasiewicz A, Zidek T, Haidinger G, Waldhör T, Vutuc C, Zacharasiewicz A, Goetz M, Pearce N. 2000. Symptoms suggestive of atopic rhinitis in children aged 6–9 years and the indoor environment. Allergy 55(10):945–950.

Zock JP, Kogevinas M, Sunyer J, Jarvis D, Toren K, Anto JM; European Community Respiratory Health Survey. 2002. Housing characteristics, reported mold exposure, and asthma in the European Community Respiratory Health Survey. Journal of Allergy and Clinical Immunology 20(3):285–292.

6

Prevention and Remediation of Damp Indoor Environments

Among the concerns that people face when dealing with indoor moisture problems are how to prevent microbial growth from starting and how to get rid of established growth safely and effectively. This chapter discusses prevention strategies, published guidelines for the removal of fungal growth (remediation), remediation protocols, and research on the effectiveness of various cleaning strategies. It also identifies weaknesses in the literature on remediation and offers suggestions for further research. The chapter does not offer guidance on which interventions are appropriate in which circumstances—this is beyond the scope of the report.

The chapter focuses on mold because most of the pertinent literature deals with mold. The observations offered here are also likely to be relevant to other indoor microbial exposures, but, because they have not been well studied, it is not possible to make definitive statements about them.

PREVENTION

The most effective way to manage mold in a building is to eliminate or limit the conditions that foster its establishment and growth. Every organism has strategies for locating a hospitable environment, obtaining water and nutrients, and reproducing. Intervention in one or more of those strategies can improve the resistance of the environment against microbial contamination.

The key to prevention in the design and operation of buildings is to limit water and nutrients. The two basic methods for accomplishing that are keeping moisture-sensitive materials dry and, when wetting is likely or unavoidable, using materials that offer a poor substrate for growth. Specifically, design and maintenance strategies must be implemented to manage

- Rainwater and groundwater, preventing liquid-water entry and accidental humidification of buildings.
- The distribution, use, and disposal of drinking, process, and wash water, making equipment and associated utilities easily accessible for maintenance and repair.
- Water vapor and surface temperatures to avoid accidental condensation.
- The wetting and drying of materials in the building and of soil in crawl spaces during construction.

Existing buildings have more limited options for water and moisture control than new construction because the systems that manage drinking, process, and wash water and that control rainwater, groundwater, water vapor, and heat flow have already been selected and installed. Flawed constituents of existing systems must be repaired, replaced, or addressed through routine operations and maintenance. Operations and maintenance procedures that reduce the likelihood of mold growth include cleaning mold-resistant materials that routinely get wet in the course of ordinary operations (floors in entryways, showers, and condensate systems or cooling coils) and quickly drying mold-prone materials that accidentally get wet through plumbing leaks, rainwater intrusion and the like.

Little scientific information on the efficacy and impact of prevention strategies is available, perhaps in part because it is easier to study problems than their absence. Moreover, little of the practical knowledge acquired and applied by design, construction, and maintenance professionals has been committed to print or subject to thorough validation; this complicates the study and dissemination of best practices. Chapters 2 and 7 address that topic and offer recommendations for research and for education of building professionals and others.

PUBLISHED GUIDANCE FOR MOLD REMEDIATION

Efforts to remediate microbial contamination involve direct intervention with building occupants, the source of the contaminant (the mold or other microbial agent), or the transport mechanism, (that is, the means by which a contaminant moves within a building environment). For example, moving people during intense remediation activities is an intervention that involves occupants, removing fungal growth and remediating the moisture problem are interventions that involve the source, depressurizing a moldy crawl space with fan-powered exhaust intervenes in the transport mechanism, and filtration and increased dilution ventilation intervene in contaminant transport by lowering airborne concentrations in general.

This section addresses similarities and differences in various published contamination-remediation guidelines, and the section that follows it is an extended discussion of the steps to be taken in remediation.

Indoor mold has historically been treated as a nuisance contaminant. Two decades ago, there was little guidance for responding to fungal contamination in buildings beyond the general instruction to clean it up. That began to change as more became known about the potential hazards of mold exposures and the practice of remediation. In 1980, allergists suggested removing mold-contaminated materials and cleaning affected areas (Kozak et al., 1980). In the same year, the U.S. Department of Agriculture published a bulletin advising people to control dampness and to treat contaminated materials with bleach (USDA, 1980). Four years later, Morey et al. (1984) recommended moisture control, improved filtration, and ventilation with outdoor air to prevent mold problems. Intervening in the moisture dynamic, cleaning contamination from hard-surface materials, and carefully discarding contaminated porous materials were suggested for dealing with existing problems. For the first time, respirators were proposed for workers performing remediation. No recommendations for containment were included.

In 1989, the Bioaerosols Committee of the American Conference of Governmental Industrial Hygienists (ACGIH) released *Guidelines for the Assessment of Bioaerosols in the Indoor Environment* (ACGIH, 1989). Those guidelines included recommendations for the design and operation of buildings and equipment and remediation of contaminated materials. Cleaning with detergent and high-efficiency particulate air (HEPA) vacuuming were suggested for removing biologic contamination, and cautious use of biocides was suggested for disinfection. For containment, the guidelines recommended that air-handling equipment be turned off during remediation. The 1992 booklet *Repairing Your Flooded Home* published by the American Red Cross and the Federal Emergency Management Agency provided guidance for drying, cleaning, and rebuilding a flood-damaged home but did not specifically address mold growth or exposure to dampness-related contaminants (ARC and FEMA, 1992). And in 1993, the Canadian Mortgage and Housing Corporation published a mold-cleanup guide for homeowners (CMHC, 1993). It recommended water and bleach cleanup, discarding some materials and using a hypochlorite-based sanitizer. Respirators and gloves were recommended during cleanup. Containment was not discussed.

While the issue was receiving more attention in both the federal and private sectors, the late 1980s also saw an increase in attention from researchers. A 1989 study discussed containment during the remediation of fungal contamination in buildings (Light et al., 1989). Containment consisted of turning off heating, ventilating, and air-conditioning (HVAC) equipment, excluding occupants from the work area, and identifying critical leakage sites and sealing them with plastic film. Contaminated materials were either to be cleaned with HEPA vacuuming, washed with a detergent–disinfectant solution, or discarded. Worker protection was not mentioned. Criteria for assessing whether the remediation effort was successful—called "clearance" criteria in many guidance documents—were also discussed.

In 1992, an American Society of Heating, Refrigerating and Air-Conditioning Engineers (ASHRAE) conference paper used a series of case studies to outline guidelines for occupant and worker protection during fungal remediation (Morey, 1992). These greatly increased the attention and detail devoted to this aspect of remediation. For a case with a high potential for dispersing spores, isolating work areas by using barriers over air leaks and HVAC openings was recommended, as was paying attention to possible bypass of leaks through ceiling and floor plenums (enclosed spaces in which air pressure is higher than outside). Airlocks and clean rooms were recommended at entries to prevent contaminant transport from the work area during entry and exit. HEPA-filtered exhaust was advised as a means to maintain the work area at a pressure 0.02 in. of water column (WC) lower than surrounding spaces. The ASHRAE paper recommended that refuse be double-bagged before removal from the work area and that HEPA vacuuming be used for cleaning. The adequacy of containment was to be documented by monitoring air-pressure relationships and collecting bioaerosol samples from occupied spaces. Air samples were to be used to document clearance after remediation activities but before containment barriers were removed.

In 1993, the New York City Department of Health (NYCDOH) convened a panel of experts to develop guidance for the assessment and remediation of *Stachybotrys atra* (*chartarum*) (NYCDOH, 1993). The resulting document included a systematic set of steps to be undertaken for investigation, including evaluation of medical issues, visual inspection, sampling, and interpretation. The second half of the document provided guidance for containment, worker protection, and training requirements for abatement personnel. Four levels of contamination were described, and identifying and eliminating the moisture source supporting mold growth was required for all four levels. Level I was for areas with less than 2 ft^2 of contaminated material, Level II for areas with 2–30 ft^2, Level III for areas with over 30 ft^2, and Level IV for the remediation of contaminated HVAC equipment. Levels I and II required respiratory protection for building-maintenance personnel with very little containment or clearance testing. Levels III and IV required full containment, including air-pressure management, isolation of HVAC equipment, and dermal and respiratory protection for workers. Air sampling was required to document containment and to provide a basis for reoccupancy.

Since the 1993 NYCDOH document was produced, a number of other guidance documents have been written, including

- *Fungal Contamination in Buildings: A Guide to Recognition and Management* (Health Canada, 1995).
- *Control of Moisture Problems Affecting Biological Indoor Air Quality* (Flannigan and Morey, 1996).
- *Bioaerosols: Assessment and Control* (ACGIH, 1999).
- *Guidelines on Assessment and Remediation of Fungi in Indoor Environments* (NYCDOH, 2000).
- *Mold Remediation in Schools and Commercial Buildings* (U.S. EPA, 2001).
- *Report of the Microbial Growth Task Force* (AIHA, 2001).

Table 6-1 compares those guidelines with regard to how they were developed, events that would trigger a fungal assessment or remediation, assessment methods, remediation activities, and prevention actions.[1]

The seven documents were each developed by a group of people with identified expertise in building and engineering issues, mycology, and occupant health assessment. Topics are not uniformly covered by the documents—for example, the ACGIH document provides extensive coverage of health effects, health assessment, and sampling, but some of the other documents do not provide information on these subjects.

The documents agree that

- Mold should not be allowed to colonize materials and furnishings in buildings.
- The underlying moisture condition supporting mold growth should be identified and eliminated. Only the International Society of Indoor Air Quality and Climate (ISIAQ) and ACGIH guidelines discuss moisture dynamics, identifying problematic moisture or remediating moisture problems. The Environmental Protection Agency (EPA) guidelines contain specific recommendations for a variety of water-damaged materials.
- The best way to remediate problematic mold growth is to remove it

---

[1]After this report was completed, the Institute of Inspection, Cleaning, and Restoration Certification published IICRC S520: Standard and Reference Guide for Professional Remediation (IICRC, 2003). This document, which was not reviewed by the committee, also addresses fungal assessment and remediation, and clearance criteria.

from materials that can be effectively cleaned and to discard materials that cannot be cleaned or are physically damaged beyond use. Managing mold growth in place is not considered by any of the documents.

• Occupants and workers must be protected from dampness-related contaminants during remediation. All the guidelines agree that some mold situations present a small enough exposure potential that cleanup does not require specific containment or worker protection but that other situations warrant full containment, air-pressure management, and full worker protection. Situations between those extremes need intermediate levels of care. Guidance for selecting appropriate containment and worker protection for different situations lacks clarity within and between documents.

• HVAC systems are special cases. But the documents disagree on how to respond to contamination in HVAC systems.

The documents are divided on the use of disinfectants. Four recommend that disinfectants be used sparingly, in appropriate locations, for specific purposes, and with caution. The original NYCDOH guidance requires the use of biocides, whereas ISIAQ suggests it for hard surfaces. Only two of the documents—those of ISIAQ and ACGIH—discuss the prevention of mold growth in buildings to any substantial degree.

The American Industrial Hygiene Association (AIHA) document differs from the others in several respects. It identifies itself as supplementary to other guidance, and it is the only document that specifically reviews other guidelines, identifying common ground, disagreements, strengths and weaknesses in the evidence, and gaps in knowledge. It also offers recommendations for best practices. The AIHA document focuses on 11 questions:

1. When should microbial growth found in occupied buildings be remediated?

2. What amounts of mold should indicate what degrees of remediation?

3. What remediation methods should be used?

4. Should biocides be used in remediation?

5. Under what circumstances should buildings be evacuated and work areas isolated?

6. How should remediation work areas be isolated?

7. How should water-damaged items be treated?

8. What quality-assurance principles should be followed to ensure that mold remediation is successful?

9. What personal protective equipment is recommended during remediation?

10. Is personal air sampling appropriate to determine worker exposure during mold remediation?

11. What medical evaluation is recommended for remediators?

TABLE 6-1   Comparison of Seven Mold-Remediation Guidance
Documents

	NYCDOH, 1993	Health Canada, 1995	Flannigan and Morey, 1996 (ISIAQ)
General	Guidance specific to *Stachybotrys atra*; earliest best-practice remediation document to give guidance on selecting containment and worker protection	Fairly comprehensive discussions with cohesive logic tree for assessment and remediation of indoor microbial contamination	In addition to remediation guidance, problem moisture sources and indoor fungal ecology receive substantial treatment
Process	Summary of recommendations from expert panel	Sections written by members of federal-provincial working group after literature review	Written by members of Task Group 1, International Society of Indoor Air Quality and Climate
ASSESSMENT			
Triggering events	Visible mold, water damage, symptoms consistent with exposure	Not specifically identified but by implication visible mold growth, accumulations of bird droppings, or evidence of fungal growth from sampling	Not specifically identified, but by implication observation of sampling that confirms colonization by mold, mites, or bacteria
Health assessment	Conditional; brief discussion	Conditional/extensive coverage	No specific discussion of assessment included
Visual inspection and building history	Required; identify extent of mold growth and water damage	Required; extensive coverage	
Intrusive inspection	Not discussed	Conditional; cautions on disturbance	
Fungal sampling	Bulk sampling to document *S. atra*; air, not routinely unless HVAC contaminated	Conditional; coverage for many methods	

ACGIH, 1999	NYCDOH, 2000	U.S. EPA, 2001	AIHA, 2001
Most extensive discussions of health effects, sampling strategies, and data analysis	Expands original scope from single species to molds in general; provides detailed guidance on assessments, containment, and worker protection	Primarily schools and commercial buildings; has specific section on planning remediation and specific remediation methods for different materials	Reviews existing guidance, basis for recommendations, information gaps, and recommendations for 11 key issues
Written by members of Bioaerosols Committee of ACGIH	Based on literature review and comments from expert review panel	Prepared by Indoor Environments Division of EPA; internal and external review process	Review of existing guidance by Microbial Growth Task Force of AIHA; minority report included
Visible fungal growth identified in remediation section; other sections give insight into medical and environmental sampling	Presence of mold, water damage, or musty odors identified in assessment section	Not specifically identified, but by implication visible mold growth	Consensus of published guidance: visible mold growth and moisture damage; hidden growth may be important but may not be immediately obvious
Conditional; extensive coverage	Conditional; brief discussion	Conditional; brief reference	Not covered
Required; extensive coverage	Required; brief discussion	Assumed; brief reference	Not specifically covered but implicit in many sections
Brief discussion; cautions on disturbance	Brief reference	Discussion of hidden mold; caution on disturbance	Includes appendix on making holes; cautions on spore release
Conditional; extensive coverage for many methods	Conditional; part of medical evaluation, suspect HVAC contamination, suspect hidden mold	Conditional; part of medical evaluation, suspect hidden mold, litigation	Discusses dust sampling and cavity sampling; other methods extensively discussed in AIHA, 1996

(continued on next page)

TABLE 6-1   continued

	NYCDOH, 1993	Health Canada, 1995	Flannigan and Morey, 1996 (ISIAQ)
Interpretation	Bulk for presence of *S. atra*; air, differential	Coverage for many methods	
Analysis	Screen laboratories for experience with indoor environmental mycology	Not covered	
REMEDIATION			
Moisture problem	identify; intervene	identify; intervene	identify; intervene
Area 1	<2 ft^2	<3.23 ft^2 (0.3 m^2)	<2.15 ft^2 (0.2 m^2)
Containment	Special containment not needed; bag refuse	Clean material before removal	Carefully remove materials
Worker protection	Full respiratory protection; 29 CFR 1910.134	Mask and gloves	No specific guidance
Training	Building maintenance with some mold-cleanup training	Trained personnel	No specific guidance
Area 2	2–30 ft^2	3.23–32 ft^2 (0.3–3.0 m^2)	2.1–32.3 ft^2 (0.2–3.0 m^2)
Containment	Bag refuse; cover adjoining surfaces with poly	Clean before removal	Bag refuse; local containment; HEPA-filtered exhaust air
Worker protection	Full respiratory protection; 29 CFR 1910.134	Half-face respirators and gloves	Proper respiratory protection
Training	Building maintenance with some mold-cleanup training	Trained personnel	Building-maintenance personnel

ACGIH, 1999	NYCDOH, 2000	U.S. EPA, 2001	AIHA, 2001
Extensive coverage for many methods	Species ID for medical; differential for hidden mold EMLAP accredited laboratories, interpretation by experienced professional	Use trained professionals; cautions on uncertainty	Dust sampling and cavity sampling
identify; intervene	identify; intervene	identify; intervene	identify; intervene
Minimal Source	$\leq$10 ft^2 Vacate work area; dust suppression, no special containment, bag refuse, damp wipe area	<10 ft^2 None required, use professional judgment	Recommends: containment based on combining innovative professional judgment with areas defined by NYCDOH (2000); worker protection based on ACGIH recommendations; health evaluation of workers advised by NYCDOH (2000) recommended
N95 mask and gloves	N95 mask, gloves, and eye; 29 CFR 1910.134 Building maintenance with some mold-cleanup training	N95 mask, gloves, and eye; use professional judgment Not covered	
Moderate	10–30 ft^2	10–100 ft^2	
Local; HEPA-filtered exhaust air	Vacate and cover work area with poly; dust suppression; bag refuse; HEPA-vacuum and damp-wipe area	Poly sheeting around area; HEPA-filtered exhaust air; block HVAC	
N95 mask, full-body covering and eye	N95 mask, gloves, and eye; 29 CFR 1910.134 Building maintenance with some mold-cleanup training	N95 mask or half-face HEPA coverall, eye Not covered	

(continued on next page)

TABLE 6-1   continued

	NYCDOH, 1993	Health Canada, 1995	Flannigan and Morey, 1996 (ISIAQ)
Area 3	>30 ft^2	>108 ft^2; area between 32 and 108 ft^2 does not seem to be directly addressed	>32 ft^2
Containment	Full; HEPA-filtered exhaust air; critical barriers, airlocks; HVAC	Full; HEPA-filtered exhaust air; critical barriers; airlocks; HVAC	Full; HEPA-filtered exhaust air; critical barriers; air locks; HVAC
Worker protection	Full-face HEPA, coverall, and eye	Full-face HEPA, coverall, and eye	Full-face HEPA, coverall, and eye implied but not specified
Training	Hazardous waste	Trained personnel	Hazardous waste
Area 4 Containment	NA	NA	NA
Worker protection Training			
HVAC Containment	Full; HEPA-filtered exhaust air; critical barriers; airlocks; HVAC	Unclear	Depends on area as above

ACGIH, 1999	NYCDOH, 2000	U.S. EPA, 2001	AIHA, 2001
Extensive; references NYCDOH, 1993 and Flannigan and Morey, 1996; does not define areas	30–100 ft^2	>100 ft^2	
Full; HEPA-filtered exhaust air; critical barriers; HVAC	Vacate work and adjacent areas; cover work area and directly adjacent areas with poly; seal HVAC openings; bag refuse; dust suppression; HEPA-vacuum and damp-wipe area; upgrade to next level of protection if dust will be raised	Full; HEPA-filtered exhaust air; critical barriers; air locks; HVAC	
N95 mask, coverall, and eye	N95 mask, gloves, and eye; 29 CFR 1910.134	Full-face HEPA, coverall, and eye	
Hazardous waste	Hazardous waste		
NA	>100 ft^2 Oversight by health and safety professional; vacate work area; full containment; HEPA-filtered exhaust air; critical barriers; airlocks; HVAC Full-face HEPA, coverall, and eye Hazardous waste	NA	
Depends on area as above; specific guidance for cooling towers included	<10 ft same as Area 2; >10 ft^2 same as Area 4	Refers to EPA document *Should You Have the Air Ducts in Your Home Cleaned?*	

(continued on next page)

TABLE 6-1   continued

	NYCDOH, 1993	Health Canada, 1995	Flannigan and Morey, 1996 (ISIAQ)
Worker protection	Full-face HEPA, coverall, and eye	Unclear	Depends on area as above
Training	Hazardous waste	Unclear	Depends on area as above
Remediating hard surfaces and semiporous	Not directly discussed	HEPA-vacuum; damp-wipe	HEPA-vacuum; damp-wipe; disinfect
Remediating porous materials	Discard contaminated absorbent materials	Discard	Discard
Biocide use	Required use of bleach to clean areas adjacent to contaminated areas; cautions on use of chlorine dioxide or ozone in HVAC systems	Discouraged or conditional; use charcoal filters in respirators	Suggested for hard surfaces
Clearance	Air monitoring for >30 ft^2 and HVAC	Not specifically covered; everything must be cleaned	Surfaces cleaned until only background fungi and bacteria remain; materials dry

ACGIH, 1999	NYCDOH, 2000	U.S. EPA, 2001	AIHA, 2001
Depends on area as above	<10 ft^2 same as Area 2; >10 ft^2 same as Area 4		
Depends on area as above; refers to NADCA and EPA;cautions against biocide use	<10 ft^2 same as Area 2; >10 ft^2 same as Area 4		
Clean; discard if physically damaged	Damp-wipe with detergent solution	Specific guidance for different materials: wet-vacuuming, damp-wiping, HEPA-vacuuming or discarding	HEPA-vacuum, damp-wipe, or scrub as needed; discard damaged materials
Discard contaminated material; wash or HEPA-vacuum materials that may harbor spores	Discard with exceptions	See above	Discard
Comprehensive discussion; discouraged or use with caution	Refers to ACGIH, 1999; recommends detergent solutions; prohibits gaseous ozone and chlorine dioxide	Discussion; discouraged or use with caution	Discouraged or use with caution
Visual; possible sampling	All areas left dry and visibly free of contamination and debris; air monitoring for areas >100 ft^2	Judgment call; water source fixed; dry and free of visible mold; reoccupying produces no complaints; if air samples have been taken, differential interpretation	Recommends documenting successful intervention in moisture source, containment, cleaning, removal, completeness, final surface dusting, and use of HEPA vacuum

(continued on next page)

TABLE 6-1   continued

	NYCDOH, 1993	Health Canada, 1995	Flannigan and Morey, 1996 (ISIAQ)
PREVENTION			
Moisture control	Not covered	Brief discussion	Substantial discussion by moisture source, climate type, envelope, and HVAC
Materials in wet spaces	Not covered	Not covered	Limit nutrients; avoid porous materials; some consideration of biostats
Operation and maintenance (O&M)	Not covered	Brief discussion	Incorporated in moisture control

NOTES:
• ID: identify.
• 29 CFR 1910.134 denotes the Occupational Health and Safety Administration (OSHA) respiratory protection standard.

A minority report in the AIHA document raises concerns about treating all molds as hazardous substances and the consequent recommendations for decontamination, worker protection, containment, and disposal.

An extensive review of the literature regarding investigation, building ecology, and, to some extent, remediation is contained in a 2002 doctoral dissertation titled *Characterizing Moisture Damaged Buildings–Environmental and Biological Monitoring* (Hyvärinen, 2002).

TASKS INVOLVED IN REMEDIATION

Responding to mold problems requires a series of actions. The order in which actions take place is sometimes important. Typically, the following actions are implemented to some extent regardless of whether a problem is small and simple or large and complex:

ACGIH, 1999	NYCDOH, 2000	U.S. EPA, 2001	AIHA, 2001
General discussion of moisture dynamics; humidity control	Not covered	Checklist of moisture-control items	Not covered
Brief discussion of nutrient and material issues	Not covered	Not covered	Not covered
Discusses ventilation, filtration, monitoring moisture events, and cleaning	Not covered	Checklist of operation and maintenance items	Not covered

- N95 class masks meet the National Institute for Occupational Safety and Health (NIOSH) standards for respiratory protective devices specified in 42 CFR Part 84.

- Take emergency actions to stop water intrusion if needed.
- Identify vulnerable populations, extent of contamination, and moisture dynamic.
- Plan and implement remediation activities.
  — Establish appropriate containment and worker and occupant protection.
  — Eliminate or limit moisture sources and dry the materials.
  — Decontaminate or remove damaged materials as appropriate.
  — Evaluate whether space has been successfully remediated.
  — Reassemble the space to prevent or limit possibility of recurrence by controlling sources of moisture and nutrients.

For small, simple problems, the entire list may be implemented by one person. For larger, more complex problems, the actions in the list may be accomplished by a series of people in different professions and trades.

Maintenance workers may find a broken pipe flooding the building and turn off the water main. A firm that specializes in water-damage emergency response may pump the basement and dehumidify the building with specialized drying equipment. A remediation-assessment consultant may determine whether fungal growth has occurred and what actions are required. Remediation contractors may set up the appropriate containment, remove some contaminated materials, and clean up materials that can be salvaged. A remodeling contractor may reconstruct damaged areas, and yet another contractor may repair the broken pipe that started the problem. The remediation-assessment consultant may provide quality assurance while work is being carried out and determine whether clearance criteria have been satisfied.

For circumstances that fall between those extremes, some combination of occupant action and professional intervention will be appropriate. In general, no single discipline brings together all the required knowledge for successful assessment and remediation.

Take Emergency Actions to Stop Water Intrusion If Needed

At times, water intrudes into a building so quickly that it must be responded to as an emergency. Broken water lines, gaping holes in the roof or walls during rainstorms, or rising water tables may cause severe wetting or structural damage. To avoid fungal growth on susceptible materials, it is important to dry them quickly. Porous materials can absorb and retain a great deal of water. If nutrients are readily available in the material itself or in collected dust, visible fungal growth can occur within a few days of wetting (Doll, 2002; Horner et al., 2001). Doll's work indicates that common construction materials may contain mold spores when they arrive from distributors and that some of the spores will germinate if exposed to relative humidity (RH) of 95% for extended periods (22–60 days); if they are wetted with liquid water, germination can occur in less than 5 days.

Interventions in continuing wetting and in drying a building out must be informed by the nature of the water source and the nature of the wet materials (IICRC, 1999). Emergency actions may include turning off the water main, repairing pipes, temporarily closing holes in roofs or walls, pumping water out of buildings, unclogging drains, vacuuming water from materials in buildings, and actively dehumidifying buildings, rooms, or cavities. Appropriate worker protection may be necessary, depending on the source of the water (for example, flood water must be considered septic) (IICRC, 1999).

Quick and effective response to water intrusion can prevent or greatly reduce fungal growth and occupant exposures to bioaerosols, further disruption to building operations, and future needs for fungal remediation.

Identify the Vulnerable Populations, Extent of Contamination, and Moisture Dynamic

The purpose of an assessment is to collect the information needed to respond appropriately to a contamination problem. People who are particularly vulnerable to dampness-related exposures should be identified, and the location and extent of contamination must be inventoried. Sources of moisture and nutrient supporting the growth must be identified.

Collect Histories of Building and Occupants

As noted in Chapter 2, the people who design, construct, maintain, and occupy a building are a valuable source of information concerning occupant health, building history, microclimates that might support fungal growth, exposure mechanisms, and other essential information that helps to form an overview of the situation of the building and the occupants. Collecting histories can provide a wealth of information that may be invaluable when planning a building remediation. The process can also help identify whether some occupants are particularly sensitive to fungal exposures—people with known mold allergies, asthma, and the like—and thus require extra protection during investigations and remediations. Interviews, questionnaires, and contact with personal physicians (when authorized by the subjects) are some approaches to collecting such information. Some standardized questionnaires have been developed, including the MM-040-EA Indoor Climate Work Environment questionnaire (Andersson, 1998) and the Stockholm Indoor Environment Questionnaire (Engvall et al., 2004).

Some of the guidance documents recommend removing people who are more sensitive to fungal contaminants from buildings during remediation activities. Building occupants who are exhibiting symptoms, have identified illness associated with fungal exposures, or have compromised immune systems may be at greater risk than the general population. The AIHA Task Force (2001) identified NYCDOH (2000) as providing a basis for making decisions regarding evacuation and containment.

Occupants may also provide clues regarding the location of mold growth. Musty odors are sometimes reported, and their timing may be predictable or erratic. Odors, for example, might occur only in an entryway during periods of heavy rain, or respiratory problems may be more frequent in an area served by a particular air handler. Someone may recall that the lower floor floods during heavy rainstorms. A member of the buildings and grounds staff may know that there is an undocumented crawl space beneath the basement, that roof drains run inside wall cavities, or that the enthalpy-control system has been disabled for 5 years. Photographs of a house under construction may show details of foundation drainage or defects in rainwater control. Failing to take advantage of the information available from architects, builders, occupants, and maintenance staffs may lead to an incomplete and ineffective remediation plan.

Ascertain Extent and Location of Fungal Contamination

Microbial contamination may consist of active growth, relic colonies that have run out of moisture or nutrients, or remnants of growth that have accumulated with other dust particles beneath furniture or in corners, cushions, or carpets. Spores and hyphal fragments deposited on hard surfaces or contained in porous materials—such as textiles, cushions, and fibrous insulation—may have originated indoors or been transported from outside. Spores are known to spread from one portion of a building to another (Morey, 1993), to be transported inside on people's clothing (Pasanen et al., 1989), and to enter with accidental or intentional ventilating air. A thorough inspection of the building is required. Because mold problems are also moisture problems, understanding how water behaves in buildings is a valuable component of the inspection.

Some remediation-guidance documents base worker protection and containment recommendations on the area of visible mold growth (Flannigan and Morey, 1996; Health Canada, 1995; NYCDOH, 1993, 2000). The ACGIH guidance divides remediation into three levels of effort distinguished by worker protection and containment but does not relate them to specific areas of contamination. Instead, it urges the use of professional judgment to determine which situations require what precautions. For remediation activities to comply with the guidance in the other documents, the area of visible mold must be identified. However, it must be remembered that although some contamination is readily apparent, other contamination may be hidden behind furnishings or wallpaper, in accessible cavities, or in spaces with no ready access.

To find fungal growth in a building, one must look for places that have both water and nutrients. Often, these are places that are likely to get wet and to take a long time to dry. Chapter 2 addresses the most common moisture dynamics. Several signs of water problems must be evaluated—not just mold, but also peeling paint, efflorescence, wet spots, wrinkled wallpaper, cracks in plaster, and warped wood. Rooms that are unavoidably wet—such as bathrooms, kitchens, spas, and pool rooms—are the most common places to find visible mold growth.

Rainwater penetration, plumbing leaks, and condensation are common moisture sources for hidden locations. Exterior walls and walls below roof penetrations are generally the most vulnerable to such sources. Oriented strand board (OSB), the backside of paper-covered gypsum board, and the topside of cellulose ceiling tile are also at risk (Doll, 2002). Often, growth is in spaces such as attics, crawl spaces, basements, and garages. Those largely unoccupied rooms are frequent sites of condensation and rainwater or plumbing leaks. The relative importance of microbial contamination in such spaces to overall exposures has not been extensively studied; available case studies (Miller et al, 2000; Pessi et al., 2002) yield inconsistent results.

As a rule of thumb, microbial contamination is most likely to be found by

- Examining attics, foundations, penetrations through walls and roofs, and exterior drainage detailing.
- Tracing water lines.
- Examining surfaces chilled by mechanical equipment or by contact with the earth (air-conditioning ducts and coils, chilled water lines, outdoor air ducts, and basements and crawl spaces).

Thorough examination can uncover mold hidden in closets, in cabinets, beneath wallpaper, under furniture, and beneath carpets. Cavities above suspended ceilings can be inspected by removing ceiling tile (unless the tile is fastened with clips as part of the fire-protection detailing). Often, access openings permit inspection of plumbing for bathtubs, utility trenches, crawl spaces, basements, and attics. However, a great deal of fungal growth in a building may be in cavities that are not open to inspection. Some organisms may cover an exposed surface with a light, nonpigmented growth (Horner et al., 2001; Miller et al., 2000).

A number of methods can be used to augment the ability to find hidden mold, for example, moisture meters, bulk or surface samples to identify organisms that are contaminating materials, and air samples of rooms, cavities, and outdoor air to identify fungal spores.

Moisture meters are useful in finding hidden damp spots and tracing leak pathways. They provide instant results that can guide an ongoing inspection. However, if a water source is intermittent—like a rainwater leak—a hidden growth site may be dry at the time of inspection. Moisture meters may also report a saturated condition where no moisture is present, if metal is part of the assembly. They must thus be used with care. Meters that use pins or electromagnetic fields (EMFs) to penetrate materials are available. Models with pins need holes in the test material to make measurements, but EMF-based meters are nonintrusive. Insulated pins allow assessment to specific depths. EMF models reflect moisture content in a depth range that varies from a half-inch to a few inches.

Hand lenses and microscopes can be used in buildings, and samples can be sent to a mycologist for analysis to determine the presence of mold on a surface (Horner et al., 2001; Miller et al., 2000). Microscopic examination of surfaces, dust samples, and tape lifts can identify hyphae, sporangia (spore cases), and spores. In the field, a stand microscope with magnifications of 20–200 can be used to examine surfaces in situ. Clear plastic tape can be used to collect samples from suspect surfaces (Swenson et al., 2003). The tape can be mounted on a slide and examined either in the field or in a laboratory to determine whether the surface is supporting or is contaminated with fungal growth. Other means of estimating the amount of fungal material on a surface in the field include chemiluminescence analysis to detect microbial enzymes (Bjurman, 1999), and measurement of mold biomass parameters such as ergosterol content or β-N-acetylhexosaminidase activity (Reeslev et al., 2003).

Fungal growth hidden in wall or ceiling cavities can be observed directly by making inspection openings or inferred from bioaerosol measurements (ACGIH, 1999; AIHA, 1996; Miller et al., 2000). It is desirable to inspect the inside of cavities directly (Miller et al., 2000) and to use tape-lift or chemiluminescence methods to assess the surface. But there are also drawbacks to making inspection openings. To be certain of detecting fungal growth in ceiling and wall cavities, one has to make several holes in the large surfaces that enclose these spaces. Making openings in walls or ceilings requires repair of materials or installation of permanent inspection ports and may release dust or fungal material into the indoor air or make a more intimate connection between the indoor air and a large fungal mass.

Although larger openings allow for more extensive inspection of a cavity, making them may release more fungal spores and construction dust than making smaller ones. Methods that reduce or confine the release of particles and facilitate immediate repair if extensive fungal contamination is revealed are preferable in opening cavities. For example, hole saws with dust-collecting shrouds can contain the dust released while a hole is being made and do not produce potentially aerosolizing vibrations as great as those made by reciprocating saws. Glove bags and temporary containments can provide more protection.

To reduce the necessity for large openings and to limit the release of fungal spores, inspectors have developed several methods to detect fungal growth and to predict where it is most likely to occur. As already noted, wall or ceiling cavities that have visible signs of moisture damage, produce increased moisture-meter readings, or yield positive moisture-assessment test results are areas where inspectors will concentrate their efforts. Likely spots for cavity troubles include finished basement walls, plumbing walls behind and floor cavities beneath bathtubs and showers, and exterior walls that have a combination of masonry cladding, interior vapor retarders, and air-conditioning.

Boroscopes[2] are sometimes used to reduce the number of particles released in the making of an opening and to decrease the damage to surfaces around wall or ceiling cavities. The opening, measuring only a fraction of an inch, releases only a small number of particles and is easily repaired. However, a boroscope does not offer a full view of the cavity and makes it difficult to conduct a thorough inspection (Miller et al., 2000).

Air samples can be drawn from wall or ceiling cavities through small openings and particles can be collected in spore traps or on culture media. For accurate analysis of spore traps or cultures, the investigator must use a microscope at the site or send samples to a laboratory. It is possible to draw air samples from wall cavities by accessing a cover plate for electric outlets, thereby obviating additional openings. The committee did not identify any research that established guidelines for interpreting bioaerosol samples drawn from cavities.

Spore-trap or culturable samples taken from indoor and outdoor air can be used to infer the presence of hidden mold. Interpretation of the data is simple in principle: examine the indoor relative to outdoor results. Relatively high indoor concentrations are evidence—but not proof—that microbial growth is occurring indoors. Organisms that appear indoors and are not likely to have entered with outdoor air are also evidence of an indoor source (ACGIH, 1999). In practice, spatial and temporal variability in spore concentrations (ACGIH, 1999), spores from settled dust disturbed by investigators (AEC, 2002), and inherent limitations of sampling methods introduce important uncertainties. Spore traps have the advantage of identifying organisms that do not compete well in culture or are not viable, but they do not allow speciation of such important genera as *Penicillium* and *Aspergillus*. Speciation is essential in determining whether spores in outdoor air are a likely source of spores found in indoor air. Culturable samples can be used to identify to the species level, but species that compete poorly may be missed.

A number of studies have compared indoor spore concentrations in buildings or rooms that have visible mold growth with those in buildings or rooms that do not have visible growth. The studies are far from conclusive. Several reported that buildings or rooms with visible fungal growth had higher spore concentrations than control buildings or rooms (Dharmage et al., 1999; Hunter et al., 1988; Hyvärinen et al., 1993, 2001a,b; Johanning et al., 1999; Klánová, 2000; Lawton et al., 1998). However, other investi-

---

[2]The boroscope is an instrument specifically designed to be inserted in wall cavities, ducting, and other otherwise inaccessible areas. It typically comprises a long thin rod with a light source at the tip and optics to direct and magnify the image.

gators found no differences between airborne spore concentrations (Dill and Niggeman, 1996; Garrett et al., 1998; Nevalainen et al., 1991; Pasanen et al., 1992; Strachen et al., 1990).

A few studies have sought to identify a relationship between fungal growth hidden in cavities and indoor bioaerosol samples. Miller et al. (2000) conducted a series of experiments to assess the extent of fungal colonization of wall cavities in 58 apartments that had suffered some water damage but had little visible mold on interior surfaces (some apartments were reported to have light mold growth in bathrooms and behind refrigerators). Air samples were taken in the center of living areas and bedrooms (four, six, or eight times depending on size of apartment). The bottom 30 cm of gypsum board was removed from the walls, the exposed wall cavities were examined, and the location and extent of observable, previously hidden mold were mapped. Samples of gypsum board were collected from areas that had visible mold and from contiguous areas (within 50 cm) that did not. A relationship was found between the area of fungal growth uncovered and the percentage of indoor air samples that yielded measurements significantly different from outdoor-air measurements. That led the authors to conclude that indoor-air sampling was potentially useful for identifying spaces where more intrusive testing (such as opening walls) to find mold infestations might be appropriate. Morey et al. (2002) studied two buildings that had extensive fungal growth in the envelope. Air sampling found many spores from the organism that was contaminating the shell in one of the buildings; in the other, air samples yielded no evidence of hidden fungi in the building. Pessi et al. (2002) compared indoor air-concentrations of actinomycetes, other bacteria, and fungi with concentrations of these contaminants in the insulation layer of an adjoining wall. The study was conducted in 50 apartment buildings in a subarctic climate. Actinomycete-contaminated insulation was associated with increased indoor-air concentrations of actinomycetes, and the moisture content of the indoor air significantly affected all measurable airborne concentrations. The authors noted that the relationship for fungi might be an artifact of the climate.

Disturbed spores may be released into the air in large numbers during inspection activities, exposing inspectors and others present to microbial materials. Little research was found on spore concentrations resulting from inspection activities. One study did report increased spore concentrations resulting from walking or vacuuming a carpet contaminated with *Penicillium chrysogenum* spores (Buttner and Stetzenbach, 1993). Activity raised spore concentrations by an order of magnitude.

Identify Moisture Dynamic

To ensure that a fungal problem will not recur after remediation, the moisture source supporting the growth must be identified, and an intervention in the moisture dynamic must be made. Floods are the easiest moisture dynamic to identify but may be the most difficult to prevent. For example, if a basement floods because storm drains are too small and cause recurring backups throughout the year, the system must be redesigned and modified—a potentially costly intervention. The basement would need to be waterproofed, and back-flow valves would have to be installed in the drainage system. The foundation might need to be structurally reinforced to withstand hydrostatic pressure caused by flood waters. Even then, a sufficiently high water level could still float the foundation. In comparison, large plumbing leaks are usually easy to identify and mend.

Many moisture problems are chronic, such as a slow leak in a plumbing line. Materials moistened by plumbing leaks remain wet until the leaks are repaired. Windows with leaks at the bottom corners wet the materials beneath them whenever they are hit by a driving rain; this wetting is more erratic because it depends on the occurrence of rain and on the direction and force of the wind. Materials wetted by intermittent sources might not always be wet. Condensation on chilled surfaces is often seasonal; surfaces are chilled by cooling equipment during the warm season and by outdoor air during the cold season. Absorptive claddings are wetted by rain only when the temperature is above freezing.

As noted elsewhere in this report, the moisture sources most difficult to identify are combinations of erratic wetting in hidden locations and multiple transport mechanisms. For example, water leaking through flashing at a roof curb may seep through the roof deck, run along the bottom side of a steel truss until it reaches the end of the bottom chord, drip onto a suspended ceiling tile, eventually run off the tile's edge, and only then become a noticeable drip.

The combination of condensation, absorptive materials, capillary suction, and temperature-driven vapor flow creates problems that are difficult to diagnose. Among the more complex of these dynamics is wind-driven rain against a brick veneer. This may create an interior pocket of moisture that heating by the sun turns to vapor. The vapor may then condense on the backside of (interior) vinyl wallpaper under the influence of air-conditioning. This dynamic can result in extensive fungal contamination without revealing a drop of moisture on the interior or exterior finish of the wall.

Identifying and solving moisture problems parallel identifying the extent and location of fungal problems. Building occupants are interviewed about problems with water, and then the building and site are thoroughly inspected. Tests and measurements with moisture meters and other instruments may also be used. Humidity and temperature measurements are useful in a number of ways. They can be used to identify microclimates that support fungal growth. For example, if the temperature and RH are 75°F (24°C) and 60% indoors, 10°F (–12°C) and 50% outdoors, and 65°F (18°C) and 99% RH in a closet on an exterior wall, the closet will be a hospitable niche for fungal growth. Temperature and RH can be used to calculate the amount of water vapor in each pound of air, and inferences can be drawn about moisture sources. In the example just mentioned, there is 0.0127 lb (5.8 g) of water per pound of dry air indoors, 0.0007 lb (0.3 g) of water per pound of dry air outdoors, and 0.011 lb (5.0 g) of water per pound of dry air in the closet. From that information, it is clear that moisture in the indoor air explains the moisture in the closet air. Although 60% RH is within general comfort conditions, it is too high for this building under these conditions. The next challenge would be to find the sources of water vapor. Common sources of water vapor in buildings include respiration, washing and drying (floors, dishes, and clothing), bathing, cooking, watering plants, damp foundations, and planned and unplanned ventilating air.

Often, finding the source of water vapor requires a little more investigative work. In the example, the planned and unplanned ventilating air would typically be drying the building; therefore, the source of the unwanted cold-weather humidity must be strong and sought elsewhere. Moisture-meter readings may find the basement walls saturated, indicating the basement as the source. Indeed, a damp basement may surpass the total of all other sources in a residential building.

Infrared thermometers and scanners can be used to find cold surfaces that may result in condensation. In the example, the dew point of the indoor air is 64.4°F (18.0°C). Any surface in the building at or below that temperature will have water condensing on it.

Challenge tests can be used to evaluate building components and plumbing fixtures for leaks. For example, a suspected rainwater leak at a window, door, flashing detail, or parapet wall can be tested with a water-spray rack, with simple gentle sprays, or by observing ponding (Endean, 1995). Two American Society for Testing and Materials (ASTM) methods specify flow rates and air-pressure differences for conducting spray tests in the field: ASTM E1105 and ASTM E514 (ASTM, 2000, 2002a). In practice, many investigators simply use a gentle spray and apply a film of water to the test surface with little inertia. The significance of air-pressure differences on the film can be assessed with using fan depressurization methods developed to test the air tightness of enclosures. Plumbing and fixtures can be tested by operating them. Sinks often leak where the basket drain seals to the sink or at the water trap in the drain line. Such leaks can be found by closing the drain, filling the sink with water, observing the drain from beneath the sink, releasing the water, and observing the trap. The same process can be used to test shower pans. Pan leaks that would otherwise go unnoticed can be detected by running the shower for an extended time and making continual moisture-content readings of the floor around the shower.

A simple and useful test can be made by masking suspected leaks. For example, if it is suspected that a skylight or a parapet wall is the source of a rainwater leak, plastic film can be temporarily installed over the suspect area. If the temporary protection prevents the leak during the next storm or during a spray test, it is taken as evidence that the leak has been found. Used in combination with the spray test, masking can distinguish between a poorly detailed flashing and a window that leaks.

Vapor-emission tests—the calcium chloride test (ASTM, 1998) and the equilibrium relative-humidity (ERH) test (ASTM, 2002b)—can be performed on concrete slabs and walls to estimate the amount of water vapor being exhaled from the surface. Historically, these tests have been used to determine whether flooring materials can be applied to a concrete floor. Too high a vapor transmission rate may cause problems for flooring materials applied to the concrete. Flooring manufacturers provide guidance on maximal emission rates. However, the committee did not find any references relating the results of either method to microbial contamination in buildings.

Plan Remediation Activities

Information collected during the assessment is used as the basis of remedial efforts. A remediation plan must protect those conducting the remediation, other building occupants, and people in the general vicinity. The guidance documents listed in Table 6-1 all specify means for intervening in the moisture dynamic and removing problem fungal material. As was noted above, implementation of a remediation involves five essential steps

- Establish appropriate containment and worker and occupant protection.
- Eliminate or limit moisture sources and dry the materials.
- Decontaminate or remove damaged materials as appropriate.
- Evaluate whether the space has been successfully remediated.
- Reassemble the space to prevent or limit the possibility of recurrence by controlling sources of moisture and nutrients.

Establish Appropriate Containment and Worker and Occupant Protection

The intention of containment and protection activities is to protect workers, building occupants, and people in the surrounding area from excessive exposure to microbial contaminants. The use of containment and negative air pressure to control the spread of contaminants has been discussed in the literature since 1989 (Light et al., 1989; Morey, 1992), many practical questions regarding when and how to use containment remain. It is known that spores may be released when contaminated materials are disturbed. Rautiala et al. (1996) report an increase of two orders of magnitude in spore concentrations during a demolition. After the demolition, concentrations returned to baseline. A later study on containment compared three methods used in two contaminated buildings (Rautiala et al., 1998). The tests used plastic film as a barrier and exhaust fans for removing air from the work area. In all cases, spore concentrations in the work zone were one to two orders of magnitude higher in the contained area than in adjacent areas. In addition to negative pressure in the work area, there are a number of possible options for containment and managing indoor-air quality during renovation. Turner (1999) describes a fungal remediation in an occupied building where negative pressure was not possible in some work areas. Two alternative methods—creating a depressurized wall cavity between the work area and the occupied space and pressurizing the occupied space—are discussed. The AIHA task force recommends the use of professional judgment and encourages innovation in isolating work areas (AIHA, 2001).

Many guidance documents link containment requirements to the area of contiguous visible mold,[3] the presence of fungal contamination in HVAC equipment, and the health status of occupants. The logic behind linking the area of contiguous visible mold to requirements for containment and worker protection is straightforward: the more mold, the greater the risk of exposure. It is easy to envision the polar extremes. Spots of mold growth on a refrigerator gasket do not require extensive containment or worker protection, but 1,000 ft^2 of mold growth on the ceiling of a ground floor because of a flood on the second floor will require containment of the work area and respiratory, eye, and skin protection for workers. However, guidance documents are unclear on the subject of mold hidden in wall or ceiling cavities and growth that is not visible to the naked eye.

The AIHA task force (2001) identifies several complicating issues (hidden mold, density of contamination, reservoirs of settled spores, and risk assessment) in deciding what levels of contamination require remediation and what levels of containment and worker protection are required. It also identifies a need for clearer guidance on these issues.

HVAC equipment is treated separately in all the guidance documents. The North American Air Duct Cleaners Association has published guidance

---

[3]The ACGIH guidance separates remediation options by level of concern but does not link them to the square footage of contamination. The ACGIH Bioaerosols Committee withdrew the numerical guidelines in 1999.

for cleaning hard-surfaced air-conveyance equipment that includes explicit containment and cleaning protocols (NADCA, 2002). Air-handling equipment is unique in that it is a contained area used to transport air throughout buildings. It also often contains processes that condense water from or intentionally add water to the air, and it may contain materials that are difficult to clean. Contaminant released inside an air-handling system is likely to be transported to other portions of the system and to other portions of the building. In hard-ducted systems, containment is relatively simple and straightforward. Hard-ducted systems convey air through ducts and fittings dedicated to the purpose rather than through building cavities formed by structural and architectural materials used as walls, ceilings, or floors. Containment and cleanup become much more complex when plenum air-return or air-supply systems—such as the plenum above a suspended ceiling—are involved: the surface areas are larger, materials with favorable nutrient content are more likely to be involved, and air leaks into building cavities and rooms outside the work area are almost certain. Those areas are enclosed by structural and architectural materials and finishes (for example, paper-covered gypsum board, roof sheathings, concrete masonry units, and ceiling tile), rather than by materials used in ductwork (such as galvanized or PVC-coated steel).

A number of papers discuss fungal contamination in HVAC systems. Batterman and Burge (1995) surveyed the literature regarding effects of HVAC characteristics on indoor-air quality, addressing microbial growth in the presence of water sources (air washers and condensate on coils and humidification systems) and poor control of indoor humidity. Dust accumulation and internal fibrous insulation were found to be contributing factors. The survey reported that few of the HVAC studies evaluating contaminant emissions have been well controlled for confounding factors. Foarde et al. (1995) found that *P. chrysogenum* could colonize samples of fiberglass duct liner, fiberglass duct board, and fiberglass insulation. Although one sample amplified the mold when exposed to high humidity alone, most of the samples required wetting or soiling. Morey (1994) presented recommendations for improving the ASHRAE 62 ventilation standard with regard to HVAC contribution to production and distribution of fungal contaminants; they included suggestions to improve moisture control in building enclosures, air handlers, and distribution systems and to ensure that materials do not provide good substrates for microbial growth.

Eliminate or Limit Moisture Sources and Dry the Materials

The relevant moisture dynamic must be identified and stopped, at least in the short term, for remediation to proceed. In some cases, such as a leak in an exposed pipe, the identification of the moisture source may be relatively easy and the intervention relatively inexpensive. At other times, however, finding the water and stopping it are more problematic. If water is condensing on the underside of a roof deck, on chilled water lines, or on the backside of gypsum board, permanent solutions may require extensive renovations. Foundations soaked by poorly managed rainwater may also require extensive modifications to the building or site. In those cases, it is best to provide short-term solutions to the wetting, using some combination of ventilation, heat, dehumidification, airflow management, and interim drainage to stop the water. It must also be remembered that drying may trigger spore release by some organisms.

Decontaminate or Remove Damaged Materials as Appropriate

Remediation should result in as much fungal contamination removal as possible. Materials that can be cleaned and are not excessively damaged can be salvaged and reused in some cases. Materials that are severely damaged by the fungi or the cleaning process are best replaced. The guidance documents all agree that smooth, hard-surface materials like ceramic tile, metals, glass, and porcelain are resistant to moisture damage and can be cleaned and salvaged. They also agree that some porous materials that are moisture-insensitive—such as concrete, masonry, stone, and fiber-cement products—may be cleaned and saved. Most agree that solid lumber, unless contaminated by wood-decaying microbes, can be cleaned and saved. Unless the contamination is minor or the value of the material great, the guidance documents recommend removing and disposing of soft porous materials that are difficult to clean or damaged by moisture.[4] Paper, books, paper-covered gypsum board, cellulose ceiling tile, and textiles are examples of these materials. Recommended cleaning methods include dry vacuuming with a HEPA vacuum cleaner, wet vacuuming, damp wiping with water or detergent mixtures, and bagging and removing.

Another factor that may be considered in selecting a containment strategy is the specific action required for removing the contamination. Gentle damp wipes or vacuuming of a hard surface with a HEPA-filtered vacuum may release only a small fraction of the contaminant being cleaned but the vibrations and effects of demolition can easily release large amounts. No studies of the comparative benefits of different cleaning and removal activities are reported in the literature.

The AIHA task force (2001) identified a number of control strategies for further research. They include blowing spores from surfaces into the air

---

[4]Conservation techniques have been developed to deal with materials—such as those with historical significance—that cannot be replaced.

where they can be removed by high-volume, HEPA-filtered fans. The need for options other than removal to manage contaminated porous materials was also identified.

The guidance documents do not specifically consider responding to mold growth in spaces such as attics, crawl spaces, and garages. It is possible that intervening in the moisture dynamic and managing the direction of airflow with fan-induced pressure differentials can be used to prevent occupant exposure at a tiny fraction of the cost of removing all the fungi from the spaces. Caution must be exercised in applying this approach.

Evaluate Whether the Space Has Been Successfully Remediated

Whenever a remediation activity is undertaken—especially when it requires containment that results in evacuation of portions or all of a building—a determination must be made that the job has been completed and that the space is suitable for reoccupation. Such determinations are necessarily subjective because there are no generally accepted health-based standards for acceptable concentrations of fungal spores, hyphae, or metabolites in the air or on surfaces (ACGIH, 1999; AIHA, 2001; Rao et al., 1996).

There is substantial variation in recommended clearance criteria between the guidance documents. The NYCDOH 2000, U.S. EPA, and AIHA guidance documents discussed above agree that the work site should be dry, clean, and free of visible mold growth. However, the NYCDOH guideline also requires air monitoring for areas that had more than 100 ft^2 of contamination. The U.S. EPA document specifically says that air samples are not required but may be useful in some circumstances. Finally, the AIHA task force guideline suggests two principal quality-assurance indicators for successful remediation: documentation that the moisture problems have been solved and physical inspection of contaminated areas to ensure that the contaminants have been removed. Air sampling and surface sampling are left as options, but the task force suggests that surface sampling for cleanliness may be more useful than biologic surface sampling. The ISIAQ guideline states that materials should be dry and surfaces cleaned until only background concentrations of mold and bacteria remain, but it does not specify how that is to be determined. The Health Canada guideline does not specifically discuss clearance but does mention that things should be left clean. And the ACGIH guideline recommends that remediated areas be clean and leaves sampling of biologics as an option. There is no specific interpretation for biologic sampling in the remediation section, but there are extensive discussions on making and interpreting biologic samples in general.

Although all the guidance documents agree that moisture problems should be fixed and materials in the building should be dry, no methods for establishing whether materials are dry are offered. Remediation failures due to regrowth of mold frequently occur, and this is of particular concern and needs to be addressed in future research. Regrowth often occurs because a faulty moisture dynamic was not mended or because a damaged area was reassembled before materials were completely dry; for example, the surface of porous materials, such as wood and concrete, may be dry while the interior remains damp.

"Clean" in the context of a clearance inspection means that the remediated area is free of residual microbial contamination. However, it is possible to ascertain that only if all potentially contaminated visible and hidden spaces have been inspected. All spaces would have to be subjected to close inspection for dust, debris, fungal contamination, and dampness. Only in this unusual case could thorough examinations and measurements be easily made. The greater the chance of hidden dampness or contamination, the more difficult it is to determine whether a remediation can be defined as successful by this criterion.

Even when visible contamination has been removed, air or surface measurements might detect mold or bacteria because fungal and other microbial material is ubiquitous. Their presence alone thus does not indicate a contamination problem, so it is difficult to set quantitative standards for evaluating when and whether a space is clean.

There is no agreement on requirements for, methods of, or interpretation of microbiologic sampling for clearance purposes. One could undertake a sampling campaign after the completion of remediation identical with that before the remediation and document whether there was a decrease in microbial contamination as a result of the remediation. Such a decrease in concentrations and microbial diversity to those of a reference building has been reported in some studies (Meklin et al., 2002). However, as discussed in Chapter 3, sampling may present an incomplete picture.

A small number of studies report decreases in symptoms experienced by occupants after remediation of moisture damage. The Savilahti et al. (2000) and Meklin et al. (2002) studies took place in Finnish schools, and both used questionnaires before and after renovation in combination with fungal sampling. In the Meklin et al. study, a comparison was made with a control building. A third study (Jarvis and Morey, 2001) looked at a new building in a hot, humid climate. Biologic sampling and questionnaires were used before and after remediation. The study found that the occurrence of illness was reduced after remediation was completed.

Reassemble the Space to Prevent or Limit the Possibility of Recurrence by Controlling Sources of Moisture and Nutrients

When portions of the building are reassembled after remediation, they must be modified so that the chance of recurring moisture damage and fungal growth is reduced. That may require

- Adding rainwater drainage elements.
- Back-venting for cladding.
- Elimination of intentional or unintentional water-vapor retarders.
- Air sealing and changes in air-handling equipment or operation to manage air-pressure relationships.
- Improvements in the dehumidification unit of the air-conditioning equipment.
- Removal of humidification equipment or controls of humidification or process-water systems.
- Replacement of materials that offer superior nutrient and substrate for fungal growth with materials that are resistant to microbial growth (ceramics, concrete products, stainless steel, and the like).
- Encapsulation of surfaces that have been dried and substantially decontaminated but cannot be completely decontaminated (for example, between floor joists and subfloors).

There is very little guidance for planning, installing, and determining acceptability of the renovation in the guidance documents. The ISIAQ and ACGIH documents provide the best discussion of these issues, but they are limited in scope.

EFFECTS OF AIR AND SURFACE CLEANING AND VENTILATION

Ventilation, air cleaning, and surface cleaning can influence exposure. Airborne spores can be removed from a building with the out-going ventilation airflow or trapped in a particle filter and thus removed from the air. Spores can also be removed from surfaces by washing or vacuuming.

Model predictions indicate that normal variations in house ventilation rates when windows are closed will have only a moderate influence on indoor airborne concentrations of fungal spores 2–10 μm in aerodynamic diameter (IOM, 2000; Chapter 10). For example, an increase of a factor of 8 in the ventilation rate from 0.25 to 2 air changes per hour would be expected to reduce airborne concentrations of 5-μm-diameter spores by 60%. The decrease in concentration of 2-μm spores would be larger (~70%); the decrease in 10 μm spores smaller (40%). In most buildings, the practical increase in ventilation rates would be considerably smaller than a factor of eight, with correspondingly smaller decreases in airborne spore concentrations.

When air enters a building through small cracks and holes, only a fraction (characterized by the penetration factor) of the particles will penetrate to the indoor air, and the remainder will deposit on the surfaces of the leaks. Considerable data indicate that the penetration factor for PM 2.5 (particle matter less than 2.5 µm) is close to 1.0 (Thatcher et al., 2001). However, the penetration factor for particles of 2–10 µm in air leaking into buildings is not well understood and varies with the size of holes through which the air leaks. In studies of particle penetration through simulated 0.5-mm-wide cracks, the penetration factor was less than 0.1 (Mosley et al., 2001). Particle penetration factors for the cracks in commercial windows were 0.6–1.0 for 5-µm particles and 0.6–0.8 for 8-µm particles (Liu and Nazaroff, 2002). Thus, the natural particle losses that occur during air infiltration provide a substantial but still uncertain amount of protection from outdoor fungal spores.

Opening windows can cause large increases in ventilation rates, depending on the weather and how often and how long the windows remain open. That ventilation will reduce exposures to indoor-generated spores. However, large increases in indoor concentrations of spores from the outdoors may occur. The rate of flow of spore-laden outdoor air into a house will increase dramatically with open windows, and few spores will be lost by deposition on surfaces (such as window sills) as the air passes through a window.

Predicted reductions in indoor airborne concentrations of spore-size particles by filtering were discussed in the 2000 Institute of Medicine report *Clearing the Air*. Reductions in spore concentrations by recirculation of air through filters in household furnace and air-conditioning systems—including filters with a much better efficiency than the common see-through furnace filter—will normally be less than 50% for 5-µm-diameter spores. Portable fan filter units can reduce spore concentrations more, but only with high rates of airflow through the filtration unit (10 room volumes per hour). Few measurement data are available for evaluating those model predictions. It seems likely, however, that normal variations in ventilation rates and filtration in buildings with closed windows will have a moderate effect on inhalation exposure to mold spores.

Surface cleaning, such as vacuuming, can remove spores, potentially preventing their resuspension and inhalation and reducing the probability of exposure dermal contact and incidental ingestion. A number of studies have been performed on surface cleaning to evaluate the reduction in total dust or lead on surfaces, fewer on the removal of dust-mite allergens, and fewer still on the removal of fungal matter. However, some data imply, but do not clearly demonstrate, that improved surface cleaning could reduce exposure to fungi. Cole et al. (1996) found that concentrations of fungi and bacteria in air correlate with concentrations of fungi and bacteria on indoor nonfloor surfaces (r = 0.6 for fungi) and correlate with concentrations on floor surfaces (no statistic provided). Intervention studies have demonstrated that improved surface cleaning can reduce the loading of uncharacterized dust, lead, and mite allergen on surfaces. Reductions of 80–90% in dust, lead, and mite allergen on surfaces were achieved by Roberts et al. (1999) after vacuuming carpets for 6–45 min/m² of carpet surface. Kildesø et al. (1998) compared nine cleaning practices and found that the quantity of dust remaining on floor surfaces where people walked varied by a factor of 2. In a cross-sectional study of schools that improved cleaning practices for floors, classrooms cleaned primarily with wet mopping had more airborne viable bacteria but less settled dust than classrooms cleaned primarily with dry methods (Smedje and Norbäck, 2001). Franke et al. (1994) reported a substantial reduction in fungal spores on surfaces after a period of deep cleaning; however, the reduction was temporary, and the benefits of the cleaning were not easily distinguishable from natural variation.

A few studies have also found that surface cleaning practices or frequency can influence airborne concentrations of particles or microorganisms. In a conference paper, Skyberg et al. (1999) described a study comparing 49 offices that received improved cleaning with 55 control offices that received superficial cleaning. The concentration of inhalable dust decreased by about one-third in the intervention offices and increased slightly in control offices (significance level not reported). In another conference paper, White and Dingle (2002) found that airborne PM 2.5 and PM 10 concentrations were decreased by about 50% (p < 0.01) after 14 weeks of intensive[5] vacuum cleaning of 19 houses, but airborne particle concentrations were not significantly changed in 17 control houses. Kemp et al. (1998) reported an 85% reduction (p < 0.04) in respirable suspended particles on two floors of an office building after improved surface cleaning (9% reduction on control floors) but initial particle concentrations were unusually high. In a cross-sectional study of classrooms, Smedje and Norbäck (2001) found that cleaning practices were associated with concentrations of airborne viable bacteria (p = 0.013 in a multivariate regression); however, no association of cleaning practices with airborne fungal concentrations was reported.

Finally, a few studies (Kemp et al., 1998; Skyberg et al., 1999; Wålinder et al., 1999) have reported significant improvements in subjective or objective health measures with improved surface cleaning or lower concentra-

---

[5]Carpets were cleaned every 2 weeks for 4 min/m² in the first cleaning, 2 min/m² in the second cleaning, and 1 min/m² in five additional cleanings. Upholstered sofas and beds were cleaned every 2 weeks for 1 min/m².

tions of dust on surfaces, which is indirect evidence of reductions in exposure to unidentified agents.

In summary, the normal variations in ventilation rates, air-filtration rates, and surface-cleaning practices of homes may under some circumstances substantially affect exposure to fungal spores and other dampness-related microbial bioaerosols. Improved surface cleaning appears to have the largest and most practical potential for bringing about large reductions in exposure; however, further research is needed to characterize its effectiveness.

FINDINGS, RECOMMENDATIONS, AND RESEARCH NEEDS

On the basis of its review of the papers, reports, and other information presented in this chapter, the committee has reached the following findings and recommendations and has identified the following research needs regarding the prevention of moisture problems and the remediation of buildings that have water damage or microbial contamination.

Findings

• The most effective way to manage a biological agent, such as mold, in a building is to eliminate or limit the conditions that foster its establishment and growth.

• There are several sources of guidance on how to respond to various indoor microbial contamination situations. However, determining when a remediation effort is warranted or when it is successful is necessarily subjective because there are no generally accepted health-based standards for acceptable concentrations of fungal spores, hyphae, or metabolites in the air or on surfaces.

• Remediation must identify and eliminate the underlying cause of dampness or water accumulation. If the underlying causes are not addressed, contamination may recur.

• Valuable information can be acquired from architects, builders, occupants, and maintenance staffs regarding health complaints, the use history of the building, moisture events, and locations of problems. Both expert assessment of the building's condition and knowledge of its history and current problems are needed to make a thorough evaluation of potential dampness-related exposures and an effective plan for remediation.

• Fungal and other microbial material is present on nearly all indoor surfaces. There is a great deal of uncertainty and variability in samples taken from indoor air and surfaces, and it may be difficult to discern which organisms are part of the natural background and which are the result of problematic contamination. However, the information gained from a careful and complete survey may aid in the evaluation of contamination sources and remediation needs.

• The potential for exposure to microbial contaminants in spaces such as attics, crawl spaces, exterior sheathing, and garages is poorly understood.

• Disturbance of contaminated material during remediation activities can release microbial particles and result in contamination of clean areas and exposure of occupants and remediation workers.

• Containment has been shown to prevent the spread of molds, bacteria, and related microbial particles to noncontaminated parts of a contaminated building. The amount of containment and worker personal protection and the determination of whether occupant evacuation is appropriate depend on the magnitude of contamination.

• Very few controlled studies have been conducted on the effectiveness of remediation actions in eliminating problematic microbial contamination in the short and long term and on the effect of remediation actions on the health of building occupants.

• Available literature addresses the management of microbial contamination when remediation is technically and economically feasible. There is no literature addressing situations where intervening in the moisture dynamic or cleaning or removing contaminated materials is not practicable.

Recommendations

• Homes and other buildings should be designed, operated, and maintained to prevent water intrusion and excessive moisture accumulation when possible. When water intrusion or moisture accumulation is discovered, the sources should be identified and eliminated as soon as practicable to reduce the possibility of problematic microbial growth and building-material degradation.

• When microbial contamination is found, it should be eliminated by means that limit the possibility of recurrence and limit exposure of occupants and persons conducting the remediation.

Research Needs

• Research is needed to characterize

— The effectiveness of remediation assessment and remediation methods in different contamination circumstances.

— The dynamics of movements of contaminants from colonies of mold and other microorganisms in spaces such as attics, crawl spaces, exterior sheathing, and garages.

— The effectiveness of various means of protection of workers and occupants during remediation activities

- Research should be performed to develop

— Methods for finding microbial contamination in HVAC systems, and in crawl spaces, attics, wall cavities, and other hidden or seldom-accessed areas.

— Building materials that, when moist, are less prone to microbial contamination.

— Standard methods of assessing the potential of new materials, designs, and construction practices for dampness problems.

— Standardized, effective protocols for cleaning up after flooding and other catastrophic water events that will minimize microbial growth.

— Methods that can distinguish between naturally-deposited spores and active microbial growth in wall cavities.

- Research should be performed to determine

— How free of microbial contamination a surface or building material must be to eliminate problematic exposure of occupants (in particular, how concentrations of microbial contamination left after remediation are related to those found on ordinary surfaces and materials in buildings where no problematic contamination is present).

— Whether and when microbial contamination that is not visible to the naked eye but is detectable through screening methods should be remediated.

— The risk of microbial contamination in the building but outside the general air circulation of the building—in crawl spaces, attics, wall cavities, building sheathing, and the like.

— The effectiveness of managing contamination in place by using negative air pressure, encapsulation, and other means of isolation.

— The best ways to address microbial contamination in situations where remediation is not technically or economically feasible.

— The best ways to open a wall or other building cavity to seek hidden contamination while controlling the release of spores, microbial fragments, and the like.

— How to measure the effectiveness and health effects of a remediation effort.

REFERENCES

ACGIH (American Conference of Governmental Industrial Hygienists). 1989. Guidelines for the Assessment of Bioaerosols in the Indoor Environment. Cincinnati, OH.
ACGIH. 1999. Bioaerosols—Assessment and Control. Cincinnati, OH.

AEC (Advanced Energy Corporation). 2002. A Field Study Comparison of the Energy and Moisture Performance Characteristics of Ventilated Versus Sealed Crawl Spaces in the South. Final Report. Raleigh, NC.

AIHA (American Industrial Hygiene Association). 1996. Field Guide for the Determination of Biological Contaminants in Environmental Samples. Fairfax, VA: AIHA Press.

AIHA. 2001. Report of the Microbial Task Force. Fairfax, VA: AIHA Press.

Andersson K. 1998. Epidemiological approach to indoor air problems. Indoor Air (Supplement 4):32–39.

ARC and FEMA (American Red Cross and Federal Emergency Management Agency). 1992. Repairing Your Flooded Home. ARC publication 4476; FEMA publication L-198, August 1992 Washington, DC. http://www.redcross.org/services/disaster/afterdis/repfhm.pdf.

ASTM (American Society of Testing and Materials). 1998. ASTM F-1869-98 Standard Test Method for Measuring Moisture Vapor Emission Rate of Concrete Subfloor Using Anhydrous Calcium Chloride. West Conshohocken, PA.

ASTM. 2000. ASTM E1105-00 Standard Test Method for Field Determination of Water Penetration of Installed Exterior Windows, Skylights, Doors, and Curtain Walls by Uniform or Cyclic Static Air Pressure Difference. West Conshohocken, PA.

ASTM. 2002a. ASTM E514-02 Standard Test Method for Water Penetration and Leakage through Masonry. West Conshohocken, PA.

ASTM. 2002b. ASTM F-2170-02 Standard Test Method for Determining Relative Humidity in Concrete Floor Slabs Using in situ Probes. West Conshohocken, PA.

Batterman SA, Burge H. 1995. HVAC Systems as Sources Affecting Indoor Air Quality: A Critical Review. HVAC&R Research 1(1):61–78.

Bjurman J. 1999. Fungal and microbial activity in external wooden panels as determined by finish, exposure, and construction techniques. International Biodeterioration and Biodegradation 43:1–5.

Buttner MP, Stetzenbach LD. 1993. Monitoring Airborne Fungal Spores in an Experimental Indoor Environment to Evaluate Sampling Methods and the Effects of Human Activity on Air Sampling. Applied and Environmental Microbiology 59(5):219–226.

CMHC (Canadian Mortgage and Housing Corporation). 1993. Clean-up Procedures for Mold in Houses. Ottawa, Canada.

Cole EC, Dulaney PD, Leese KE, Hall RM, Foarde KK, Franke DL, Myers EM, Berry MA. 1996. Biopollutant sampling and analysis of indoor surface dusts: characterization of potential sources and sinks. In Characterizing Sources of Indoor Air Pollution and Related Sink Effects, ASTM STP 1287:153–165. Bruce A Tichenor, ed. West Conshohocken, PA: American Society for Testing and Materials.

Dharmage S, Bailey M, Raven J, Mitakakis T, Thien F, Forbes A, Guest D, Abramson M, Walters EH. 1999. Prevalence and residential determinants of fungi within homes in Melbourne, Australia. Clinical and Experimental Allergy 29(11):1481–1489.

Dill I, Niggeman JD. 1996. Domestic fungal viable propagules and sensitization in children with IgE mediated allergic disease. Pediatric Allergy and Immunology 7(3):151–155.

Doll SC. 2002. Determination of Limiting Conditions for Fungal Growth in the Built Environment. Doctoral Dissertation, Harvard School of Public Health, Boston, MA.

Endean KF. 1995. Investigating Rainwater Penetration of Modern Buildings. Gower, Aldershot, England.

Engvall K, Norrby C, Sandstedt E. 2004. The Stockholm Indoor Environment Questionnaire: a sociologically based tool for the assessment of indoor environment and health in dwellings. Indoor Air 14(1):24–33.

Flannigan B, Morey P, eds. 1996. Control of moisture problems affecting biological indoor air quality. ISIAQ Guideline TFI-1996. International Society of Indoor Air Quality and Climate, Milan, Italy.

Foarde KK, VanOsdell DW, Chang JCS. 1995. Evaluation of fungal growth on fiberglass duct materials for various moisture, soil, use and temperature conditions. Indoor Air 6(2): 83–92.

Franke DL, Cole EC, Berry MA. 1994. Deep cleaning—a first step to improved environmental quality. Report 94-TP48.06, Proceedings of the Air and Waste Management Association, 87th Annual Meeting and Exhibition, Cincinnati, OH. Pittsburgh, PA: AWMA.

Garrett MH, Rayment PR, Hooper MA, Abramson MJ, Hooper BM. 1998. Indoor airborne fungal spores, house dampness and associations with environmental factors and respiratory health in children. Clinical and Experimental Allergy 28(4):459–467.

Health Canada. 1995. Fungal Contamination in Buildings: A Guide to Recognition and Management. Ottawa, Ontario; Health Canada.

Horner W, Morey PR, Ligman BK. 2001. How Quickly Must Gypsum Board and Ceiling Tile Be Dried to Preclude Mold Growth After a Water Accident? Proceedings IAQ 2001. American Society of Heating Refrigeration and Air Conditioning Engineers, Atlanta, GA.

Hunter CA, Grant C, Flannigan B, Bravery AF. 1988. Mould in buildings: the air spora of domestic dwellings. International Biodeterioration 24:81–101.

Hyvärinen A. 2002. Characterizing moisture damaged buildings—environmental and biological monitoring. Academic Dissertation. Department of Environmental Sciences, University of Kuopio, Kuopio, Finland, and the National Public Health Institute.

Hyvärinen A, Reponen T, Husman T, Ruuskanen J, Nevalainen A. 1993. Characterizing Mold Problem Buildings—Concentrations and Flora of Viable Fungi. Indoor Air 3: 337–343.

Hyvärinen A, Reponen T, Husman T, Nevalainen A. 2001a. Comparison of the indoor air quality in mould damaged and reference buildings in a subarctic climate. Central European Journal of Public Health 9(3):133–139.

Hyvärinen A, Vahteristo M, Meklin T, Jantunen A, Nevalainen A. 2001b. Temporal and spatial variation of fungal concentrations in indoor air. Aerosol Science and Technology 35:688–695.

IICRC (Institute of Inspection, Cleaning and Restoration Certification). 1999. IICRC S500: Standard and Reference Guide for Professional Water Damage Restoration. Vancouver, WA: IICRC.

IICRC. 2003. IICRC S520: Standard and Reference Guide for Professional Remediation. Rockville, MD: Indoor Air Quality Association.

IOM (Institute of Medicine). 2000. Clearing the Air: Asthma and Indoor Air Exposures. Washington, DC: National Academy Press.

Jarvis JQ, Morey PR. 2001. Allergic respiratory disease and fungal remediation in a building in a subtropical climate. Applied Occupational and Environmental Hygiene 16(3):380–388.

Johanning E, Lansbergis P, Gareis M, Yang CS, Olmstead E. 1999. Clinical experience and results of a sentinel health investigation related to indoor fungal exposure. Environmental Health Perspectives 107(3):489–494.

Kemp PC, Dingle P, Neumeister HG. 1998. Particulate-matter intervention study: a causal factor of building related symptoms in an older building. Indoor Air 8:153–171.

Kildesø J, Tornvig L, Skov P, Schneider T. 1998. An intervention study of the effect of improved cleaning methods on the concentration and composition of dust. Indoor Air 8:12–22.

Klánová K. 2000. The Concentrations of Mixed Populations of Fungi in Indoor Air: Rooms with and without Mould Problems; Rooms with and without Health Complaints. Central European Journal of Public Health 8(1):59–61.

Kozak PP, Gallup J, Cummings LH, Gillman SA. 1980. In: Currently Available Methods for Home Surveys. Annals of Allergy 45:167.

Lawton MD, Dales RE, White J. 1998. The Influence of House Characteristics in a Canadian Community on Microbiological Contamination. Indoor Air 8:2–11.

Light EN, Coco JA, Bennett AC, Long KL. 1989. Abatement of *Aspergillus niger* contamination in a library. Proceedings of IAQ 1989. Atlanta, GA: ASHRAE.

Liu DL, Nazaroff WW. 2002. Particle penetration through windows. Proceedings of Indoor Air 8:862–867.

Meklin T, Husman T, Pekkanen J, Hyvärinen A, Hirvonen M-R, Nevalainen A. 2002. Effects of moisture damage repair on microbial exposure and health effects in schools. Proceedings of Indoor Air 2002—The 9th International Conference on Indoor Air Quality and Climate, Monterey, CA.

Miller JD, Haisley PD, Reinhardt JH. 2000. Air sampling results in relation to extent of fungal colonization of building materials in some water-damaged buildings. Indoor Air 10(3):146–151.

Morey PR. 1992. Microbiological Contamination in Buildings: Precautions Against Remediation. Proceedings of IAQ92. Atlanta, GA: ASHRAE.

Morey PR. 1993. Microbial events after a fire in a high-rise building. Indoor Air 3:354–360.

Morey PR. 1994. Suggested Guidance on Prevention of Microbial Contamination for the Next Revision of ASHRAE Standard 62. Proceeding for IAQ94. Atlanta, GA: American Society of Heating Air Conditioning and Refrigeration Engineers.

Morey PR, Hodgson MJ, Sorenson WG, Kullman GJ, Rhodes WW. 1984. Environmental studies in moldy office buildings: biological agents, sources and preventative measures. Annals of the American Conference of Governmental Industrial Hygienists 10:21–34.

Morey PR, Ligman B, Jarvis J. 2002. 2C2p1 Hidden mold sometimes enters the indoor air. International Society of Indoor Air Quality and Climate. Proceedings from IAQ 2002, Monterey, CA.

Mosley RB, Greenwell DJ, Sparks LE, Guo Z, Tucker WG, Fortmann R, Whitfield C. 2001. Penetration of ambient fine particles into the indoor environment. Aerosol Science and Technology 34:127–136.

NADCA (North American Air Duct Cleaners Association). 2002. ACR2002 Assessment, Cleaning and Restoration of HVAC systems. Washington, DC.

NAS (National Academy of Sciences). 2000. Clearing the air: Asthma and indoor air exposure. Institute of Medicine. Washington, DC: National Academy Press.

Nevalainen A, Pasanen A-L, Niininen M, Reponen T, Kalliokoski P. 1991. The indoor air quality in Finnish homes with mold problems. Environment International 17:299–302.

NYCDOH (New York City Department of Health). 1993. Assessment and Remediation of *Stachybotrys Atra* in Indoor Environments, New York City Department of Health.

NYCDOH. 2000. Guidelines on Assessment and Remediation of Fungi in Indoor Environments. New York City Department of Health.

Pasanen AL, Kalliokoski P, Pasanen P, Salmi T, Tossavainen A. 1989. Fungi carried from farmers' work into farm homes. American Industrial Hygiene Association Journal 50(12):631–633.

Pasanen AL, Niininen M, Kalliokoski P, Nevalainen A, Jantunen MJ. 1992. Airborne *cladosporium* and other fungi in damp versus reference residences. Atmospheric Environment 26B(1):121–124.

Pessi A-M, Suonketo J, Pentti M, Kurkilahti M, Peltola K, Rantio-Lehtimäki A. 2002. Microbial growth inside external walls as an indoor air biocontamination source. Applied and Environmental Microbiology 68(2):963–967.

Rao C, Burge H, Chang JCS. 1996. Review of Quantitative Standards and Guidelines for Fungi in Indoor Air. Journal of the Air and Waste Management Association 46:899–908.

Rautiala S, Reponen T, Hyvärinen A, Nevalainen A, Husman T, Vehvilainen A, Kalliokoski P. 1996. Exposure to airborne microbes during the repair of moldy buildings. American Industrial Hygiene Association Journal 57:279–284.

Rautiala S, Reponen T, Nevalainen A, Husman T, Kalliokoski P. 1998. Control of exposure to airborne viable microorganisms during remediation of moldy buildings; report of three case studies. American Industrial Hygiene Association Journal 59(7):455–460.

Reeslev M, Miller M, Nielsen KF. 2003. Quantifying mold biomass on gypsum board: comparison of ergosterol and beta-N-acetylhexosaminidase as mold biomass parameters. Applied and Environmental Microbiology 69(7):3996–3998.

Roberts JW, Clifford WS, Glass G, Hummer PG. 1999. Reducing dust, lead, dust mites, bacteria, and fungi in carpets by vacuuming. Archives of Environmental Contamination and Toxicology 36:477–484.

Savilahti R, Uitti J, Laippala P, Husman T, Roto P. 2000. Respiratory morbidity among children following renovation of a water-damaged school. Archives of Environmental Health 55(6):405–410.

Skyberg K, Skulberg KR, Kruse K, Huser PO, Djupesland P. 1999. Dust reduction relieves nasal congestion—a controlled intervention study on the effect of office cleaning, using acoustic rhinometry. Proceedings of Indoor Air 1:153–154. London: Construction Research Communications, Ltd.

Smedje G, Norbäck D. 2001. Irritants and allergens at school in relation to furnishings and cleaning. Indoor Air 11(2):127–133.

Strachan DP, Flannigan B, McCabe EM, McGarry F. 1990. Quantification of airborne moulds in homes of children with and without wheeze. Thorax 45:382–387.

Swenson L, Geer W, Krause M, Robbins C. 2003. Tape-Lift Survey of Settled Dust in Non-Water-Damaged Homes. AIHA (American Industrial Hygiene Association) 2003 Poster papers.

Thatcher TL, McKone TE, Fisk WJ, Sohn MD, Delp WW, Riley WJ, Sextro RG. 2001. Factors affecting the concentration of outdoor particles indoors (COPI): identification of data needs and existing data. LBNL-49321. Lawrence Berkeley National Laboratory Report, Berkeley, CA.

Turner W. 1999. Avoiding problems during renovation. In: A Practical Guide to Ventilation Practices & Systems for Existing Buildings. Supplement to the April issue of Contracting Business and the May issue of Heating/Piping/Air Conditioning Engineering. pp. 16–19.

USDA (United States Department of Agriculture). 1980. How to Prevent and Remove Mildew. Home and Garden Bulletin 68, Washington, DC.

U.S. EPA (United States Environmental Protection Agency). 2001. Mold Remediation in Schools and Commercial Buildings. Washington, DC.

Wålinder R, Norbäck D, Wieslander G, Smedje G, Erwall C, Venge P. 1999. Nasal patency and lavage biomarkers in relation to settled dust and cleaning routines in schools. Scandinavian Journal of Work, Environment and Health 25(2):137–143.

White K, Dingle P. 2002. The effect of intensive vacuuming on indoor PM mass concentration. Proceedings of the 9th International Conference on Indoor Air Quality and Climate, Monterey, CA. 3:92–97.

7

The Public Health Response

Studies reviewed in this report indicate that:

- Dampness is prevalent in residential housing in a wide array of climates (Chapter 2);
- Sufficient evidence of an association exists between signs of dampness and upper respiratory tract symptoms, cough, wheeze, and asthma symptoms in sensitized persons (Chapter 5);
- Sufficient evidence of an association exists between signs of mold and upper respiratory tract symptoms, cough, wheeze, asthma symptoms in sensitized persons and hypersensitivity pneumonitis in susceptible persons (Chapter 5).

The committee concludes, on the basis of this information and other findings presented in Chapters 2 through 6, that excessive indoor dampness is a public health problem.

This chapter draws together findings and recommendations presented in earlier parts of the report and places them in the context of the mission of public health to "[fulfill] society's interest in assuring conditions in which people can be healthy" and its aim "to generate organized community effort to address the public interest in health by applying scientific and technical knowledge to prevent disease and promote health" (IOM, 1988). It addresses the public health interest in housing, barriers to the adoption of dampness prevention and reduction measures, and public health approaches to addressing the problems created by damp indoor environments. The chapter also operationalizes some of the findings, recommendations and research needs presented in earlier chapters by suggesting specific actions and actors to implement them.

PUBLIC HEALTH AND HOUSING

The influence of housing and the workplace on human health has long been an element of public-health action and research (Susser, 1973). With the industrialization of the United States, responsibility for workplace, housing, and public-health improvement efforts was clearly distinguished at the local and national levels (Duffy, 1974; Galishoff, 1988; Melosi, 2000; Rosner, 1995; Veiller, 1921). The sanitary-reform movement during the 1800s sought locally and federally to rectify the hazardous effects of overcrowding, insufficient light and air, impure water, and the like associated with inadequate housing. Similarly, the mechanisms through which housing promotes health or disease became the focus of considerable research related to public health (Dedman et al., 2001; Dunn and Hayes, 2000; Matte and Jacobs, 2000). The societal obligation to ensure safe and healthy housing and workplaces has long been evident in building codes and zoning policies, which, according to Freeman (2002), are

> prima facie evidence that America has deemed a certain standard of housing a basic requirement of civilized society. If this were not so, we would allow the poor and homeless to build shantytowns, as is done in many cities of the Third World.

The environments in which we live and work, then, are widely accepted as determinants of our health; and, as this report delineates, damp buildings may pose a health risk.

Research also indicates that damp buildings have economic implications. Nguyen and colleagues (1998) examined direct costs (in inpatient care, drugs, physicians, and nursing and clinic services) and indirect costs (in lost work days and duration of disability) of asthma and other respiratory diseases associated with the presence of moisture and mold in residences in Finland. The authors estimated the cost of asthma associated with moisture in buildings in 1996 at 137.5 million Finnish markkaa, which translates to about U.S.$6.06 per person.[1] The corresponding per capita cost associated with mold in buildings was about U.S.$3.20. Such estimates are necessarily based on simplifying assumptions and are subject to substantial uncertainty, but they highlight the fact that the effect of indoor dampness on health has an economic dimension and that cost-effectiveness

---

[1]The calculation is based on a July 1996 population estimate of 5,105,230 and a January 1996 exchange rate of 4.4425 markkaa per U.S. dollar (CIA, 1997).

studies are needed to assess the savings that damp-building prevention or remediation might generate by reducing morbidity.

As noted in Chapter 2, dampness is more likely to be found in buildings that are older, lack central heating, are poorly insulated (and hence subject to cold and damp conditions), and are overcrowded. Buildings with those characteristics are most evident in low-income communities (Krieger and Higgins, 2002).

BARRIERS TO THE ADOPTION OF DAMPNESS PREVENTION AND REDUCTION MEASURES

Chapter 6 identifies many technical measures and practices that could prevent or reduce problematic indoor dampness. However, social and institutional barriers hinder their widespread adoption.

One important barrier is poverty. Historically, the distribution of poor housing stock in the United States—as in many other nations—has been largely associated with socioeconomic status (Evans and Kantrowitz, 2002) and ethnicity (Lawrence and Martin, 2001). Ostensibly, the population at greatest risk of exposure to dampness-related health problems in the United States is the population that is the most poorly housed. Census statistics indicate that the poor are more than three times as likely (22% versus 7%) to have substandard-quality housing (Evans and Kantrowitz, 2002) and that blacks and low-income people are more likely than the general population to be in housing with severe physical problems (Krieger and Higgins, 2002).

Given the costs of maintaining a clean, dry, well-heated, and properly-ventilated home, it should not be surprising that low-income families are more likely to have substandard housing and to live in the kind of damp interiors that may be associated with health problems. Children in such families may bear an additional burden because they are more likely to be in school buildings that have environmental problems: poor plumbing, inadequate heating, and poor indoor air quality (Evans and Kantrowitz, 2002). Reviewing data accumulated in the second National Health and Nutrition Examination Survey (NHANES II) and the Harvard Six Cities Study, Eggleston (2000) concluded that ethnicity, poverty, and residence combined to influence asthma prevalence in inner-city children in ways that could not be easily disentangled.

Economic factors may encourage poor building practices. Combinations of pressure to build quickly and cheaply can result in poorly constructed buildings that are more likely to have water leaks. Under ordinary circumstances, the market works to sift out builders that produce shoddy construction. However, in low-income neighborhoods—where options are limited because there is a shortage of affordable housing—and in other circumstances in which demand outstrips supply, the market may not penalize poor workmanship effectively (Parker, 1994).

Poverty combined with the lack of affordable housing may also create incentives to forgo or limit investment in maintenance that might help to prevent moisture problems. Landlords have little incentive to spend money on repair when there is a surplus of people ready to accept any kind of low-rent shelter (Ehrenreich, 2001). As already noted, those pressures also result in overcrowding, which can lead to excessive indoor moisture and condensation problems (Markus, 1993), which in turn promote mold and bacterial growth.

There are also other barriers to the implementation of dampness prevention or reduction measures. The 2000 Institute of Medicine (IOM) report *Clearing the Air: Asthma and Indoor Air Exposures* noted that "the relevant features of building design, operation, and maintenance may be determined substantially by speculative builders or other decision-makers who are substantially unaffected by future moisture problems" (IOM, 2000). Such market imperfections isolate decision-makers from the consequences of their choices and may thus lead to socially undesirable outcomes. Insufficient awareness and training about dampness, its prevention, and its consequences and the lack of a clear definition of the environmental conditions and practices that are harmful to health also hinder actions. The following pages discuss those impediments and means of addressing them.

PUBLIC HEALTH APPROACHES TO
DAMP INDOOR ENVIRONMENTS

If excessive indoor dampness is a public health problem, then an appropriate public health goal should be to prevent or reduce the incidence of potentially problematic damp indoor environments, that is, environments that may be associated with undesirable health effects, particularly in vulnerable populations. However, there are serious challenges associated with achieving that goal. As the literature reviewed in this report indicates, there is insufficient information on which to base quantitative recommendations for either the appropriate level of dampness reduction or the "safe" level of exposure to dampness-related agents. The relationship between dampness or particular dampness-related agents and health effects is sometimes unclear and in many cases indirect. Questions of exposure and dose have not, by and large, been resolved (see Chapters 3 and 4). An additional challenge is posed by the fact that it is not possible to objectively rank dampness-related health problems within the larger context of threats to the public's health. As the report notes, there is insufficient information available to confidently quantify the overall magnitude of the risk resulting from exposures in damp indoor environments. Even if those data were available, the choice of metric used for ranking—the number of illnesses or the cost of lost work and school days, for example—would greatly influence the result.

What then can be done through public health mechanisms to prevent or reduce the incidence of damp indoor environments? At least seven areas of endeavor deserve discussion with relation to the agents and exposures examined in this report:[2]

- Assessment and monitoring of indoor environments at risk for problematic dampness.
- Modification of regulations, building codes, and building-related contracts to promote healthy indoor environments; and enforcement of existing rules.
- Creation of incentives to construct and maintain healthy indoor environments.
- Development, dissemination, and implementation of guidelines for the prevention of dampness-related problems.
- Public-health-oriented research and demonstration projects to evaluate the short-term and long-term effectiveness of intervention strategies.
- Education and training of building occupants, health professionals, and people involved in the design, construction, management, and maintenance of buildings to improve efforts to avoid or reduce dampness and dampness-related health risks.
- Collaborations among stakeholders to achieve healthier indoor environments.

The separate areas are discussed in greater detail below.

Assessment and Monitoring

Although poor housing and health problems are especially evident in low-income populations, the problems created by indoor dampness are by no means limited to the homes of poor people, and they extend beyond homes to workplaces, schools, and other commercial and public buildings. If interventions are to be effective, an accepted first step is to establish mechanisms for identifying existing or potential problems and their determinants. At least one local health jurisdiction in the United States has

---

[2]A more general examination of housing and health is outside the scope of this report. However, house dust mites, respiratory viruses, and cockroaches—all of which are associated with damp conditions—may have an important effect on occupant health (IOM, 1993, 2000). The literature also addresses how temperature acts as a contributory and independent risk factor for increased morbidity and mortality in damp indoor environments, especially among the elderly and poor (Evans et al., 2000; Healy, 2003; Mercer, 2003).

systematically collected and analyzed data on the adequacy of housing (Marsh, 1982), but there appear to be few, if any, current population-based, systematic efforts to identify and anticipate trouble spots (Krieger and Higgins, 2002). Some general population surveys that include collection of data on dampness and specific dampness-related exposures have been conducted.

Comprehensive data on the prevalence of dampness in residences are collected as part of the biennial American Housing Survey (U.S. Census Bureau, various years), which gathers information on the prevalence of water leaks with outdoor and indoor sources. The number of households surveyed has varied between roughly 88,000 and 106,000 over the years 1985–2001. Data indicate that over that time span there was little difference between the overall incidence of outdoor leaks and the incidence in homes occupied by blacks, the elderly (65 years old or older) or those below the poverty level. Leaks with indoor sources were more likely in homes occupied by blacks and less likely in the homes of the elderly than in the average home. It should be remembered that the data do not record how able the occupants were to respond to instances of leakage—a major factor in whether a particular incident becomes problematic.

As noted in Chapter 2, a primary challenge in formulating a public-health strategy in response to indoor dampness is that there is no generally accepted definition of dampness or what constitutes a "dampness problem" and no generally-accepted metric for characterizing dampness. Studies of specific populations conducted as part of research projects (summarized in Table 2-1) show a wide range of estimates of the number of homes with dampness problems or water damage. In recognition of that diversity, the committee recommends that precise, agreed-on definitions of "dampness" be developed to facilitate greater uniformity (and thus comparability) in the data collected by researchers and the actions taken by those involved in prevention and remediation. That will permit important information to be gathered about mechanisms by which dampness and dampness-related effects and exposures affect occupant health.

Efforts to survey the prevalence of indoor allergens and other environmental health risk factors comprehensively have also been carried out. The National Survey of Lead and Allergens in Housing (NSLAH) was designed to assess the potential exposure of children to a variety of agents, including bacterial endotoxins and allergens of the fungus *Alternaria alternata* (Vojta et al., 2002). A companion study—the First National Environmental Health Survey of Child Care Centers—used methods similar to those of NSLAH to evaluate exposures in licensed day-care centers (Viet et al., 2003). Both efforts collected qualitative information on the presence of dampness and mold. Publications describing the findings of those studies were in preparation or in press when the present report was completed.

The committee did not identify any current surveillance of mold or other dampness-related exposures in homes.[3] A Connecticut law, however, establishes requirements for assessment and monitoring efforts in the state's public schools: Public Act 03-220, "An Act Concerning Indoor Air Quality in Schools," requires boards of education to provide for a uniform program of inspection and evaluation of indoor air quality in schools, citing the Environmental Protection Agency (EPA) Indoor Air Quality Tools for Schools Program (U.S. EPA, 2000) as an example. The program includes review, inspection, or evaluation of the "potential for exposure to microbiological airborne particles, including, but not limited to, fungi, mold and bacteria"; moisture incursion; and a number of other building and maintenance factors related to indoor air quality. Boards of education are required to perform an evaluation before January 1, 2008, and every 5 years thereafter of every school building that is or has been constructed, extended, renovated, or replaced on or after January 1, 2003. No information on the implementation of the law was available at the time the present report was completed.

Despite its intuitive appeal, there is reason to have modest expectations about the extent to which general surveillance for indoor microbial agents would inform public-health decision-making at this point. As was mentioned in Chapter 3, no exposure standards have been established for molds or other dampness-related agents, although some occupational-exposure limitations were under discussion at the time this report was completed. Among the factors hindering the development of standards are uncertainties over which dampness-related exposures and at what exposure levels may be harmful and limitations associated with all the established means of exposure measurement. Qualitative indicators of the presence of mold also have limitations. The presence of visible mold—even toxigenic genera—is not an absolute indicator that health problems will result. The absence of visible mold also is not an absolute indicator that a building is free of infestation; there may be hidden sources.

The committee recommends (Chapter 2) that the determinants of dampness problems in buildings be studied to ascertain where to focus intervention efforts and health-effects research. The present state of the science, however, is insufficient to support a general assessment and monitoring effort for mold or other dampness-related agents for public-health policy purposes.

---

[3]Periodic inspections of rental properties and public housing for mold were proposed as part of the U.S. Toxic Mold Safety and Protection Act of 2003 (the Melina Bill); HR 1268, 108th Congress, 1st session. However, the version of this legislation introduced on March 13, 2003, did not include provisions for compiling the information gathered in these inspections.

Modification of Regulations, Building Codes, and Building-Related Contracts

As detailed in Chapter 6, the use of suitable materials, appropriate design and construction techniques, and common-sense maintenance practices can be expected to prevent or dramatically reduce building dampness and problematic water accumulation. Existing local, state, and national codes and methods of code enforcement may not, however, be sufficient to ensure that good practices are implemented. Indeed, some existing codes may inadvertently promote dampness. Chapter 2 notes, for example, that most codes require passive or active ventilation of crawl spaces; the entry of warm, humid outdoor air into ventilated crawl spaces, which are often cooler than outdoors, is a moisture source for the crawl space.

Some official entities have introduced code provisions that specifically address building dampness. In 2002, the California Occupational Safety and Health Standards Board promulgated the following new section in its "General Industry Safety Orders":

> (g) When exterior water intrusion, leakage from interior water sources, or other uncontrolled accumulation of water occurs, the intrusion, leakage or accumulation shall be corrected because of the potential for these conditions to cause the growth of mold. (Chapter 4, Subchapter 7, Article 9, §3362)

Numerous codes address water and moisture control and ventilation in general (Alliance for Healthy Homes, 2003). However, many of them "are based primarily on practical experience within the building sector or on non-health-related criteria such as perceived acceptability of air (for example, immediate perception of odor or irritation)" (Mendell et al., 2002). The effectiveness and cost effectiveness of codes with regard to health-risk reduction have not been systematically studied. The committee thus recommends, on the basis of its review of the evidence presented in Chapter 2, that current building codes be reviewed and modified as necessary to reduce dampness problems. It cannot draw any informed conclusions about code enforcement, but common sense suggests that more-rigorous enforcement—especially in low-income housing—may yield health benefits for residents.

Another strategy that might be considered is changes in contracts—lease agreements, professional liability-insurance terms, contracts between builders and owners, maintenance contracts, and the like—that would promote building design, construction, operation, and maintenance practices that reduce the potential for dampness problems or that would clarify the responsibilities of parties. Model contract language could be formulated by consensus groups for possible use by individuals or in communities.

Economic and Other Incentives

Action on the part of residents living in damp environments, builders, building owners, and other stakeholders will be required to change the present level of effort applied to preventing or reducing building dampness. Economic incentives are one well-established means of achieving such policy goals. Although specific examples of incentives for policy change related to the prevention of damp environments could not be located, there is precedent for using them in fire prevention and energy efficiency. Mills (2003) cites several examples of premium reductions and other incentives for actions and training that promote fuel conservation and for safety measures that result in cost savings to the insured and insurers. Such strategies depend on actuarially sound methods for quantifying problems and assessing expected benefits—a feature that poses considerable challenges when applied to the assessment of health effects of damp buildings. However, estimates of the health and productivity gains resulting from improvements in indoor environmental characteristics in general have been generated, and these suggest that the benefits may be considerable (Fisk, 2000).

Chapter 2 indicates that research is needed to determine the societal cost of dampness problems and to quantify the economic effects of design, construction, and maintenance practices that prevent or limit dampness problems. Such data would facilitate more informed evaluation of the priority that should be assigned to dampness interventions in the wide spectrum of housing-related issues. Inasmuch as housing stock, climatic conditions, and other factors vary across the country, this research may best be conducted on a state level (via departments of health or departments of social services) or a regional level, perhaps with a common protocol and funding. Climate, geography, and building type all influence indoor moisture levels and shape which problems are most likely to occur and which interventions may be most effective.

Incentives that address dampness might take several forms. The committee did not address the topic in detail, but it offers the following as examples of studies or experiments that might be undertaken to assess their effectiveness in reducing dampness problems:

- Governments could provide tax incentives, low-interest loans, streamlined application procedures, or other means to facilitate dampness-remediation efforts. For example, a provision of Connecticut Public Act 03-220 allows the state commissioner of education to approve applications for grants in excess of $100,000 for projects to remedy a "certified school indoor air quality emergency" (as determined by the state Department of Public Health) without seeking legislative approval.

• Health departments could be given permission to levy fines if water-leak problems in rental properties are not corrected after a specified period.
• Those responsible for the maintenance of public housing, office buildings, or schools could receive bonuses for meeting a defined set of goals for the prevention or reduction of dampness-related conditions (such as leaky roofs) or problems. Such programs could first be pilot-tested in government-owned buildings.

Guidelines for the Prevention of Dampness-Related Problems

Chapter 6 discusses several sets of guidelines for the assessment and management of mold-remediation activities that were developed by various government agencies and professional organizations. However, there is a lack of analogous guidance on preventing—or, more realistically, limiting the opportunity for—the conditions that might precipitate the need for such remediation. Guidelines are typically easier to develop than regulations or other more formal instruments but can still have great effect if they earn status as professional standards of care.

Prevention is a foundation principle in public health, and the committee believes that there is a need to develop and disseminate guidelines on building design, construction, operation, and maintenance for prevention of problematic damp indoor environments. Ideally, development should take place at the national level to promote their widespread adoption and help to avoid the proliferation of multiple and possibly conflicting sets of advice. Stakeholder groups should play an active role in providing input for the guidelines, but they should not be the organizing or sanctioning body for the effort in order to promote its credibility and general acceptance. The committee suggests that any effort to develop guidelines for prevention of problematic damp indoor environments take the following considerations into account:

• The guidelines should be formulated with multidisciplinary input and with input from a wide array of stakeholders.[4]
• The costs of implementing actions should be evaluated and their expected benefits identified.
• The guidelines should account for how differences in climate, geography, building type, and building age influence vulnerabilities to dampness and the best approaches to prevention.

---

[4]In 1995, participants in a workshop on indoor air convened by the American Thoracic Society decried the lack of venues where professionals working in the field could interact. They recommended that the Society and the American Lung Association "take the lead in conducting regular interdisciplinary workshops that promote in-depth discussion of key and timely issues" (ATS, 1997).

- To be effective, the guidelines should go beyond simple prescriptive application of available science and technology; professional judgment should be allowed.
- Draft guidelines should be subject to external review.

Because the prevention of problematic dampness is just one of a set of interrelated factors that can affect the indoor environment and the health of its occupants, consideration should also be given to developing the guidelines in the context of a wider set of principles that guide the creation and maintenance of healthy buildings.

Public-Health-Oriented Research and Demonstration

Prevention and control of building dampness are hampered by the lack of evidence regarding the effectiveness of various interventions. *Clearing the Air* (IOM, 2000), for example, acknowledged the logic of reducing dampness as a method for reducing asthma symptoms related to indoor dampness but noted that "no intervention studies clearly document that any form of dampness control works effectively to reduce symptoms or to reduce chances of asthma development." Additional support for that observation is provided in Chapters 2 and 6 of the present report.

Although there are references in the public-health literature to household-level housing interventions to improve health status, randomized controlled trials are comparatively rare. A review of studies of the relationship between housing interventions and improvements in health identified only 18 (11 prospective and seven retrospective) that evaluated effects on health, illness, and a variety of social measures (Thomson et al., 2001). Only six of the prospective studies and three of the retrospective studies included a control group. The researchers found that many studies showed improvements in self-reported health or reductions in symptoms after the intervention, but "small study populations and the lack of controlling for confounders limit the generalisability of these findings" (Thomson et al., 2001).

A later review by Saegert et al. (2003) examined a wider array of intervention studies but restricted the analysis to investigations conducted in the United States in 1990–2001. They found 72 studies; 21 (29%) addressed asthma triggers or air-quality hazards (including moisture or mold). Among the characteristics noted was that 85% of the interventions were one-time efforts—a single training program, cleaning or remediation. The type of intervention performed was almost evenly split among environmental improvements (31%), participant education (32%), or both (35%). Although 81% of the studies reported the interventions to be successful, only 51% indicated that the measured improvement was sustained. The authors observed that some common factors were associated with success in intervention studies:

- "Technological interventions appear most successful when the technology is effective, cheap, and durable and requires little effort to maintain or use. Such interventions are especially effective if accompanied by behavioral or knowledge training, and if hazard amelioration can be successfully accomplished through individual-level efforts alone."
- "Involving people more deeply in the solution of health problems, especially by home visits, appears to be especially effective and can improve multiple health outcomes."

The Seattle-King County Healthy Homes Project (Krieger et al., 2000, 2002, 2003) is among recent demonstration and research efforts that included a moisture or mold component. The project targeted 274 low-income households in the Seattle area that included a child with diagnosed asthma. Participants were randomly assigned to a "high-intensity" group (n = 138), which received comprehensive intervention services, or a "low-intensity" group (n = 136), which received a single home visit and limited services. (The low-intensity group received additional services at the completion of the 1-year study period.) For the high-intensity group, community health workers provided home assessments of several potential allergens associated with asthma and other risk factors (including dust, house dust mites, cockroaches, environmental tobacco smoke, rodents, and pesticides), followup education on how to prevent their occurrence or limit their effects, and in some circumstances active interventions to change the indoor environment. The mold and moisture portion of the assessment found that 77% of participant homes had "moisture problems" and over 20% had signs of water intrusion (Krieger et al., 2002). Several possible interventions were identified, including education on moisture sources and barriers, provision of cleaning materials, replacement of moldy shower curtains, installation or inspection and cleaning of ventilation fans, plugging of holes leading to the interior, and installation of vapor barriers. However, the authors noted that implementation of some of the more rigorous building interventions was beyond the resources of the study and that participants did not have the means to perform them independently. At the end of the study period, children in the high-intensity group experienced a greater decrease in asthma-symptom days (4.7; 95% CI, 3.6–5.9; vs 3.9; 2.6–5.2) and a statistically significant decrease in urgent health-services use (15%; 6.3–23.6%; vs –3.8%; –13.1–5.4%) compared with the low-intensity children (Krieger et al., 2003). That translated to an estimated $6,301–8,854 savings in urgent-care costs over a 2-month period relative to the low-intensity group. "Excessive moisture" in the home was reported to have decreased significantly in the high-intensity group,[5] whereas no significant changes occurred in the low-intensity group; no changes in the presence of mold were observed in either group. The degree to which the improvements were a consequence of the dampness-related portion of the intervention program is not known.

The Seattle effort and similar ones conducted as part of the National Cooperative Inner-City Asthma Study (Crain et al., 2002; Sullivan et al., 2002) and in Detroit (Parker et al., 2003) show that interventions at both the household and community levels can be mounted and targeted at a specific disease for susceptible people. Local health jurisdictions—such as Boston, Cleveland, and New York City—have developed comparable community wide "Healthy Homes" initiatives related to asthma and chemical hazards (Krieger and Higgins, 2002).

The Department of Housing and Urban Development (HUD) has provided grants to the Cuyahoga County (Ohio) Department of Development and the Illinois Department of Health to fund targeted mold and moisture interventions under its Healthy Homes Initiative (HUD, 2003a). The programs include education; environmental, biologic, and medical monitoring; and in some cases remediation. The projects are going on now, and no results had been published as of late 2003.

Virtually no data are available to compare effects of different interventions and intervention strategies. There is thus insufficient information to draw conclusions on the benefits of specific public-health-oriented housing interventions specifically related to moisture or mold. Existing research on exposure to indoor environmental agents in general suggests that targeted, intensive interventions may yield health benefits. And, as documented in Chapter 6, there is universal agreement that prompt remediation of water intrusion, leaks, spills, and standing water substantially reduces the potential for growth of dampness-related microbial agents and dampness-related degradation of building materials and furnishings.

The committee recommends that carefully designed and controlled longitudinal research be undertaken to assess the effects of population-based housing interventions on dampness and to identify effective and efficient strategies. As part of such studies, attention should be paid to definitions of dampness and to measures of effect; and the extent to which interventions are associated with decreased occurrence of specific negative health conditions should be assessed when possible.

---

[5]Floor dust loading, roach activity, and a composite measure of exposure to asthma triggers also decreased significantly in the high intensity group. It must be remembered that all of these environmental factors are thought to affect asthma outcomes and that the high intervention group also changed several behaviors thought to influence asthma exacerbation. The authors did not attribute the improvement in health outcomes to any one factor.

Education and Training

If prevention of potentially problematic damp indoor environments is to be achieved, public education and training of professionals will be integral parts of the solution.

Education and outreach to citizens—especially those in vulnerable communities—could have a large role in preventing or limiting the future effects of damp indoor environments. The research and demonstration projects referred to above that included education and training components are examples of how this might be accomplished in targeted, intensive interventions. Most efforts undertaken to date, however, have been simply the provision of information. Several federal agencies—including EPA (EPA, 2002), the Federal Emergency Management Agency (FEMA, 2003), and HUD and the U.S. Department of Agriculture (Healthy Homes Partnership, 2003)—have, for example, published guidance on identifying and remediating problematic dampness and mold in the home. Some states, including Minnesota (MDH, 2003), and such cities as New York (NYCDOH, 2002) also provide dampness-remediation information on their health-department Web sites. However, the committee did not identify any assessments of public awareness of such information, its distribution outside households with Internet access, or its diffusion to individuals concerned with dampness or mold problems.

Chapter 2 notes that although technical information on controlling moisture in residences and larger buildings has been developed and published (Lstiburek, 2001, 2002; Lstiburek and Carmody, 1996; Rose, 1997), anecdotal experience suggests that architects, engineers, facility managers, and contractors in the building trades often do not apply it when designing, constructing, or maintaining buildings. The chapter recommends that these building professionals receive better training on how and why dampness problems occur and on their prevention. The committee specifically recommends that a curriculum be developed for the training of building inspectors so that they can identify and require correction of common construction errors that lead to dampness problems; the curriculum should be disseminated broadly.

At the time this report was completed (late 2003), the mold assessment and remediation industry in the United States was largely nonregulated; laws regarding specific standards for education and training of practitioners were just coming into effect in Louisiana (Act 880; effective August 15, 2003) and Texas (HB 329; effective September 1, 2003). In the absence of government standards, a number of groups have created "certifications" with widely varied requirements for instruction, testing, and continuing education. The committee did not undertake to evaluate them.

Many allergy and pulmonary-medicine specialists (especially those practicing in areas with large agricultural populations) have experience in dealing with patients who have experienced adverse health effects due to exposure to mold, endotoxin, or organic dust. In general, however, physicians and other health-care providers are not well educated in the diagnosis and treatment of environmental health-related problems (IOM, 1995). Recognizing that, the IOM Committee on Curriculum Development in Environmental Medicine (IOM, 1995) identified five competence objectives related to the clinical management of patients, stating that graduating medical students should

- Understand the influence of the environment and environmental agents on human health based on knowledge of relevant epidemiologic, toxicologic, and exposure factors.
- Be able to recognize the signs, symptoms, diseases, and sources of exposure relating to common environmental agents and conditions.
- Be able to elicit an appropriately detailed environmental exposure history, including a work history, from all patients.
- Be able to identify and access the informational, clinical, and other resources available to help address patient and community environmental health problems and concerns.
- Be able to discuss environmental risks with their patients and provide understandable information about risk-reduction strategies in ways that exhibit sensitivity to patients' health beliefs and concerns.

Such proficiency clearly would be of benefit in addressing health issues that might be related to indoor dampness. The committee is aware that guidance to physicians on the recognition and management of health effects related to indoor mold exposure was being developed when this report was completed, but there were no publicly available documents for its review.

Public health professionals, particularly those who work in environmental health, will increasingly be required to assist in drafting recommendations for the prevention of excessive building dampness and the implementation of interventions to eliminate microbial contamination in affected buildings. It follows that their training should develop competence in the recognition of and appropriate response to problematic indoor dampness in workplace settings and in housing. Public health departments should assist in the development of educational campaigns to alert members of the general public to the risks associated with indoor dampness.

Environmental health is identified as a core component of public health education (IOM, 2003), and some courses related to indoor dampness and mold have been accredited for continuing education for public-health and industrial-hygiene professionals. Awareness of the possible health implications of damp indoor spaces in the public-health community has not, however, been examined.

Education is a common component of public-health interventions because it is relatively inexpensive and, at least in some circumstances, has been shown to be effective. A great deal of information is at least potentially available to citizens and to the various health and building professionals whose work is related to indoor environments, but little research has been conducted to evaluate whether it is reaching the right people or affecting their responses to possibly problematic dampness in buildings. The committee recommends that such research be conducted. Those formulating public-health interventions should examine asthma-education efforts, which may yield clues to effective strategies for communicating information on environmental health risks and effecting favorable changes in subjects' behavior. Any efforts undertaken need to include rigorous assessments of their short-term and long-term impacts. Six aspects of such education efforts merit attention:

- Do they raise awareness of potentially problematic conditions and exposures?
- Are they culturally and linguistically appropriate?
- Are they reaching vulnerable populations—those that are at high risk for dampness problems or adverse health outcomes?
- Do they effect changes in behavior that result in decreases in exposure or risk?
- Are changes sustained?
- Do the efforts address the possible role of the media and local and community-based organizations or institutions (such as schools and churches) in raising public awareness?

Collaboration

Implementation of the recommendations in this report will require collaboration with and among stakeholders in public health. Health departments, housing authorities, policy-makers, insurers, community-based organizations, and voluntary agencies will need to coordinate efforts to advocate for and effect the research efforts and changes in policy proposed here.

At least two collaborative approaches seem reasonable for introducing or furthering the prevention and control of damp indoor environments. The first is to integrate, or better integrate, dampness considerations into current efforts. HUD's Healthy Homes Initiative—which focuses on "researching and demonstrating low-cost, effective home hazard assessment and intervention methods, as well as on public education that stresses ways in which communities can mitigate housing-related hazards" (HUD, 2003b)—is one vehicle for promoting strategies that emphasize prevention of dampness. Another is the many respiratory-health-related, communitywide coalitions in the United States. A 2000 survey identified 63 asthma coalitions of varied size and resources (ACCP, 2000), and there were about 150 such organizations across the United States at the end of 2003 (Allies Against Asthma, 2003). Chapter 5 notes that the scientific literature supports an association between dampness and asthma exacerbation; it is appropriate to integrate dampness initiatives into coalition-led programs that do not have them.

A second approach is to develop new communitywide partnerships to address the specific prevention of potentially problematic damp indoor environments. Some aspects of dampness and the exposures that it promotes are unrelated to asthma or respiratory disease. It is possible and perhaps desirable to form collaborations that bring the stakeholders (including occupants and tenant organizations) together and focus them on effective actions in their mutual interest. Mobilization of stakeholders and communitywide approaches may be particularly important in low-income areas, where resources are scarce. The new partnerships may attract stakeholders not usually evident in community health-related endeavors—for example, building professionals—who can lend new perspectives to efforts to affect change.

FINDINGS, RECOMMENDATIONS, AND RESEARCH NEEDS

On the basis of the review of the papers, reports, and other information presented in this chapter, the committee has reached the following findings and recommendations and has identified the following research needs regarding the public-health dimension of damp indoor environments. Preceding chapters of this report provide the foundation for some the recommendations and offer additional observations on health, building, and prevention and remediation issues.

Findings

• Excessive indoor dampness is a public-health problem: dampness is prevalent in residential housing in a wide array of climates; sufficient evidence of an association exists between signs of dampness and upper respiratory tract symptoms, cough, wheeze, and asthma symptoms in sensitized persons; and sufficient evidence of an association exists between signs of mold and upper respiratory tract symptoms, cough, wheeze, asthma symptoms in sensitized persons and hypersensitivity pneumonitis in susceptible persons.

• In the absence of a generally accepted definition of dampness or what constitutes a "dampness problem," the advice offered in remediation guidelines developed by government and well-established professional associations can serve as references for when and what level of response is appropriate.

- Indoor dampness has an economic dimension. Economic factors may encourage poor building design and construction practices; they may also create incentives to forgo or limit investment in maintenance and other measures that might help to prevent or reduce moisture problems. Cost-effectiveness studies are needed to assess the savings that damp-building prevention or remediation might generate.

- An appropriate public health goal should be to prevent or reduce the incidence of potentially problematic damp indoor environments, that is, environments that may be associated with undesirable health effects, particularly in vulnerable populations. However, there are serious challenges associated with achieving that goal, given the lack of information on key scientific questions regarding the health effects of dampness-related agents, and questions of exposure and dose.

Recommendations and Research Needs

- CDC, other public-health-related, and building-management-related funders should provide new or continuing support for research and demonstration projects that address the potential and relative benefit of various strategies for the prevention or reduction of damp indoor environments, including data acquisition through assessment and monitoring, building code modification or enhanced enforcement, contract language changes, economic and other incentives, and education and training. These projects should include assessments of the economic effects of preventing building dampness and repairing damp buildings and should evaluate the savings generated from reductions in morbidity and gains in the useful life of structures and their components associated with such interventions.

- Carefully designed and controlled longitudinal research should be undertaken to assess the effects of population-based housing interventions on dampness and to identify effective and efficient strategies. As part of such studies, attention should be paid to definitions of dampness and to measures of effect; and the extent to which interventions are associated with decreased occurrence of specific negative health conditions should be assessed when possible.

- Government agencies with housing-management responsibility should evaluate the benefit of adopting economic-incentive programs designed to reward actions that prevent or reduce building dampness. Ideally, these should be coupled with independent assessments of effectiveness.

- HUD or another appropriate government agency with responsibility for building issues should provide support for the development and dissemination of consensus guidelines on building design, construction, operation, and maintenance for prevention of dampness problems. Development of the guidelines should take place at the national level and should be under the aegis of either a government body or an independent nongovernment organization that is not affiliated with the stakeholders on the issue.

• CDC and other public-health-related funders should provide new or continuing support for research and demonstration projects that:

— Develop communication instruments to disseminate information derived from the scientific evidence base regarding indoor dampness, mold and other dampness-related exposures, and health outcomes to address public concerns about the risk from dampness-related exposures, indoor conditions, and causes of ill health.

— Foster education and training for clinicians and public-health professionals on the potential health implications of damp indoor environments.

• Government and private entities with building design, construction, and management interests should provide new or continuing support for research and demonstration projects that develop education and training for building professionals (architects, home builders, facility managers and maintenance staff, code officials, and insurers) on how and why dampness problems occur and how to prevent them.

• Those formulating the education and training programs discussed above should include means of evaluating whether their programs are reaching relevant persons and, ideally, whether they materially affect the occurrence of moisture or microbial contamination in buildings or occupant health.

REFERENCES

ACCP (American College of Chest Physicians). 2000. A Descriptive Study of Asthma Coalitions. http://www.chestnet.org/education/physician/asthma/study/. accessed October 28, 2003.

Alliance for Healthy Homes. 2003. Housing and Building Codes. http://www.afhh.org/aa/aa_housing_codes.htm. accessed October 24, 2003.

Allies Against Asthma. 2003. Allies Against Asthma—coalition connections. http://www.asthma.umich.edu/coalition_connections/map.html.

ATS (American Thoracic Society). 1997. Achieving Healthy Indoor Air. Report of the ATS Workshop: Santa Fe, New Mexico, November 16–19, 1995. American Journal of Respiratory and Critical Care Medicine 156(3):S33–S64.

CIA (Central Intelligence Agency). 1997. The World Factbook 1996.

Crain EF, Walter M, O'Connor GT, Mitchell H, Gruchalla RS, Kattan M, Malindzak GS, Enright P, Evans R 3rd, Morgan W, Stout JW. 2002. Home and allergic characteristics of children with asthma in seven U.S. urban communities and design of an environmental intervention: the Inner-City Asthma Study. Environmental Health Perspectives 110(9):939–945.

Dedman DJ, Gunnell D, Davey Smith G, Frankel S. 2001. Childhood housing conditions and later mortality in the Boyd Orr cohort. Journal of Epidemiology and Community Health 55:10–15.

Duffy J. 1974. A History of Public Health in New York City 1866–1966. New York: Russell Sage Foundation.

Dunn JR, Hayes MV. 2000. Social inequality, population health, and housing: a study of two Vancouver neighborhoods. Social Science and Medicine 51:563–587.

Eggleston PA. 2000. Environmental causes of asthma in inner city children; the National Cooperative Inner City Asthma Study. Clinical Reviews in Allergy & Immunology 18(3):311–324.

Ehrenreich B. 2001. Nickel and Dimed: On (Not) Getting By in America. New York: Owl Books. Henry Holt and Company.

Evans GW, Kantrowitz E. 2002. Socioeconomic status and health: the potential role of environmental risk exposure. Annual Review of Public Health 23:303–331.

Evans J, Hyndman S, Stewart-Brown S, Smith D, Petersen S. 2000. An epidemiological study of the relative importance of damp housing in relation to adult health. Journal of Epidemiology and Community Health 54(9):677–686.

FEMA (Federal Energy Management Agency). 2003. Dealing with Mold & Mildew in Your Flood Damaged Home. http://www.fema.gov/pdf/reg-x/mold_mildew.pdf.

Fisk WJ. 2000. Health and productivity gains from better indoor environments and their relationship with building energy efficiency. Annual Review of Energy and the Environment 25(1):537–566.

Freeman L. 2002. America's affordable housing crisis: a contract unfulfilled. American Journal of Public Health 92:709–712.

Galishoff S. 1988. Newark: The Nation's Unhealthiest City 1832–1895. New Brunswick, NJ: Rutgers University Press.

Healthy Homes Partnership. 2003. Help Yourself to a Healthy Home. http://www.hud.gov/offices/lead/healthyhomes/healthyhomebook.pdf. accessed October 28, 2003.

Healy JD. 2003. Excess winter mortality in Europe: a cross country analysis identifying key risk factors. Journal of Epidemiology and Community Health 57(10):784–789.

HUD (Department of Housing and Urban Development). 2003a. HHI Mold and Moisture Project Summaries. http://www.hud.gov/offices/lead/hhi/mm.cfm. accessed October 28, 2003.

HUD. 2003b. Healthy Homes Program. http://www.hud.gov/offices/lead/hhi/index.cfm. accessed October 28, 2003.

IOM (Institute of Medicine). 1988. The Future of Public Health. Washington, DC: National Academy Press.

IOM. 1993. Indoor Allergens: Assessing and Controlling Adverse Health Effects. Washington, DC: National Academy Press.

IOM. 1995. Environmental Medicine: Integrating a Missing Element into Medical Education. Washington, DC: National Academy Press.

IOM. 2000. Clearing the Air: Asthma and Indoor Air Exposures. Washington, DC: National Academy Press.

IOM. 2003. Who Will Keep the Public Healthy? Educating Public Health Professionals for the 21st Century. Washington, DC: The National Academies Press.

Krieger J, Higgins DL. 2002. Housing and health: time again for public health action. American Journal of Public Health 92:758–768.

Krieger JW, Song L, Takaro TK, Stout J. 2000. Asthma and the home environment of low-income urban children: preliminary findings from the Seattle-King County health homes project. Journal of Urban Health 77(1):50–67.

Krieger J, Takaro TK, Allen C, Song L, Weaver M, Chai S, Dickey P. 2002. The Seattle-King County healthy homes project: implementation of a comprehensive approach to improving indoor environmental quality for low-income children with asthma. Environmental Health Perspectives Supplements 110(2):311–322.

Krieger JW, Takaro TK, Song L, Weaver M. 2003. The Seattle-King County healthy homes project: a randomized, controlled trial of a community health worker intervention to decrease exposure to indoor asthma triggers among low-income children. Presented at the 2003 annual meeting of the American Thoracic Society, Seattle, WA.

Lawrence R, Martin D. 2001. Moulds, moisture and microbial contamination of First Nations housing in British Columbia, Canada. International Journal of Circumpolar Health 60:150–156.

Lstiburek J. 2001. Moisture, building enclosures and mold. Part 1 of 2. HPAC Engineering. Dec.:22–26.

Lstiburek J. 2002. Moisture, building enclosures and mold. Part 2 of 2. HPAC Engineering. Jan.:77–81.

Lstiburek J, Carmody J. 1996. Moisture Control Handbook: Principles and Practices for Residential and Small Commercial Buildings. New York: John Wiley and Sons.

Markus TA. 1993. Cold, condensation and housing poverty. In: Burridge R, Ormandy D, eds. Unhealthy Housing: Research, Remedies and Reform. New York: Spon Press. pp. 141–167.

Marsh BT. 1982. Housing and health: the role of the environmental health practitioner. Journal of Environmental Health 45(3):123–128.

Matte TD, Jacobs DE. 2000. Housing and health—current issues and implications for research and programs. Journal of Urban Health 77:7–25.

MDH (Minnesota Department of Health). 2003. Recommended Best Practices for Mold Remediation in Minnesota Schools. Environmental Health Division, Indoor Air Unit. http://www.health.state.mn.us/divs/eh/indoorair/schools/remediation.pdf.

Melosi MV. 2000. The Sanitary City: Urban Infrastructure in America from Colonial Times to the Present. Baltimore, MD: Johns Hopkins University Press.

Mendell MJ, Fisk WJ, Kreiss K, Levin H, Alexander D, Cain WS, Girman JR, Hines CJ, Jensen PA, Milton DK, Rexroat LP, Wallingford KM. 2002. Improving the health of workers in indoor environments: priority research needs for a national occupational research agenda. American Journal of Public Health 92(9):1430–1440.

Mercer JB. 2003. Cold—an underrated risk factor for health. Environmental Research 92(1):8–13.

Mills E. 2003. The insurance and risk management industries: new players in the delivery of energy-efficient and renewable energy products and services. Energy Policy 31:1257–1272.

Nguyen TTL, Pentikäinen T, Rissanen P, Vahteristo M, Husman T, Nevalainen A. 1998. Health related costs of moisture and mold in dwellings. Publication of the National Public Health Institute, B13, Kuopio University Printing Office, Finland.

NYCDOH (New York City Department of Health). 2002. Guidelines on Assessment and Remediation of Fungi in Indoor Environments. New York City Department of Health & Mental Hygiene, Bureau of Environmental & Occupational Disease Epidemiology. http://www.ci.nyc.ny.us/html/doh/html/epi/moldrpt1.html.

Parker EA, Israel BA, Williams M, Brakefield-Caldwell W, Lewis TC, Robins T, Ramirez E, Rowe Z, Keeler G. 2003. Community Action Against Asthma: examining the partnership process of a community-based participatory research project. Journal of General Internal Medicine 18(7):558–567.

Parker J. 1994. Building codes: the failure of public policy to institutionalize good practice. Environmental and Urban Issues 21(4):21–26.

Rose W. 1997. Control of moisture in the modern building envelope: the history of the vapor barrier in the United States 1923–1952. APT Bulletin 18(4):13–19.

Rosner D. 1995. Hives of Sickness: Public Health and Epidemics in New York City. New Brunswick, NJ: Rutgers University Press.

Saegert SC, Klitzman S, Freudenberg N, Cooperman-Mroczek J, Nassar S. 2003. Healthy housing: a structured review of published evaluations of US interventions to improve health by modifying housing in the United States, 1990–2001. American Journal of Public Health 93(9):1471–1477.

Sullivan SD, Weiss KB, Lynn H, Mitchell H, Kattan M, Gergen PJ, Evans R; National Cooperative Inner-City Asthma Study (NCICAS) Investigators. 2002. The cost-effectiveness of an inner-city asthma intervention for children. Journal of Allergy and Clinical Immunology 110(4):576–581.

Susser M. 1973. Causal Thinking in the Health Sciences: Concepts and Strategies in Epidemiology. New York: Oxford Press.

Thomson H, Petticrew M, Morrison, D. 2001. Health effects of housing improvement: systematic review of intervention studies. British Medical Journal 323:187–190.

U.S. Census Bureau. 1985 and onward, biennially. The American Housing Survey. Bureau of the Census, for the Department of Housing and Urban Development. http://www.census.gov/hhes/www/ahs.html.

U.S. EPA (US Environmental Protection Agency). 2000. Indoor Air Quality (IAQ) Tools for Schools Kit, Second Edition. Office of Air and Radiation, Office of Radiation and Indoor Air, Indoor Environments Division. EPA 402-K-95-001. http://www.epa.gov/iaq/schools/toolkit.html.

U.S. EPA. 2002. A Brief Guide to Mold, Moisture, and your Home. Office of Air and Radiation, Indoor Environments Division. Washington, DC: EPA. EPA 402-K-02-003. http://www.epa.gov/iaq/molds/images/moldguide.pdf.

Veiller L. 1921. Housing as a factor in health progress in the past fifty years. In: MP Ravenel (ed.): A Half Century of Public Health. New York: American Public Health Association. pp. 323–334.

Viet SM, Rogers J, Marker, Fraser A. Bailey M. 2003. First National Environmental Health Survey of Child Care Centers—Final Report—Volume II: Analysis of Allergen Levels on Floors. Prepared for the Office of Healthy Homes and Lead Hazard Control, U.S. Department of Housing and Urban Development.

Vojta PJ, Friedman W, Marker DA, Clickner R, Rogers JW, Viet SM, Muilenberg ML, Thorne PS, Arbes SJ Jr, Zeldin DC. 2002. First National Survey of Lead and Allergens in Housing: survey design and methods for the allergen and endotoxin components. Environmental Health Perspectives 110(5):527–532.

Appendix A

Workshop Presentations and Speakers

WORKSHOP ONE

March 26, 2002

The Foundry Building
1055 Thomas Jefferson St. NW
Washington, DC

Review of Acute Idiopathic Pulmonary Hemorrhage in Infants
Clive M. Brown, MBBS, MPH, and Stephen Redd, MD, National Center
for Environmental Health, Centers for Disease Control and Prevention

Overview of EPA Mold-Related Activities
Laura S. Kolb, MPH, Indoor Environments Division, EPA

Mold and Moisture Control Activities at HUD
Peter Ashley, DrPH, Office of Healthy Homes and Lead Hazard Control,
HUD

Overview of Mold-Related Research Activities in Finland
Aino Nevalainen, PhD, Head of the Laboratory of Environmental
Microbiology, National Public Health Institute, Finland

Toxigenic Fungi: Are they significant environmental health threats?
Dorr G. Dearborn, PhD, MD, Rainbow Babies and Children's Hospital,
Cleveland, Ohio

Damp Indoor Spaces Investigation and Remediation
Terry M. Brennan, MS, Camroden Associates, Inc., Westmoreland, New York

"Toxic Mold": History and Measurement
John H. Haines, PhD, New York State Museum, New York State Biological Survey, Albany, New York

WORKSHOP TWO

June 17, 2002

**Keck Center of the National Academies
500 Fifth Street NW
Washington, DC**

Indoor Damp Spaces Epidemiology
Bert Brunekeef, PhD, Professor of Environmental Epidemiology, Institute of Risk Assessment Sciences, University of Utrecht, Utrecht, The Netherlands

Experimental Methods to Assess the Pulmonary Effects of Fungi
Carol Y. Rao, ScD, National Institute for Occupational Safety and Health, Division of Respiratory Disease Studies, Field Studies Branch, Morgantown, West Virginia

Consultant's Viewpoint on Mold
John A. Tiffany, MS, President, Industrial Hygienist, Tiffany-Bader Environmental, Inc., Lawrenceville, New York

WORKSHOP THREE

October 8, 2002

**J. Erik Johnson Woods Hole Center
314 Quissett Ave.
Woods Hole, MA**

Clinical Issues
Eckardt Johanning, MD, MSc, Occupational and Environmental Life Science, Fungal Research Group, Inc., Albany, New York

Toxicology
Joseph Brain, SD, Cecil K. and Philip Drinker Professor of Environmental
Physiology, Department of Environmental Health, Harvard School of
Public Health

Limiting Conditions for Fungal Growth on Building Materials
Susan C. Doll, ScD, Department of Environmental Health, Harvard
School of Public Health

Appendix B

Committee, Consultant, and Staff Biographies

COMMITTEE AND CONSULTANT BIOGRAPHIES

Noreen M. Clark, PhD (Chair), is dean of the University of Michigan School of Public Health and the Marshall H. Becker Professor of Public Health and Professor of Pediatrics at the University. She was formerly chair of the school's Department of Health Behavior and Health Education. Her re-search specialty concerns the social and behavioral aspects of chronic-disease management, including asthma and other respiratory diseases with environmental-exposure components. She has also conducted large-scale trials of behavioral and educational interventions in clinical and community settings aimed at improving disease management by patients, families, and health-care providers. Outcomes of interest in those evaluations have included patient health status, quality of life, and health-care use. Her studies have also included assessments of partnerships and coalitions designed to enhance community wide management of disease. Dr. Clark is a member of the Institute of Medicine.

Harriet M. Ammann, PhD, DABT, is senior toxicologist for the Washington (state) Department of Ecology Air Quality Program. She is also adjunct associate professor in the Department of Environmental Health of the University of Washington's School of Occupational Health and Community Medicine. In her work for the state of Washington, Dr. Ammann provides support to a variety of environmental-health programs, including ambient-and indoor-air programs. She has participated in evaluations of schools and public buildings with air-quality problems, and has presented on toxic effects of air contaminants (indoor and outdoor), effects on sensitive populations, and other health issues throughout the state. Through her work, she has developed an interest in the toxicology of mold as an indoor air contaminant and has published and presented on mold toxicity related to human health. Dr. Ammann is involved in several professional organizations, serving on the American Lung Association's National Technical Committee for the Building Indoor Air Quality Program. She previously served as vice-chair of the Bioaerosols Committee of the American Conference of Governmental Industrial Hygienists.

Terry Brennan, MS, is a building scientist and educator. He owns a small company, Camroden Associates, that provides forensic, analytic, research, and training services to the building community, the research community, the public-health community, and owners and occupants of buildings. Mr. Brennan is on the editorial boards of *Environmental Building News* and *Heating Piping and Air Conditioning Magazine.* His recent work includes healthy-building training with Joe Lstiburek of Building Science Corporation for the Boston Region HUD Healthy Homes project, analysis of fungal dynamics in crawl space homes in North Carolina for AEC's Crawl space Characterization Pilot Study of sealed and vented crawl spaces, and teaching moisture and mold workshops in the Pacific Northwest for Washington State University and the University of Alaska. He is a member of the ASHRAE 62.2 Ventilation and Air Quality Committee.

Bert Brunekreef, PhD, is professor of environmental and occupational health at the Institute for Risk Assessment Sciences at the University of Utrecht, The Netherlands. He obtained his PhD in environmental epidemiology from Wageningen University and did postdoctoral work at the Harvard School of Public Health, where he studied the health effects of living in damp homes. Dr. Brunekreef has served as coordinator of a European Union-funded study on acute effects of air pollution on the airways of asthmatic children conducted by 14 research centers in Europe. He spent several years as a councilor for the International Society for Environmental Epidemiology and was its president in 2000-2001. Dr. Brunekreef is a coauthor of over 180 peer-reviewed journal articles in environmental epidemiology and exposure assessment.

Jeroen Douwes, PhD, is associate director of the Massey University Centre for Public Health Research (CPHR), Wellington, New Zealand. He worked jointly at the CPHR and the Institute for Risk Assessment Sciences (University of Utrecht, The Netherlands) for several years and is now based permanently in New Zealand. Dr. Douwes is engaged in a large number of national and international studies on respiratory health in relation to indoor and occupational microbial exposures. His current research includes environmental factors that protect against the development of allergies and asthma and the role of nonallergic airway inflammation in asthma. He is also a member of the World Health Organization (WHO) Committee on the Health Guidelines for Biological Agents in the Indoor Environment.

Peyton A. Eggleston, MD, is professor of pediatrics and professor of environment health sciences at the Johns Hopkins University, where he maintains both a clinical practice and a research effort. He is board-certified in allergy and immunology and in pediatrics. Dr. Eggleston is principal investigator of the Johns Hopkins University Center for the Asthmatic Child in the Urban Environment, one of EPA's national centers for environmental research. His research focus is environmental allergens—their role in respiratory diseases (in particular, asthma), risk factors for sensitization, means of avoidance, and methods and effectiveness of indoor environmental control.

William J. Fisk, MS, PE, is senior staff scientist and head of the Indoor Environment Department (IED) in the Environmental Energy Technologies Division of the Lawrence Berkeley National Laboratory. He also serves as group leader for IED's Ventilation and Indoor Air Quality Control Technologies Group. Mr. Fisk's primary research interests include indoor air-pollutant exposure, indoor air-quality control technologies, and indoor environmental quality and health; he has published and consulted extensively in these fields. He is on the editorial board of the journal *Indoor Air: International Journal of Indoor Air Quality and Climate*. In 1999, he was elected a member of the International Academy of Indoor Air Sciences.

Robert E. Fullilove III, EdD, is associate dean for community and minority affairs and associate professor of clinical public health in sociomedical sciences at the Columbia University Mailman School of Public Health. He codirects the Community Research Group at the New York State Psychiatric Institute and Columbia University. His research includes work on housing and health and the effects of public-health interventions on minority and low-socioeconomic-status communities. Dr. Fullilove served on the Board on Health Promotion and Disease Prevention at the Institute of Medicine in 1995–2001. He also serves on the editorial board of the *Journal of Public Health Policy*.

Judith Guernsey, PhD, is associate professor in the Department of Community Health and Epidemiology and holds a cross appointment in the Institute for Resource and Environmental Studies at Dalhousie University in Halifax, Nova Scotia. Her research interests include epidemiology and population health interventions in occupational and environmental health, with particular focus on bioaerosols in workplace and community settings. Dr. Guernsey directs the Atlantic Provinces Centre for Research in Rural Environmental Health. She is a peer reviewer for the Canadian Institute for Health Research, National Cancer Institute of Canada, the *Canadian Journal of Public Health,* and the *American Journal of Industrial Medicine.* Dr. Guernsey is a member of the Occupational Epidemiology Committee and the International Affairs Committee of the American Industrial Hygiene Association, the International Society of Environmental Epidemiology, and the Canadian Public Health Association.

Aino Nevalainen, PhD, is head of the Laboratory of Environmental Microbiology at the National Public Health Institute and a senior researcher at the Academy of Finland. Her research interests include microbial growth on building materials, health effects of exposure to building-related mold and moisture exposure, and the microbiologic quality of indoor air. Dr. Nevalainen serves as a member of the Steering Committee for the WHO Committee on the Health Guidelines for Biological Agents in the Indoor Environment; she was also the international advisor for the International Conferences for Indoor Air Quality and Climate and an evaluator for the Danish Guidelines for Indoor Air Quality. She has written numerous peer-reviewed research articles on the characterization and remediation of damp indoor spaces and has been an invited speaker at scientific conferences worldwide.

Susanna G. Von Essen, MD, is professor in the Department of Internal Medicine, Section of Pulmonary and Critical Care Medicine, at the University of Nebraska Medical Center. Her research focuses on the respiratory health of farmers—a population with extensive exposure to bioaerosols—and other rural-health issues. In addition to having a clinical practice, she is project director of a 5-year grant from the National Institute of Environmental Health Sciences that is designed to enhance education in environmental and occupational medicine for primary-care providers. Dr. Von Essen studied parasitology in Freiburg, Germany, on a Fulbright scholarship. She completed her internal-medicine residency and pulmonary training at UNMC, where she has been on the faculty since 1988.

STAFF BIOGRAPHIES

Rose Marie Martinez, ScD, is director of the Institute of Medicine (IOM) Board on Health Promotion and Disease Prevention. Before coming to IOM, she was a senior health researcher at Mathematica Policy Research, where she focused on health-workforce issues, access to care for vulnerable populations, managed care, and public-health issues. Dr. Martinez is a former assistant director for health financing and policy with the U.S. General Accounting Office, where she led evaluations and policy analysis in national and public health issues.

David A. Butler, PhD, is a senior program officer in the Institute of Medicine (IOM) Board on Health Promotion and Disease Prevention. He received his BS and MS in engineering from the University of Rochester and PhD in public-policy analysis from Carnegie-Mellon University. Before joining IOM, Dr. Butler served as an analyst for the U.S. Congress Office of Technology Assessment and was a research associate in the Department of Environmental Health at the Harvard School of Public Health. He has directed several IOM studies on environmental-health and risk-assessment topics, resulting in the reports *Veterans and Agent Orange: Update 1998; Veterans and Agent Orange: Update 2000; Clearing the Air: Asthma and Indoor Air Exposures;* and *Escherichia coli O157:H7 in Ground Beef: Review of a Risk Assessment.*

Jennifer A. Cohen is a research associate in the Institute of Medicine (IOM) Board on Health Promotion and Disease Prevention. She received her undergraduate degree in art history from the University of Maryland. She has also been involved with the IOM committees that produced *Clearing the Air: Asthma and Indoor Air Exposures; Escherichia coli O157:H7 in Ground Beef: Review of a Risk Assessment; Organ Procurement and Transplantation; Veterans and Agent Orange: Herbicide/Dioxin Exposure and Type 2 Diabetes; Veterans and Agent Orange: Update 2000; Veterans and Agent Orange: Herbicide/Dioxin Exposure and Acute Myelogenous Leukemia in the Children of Vietnam Veterans; Veterans and Agent Orange: Update 2002; Characterizing Exposure of Veterans to Agent Orange and Other Herbicides Used in Vietnam: Final Report;* and *Veterans and Agent Orange: Length of Presumptive Period for Association Between Exposure and Respiratory Cancer.*

Joe A. Esparza is a senior project assistant in the Institute of Medicine (IOM) Board on Health Promotion and Disease Prevention. He attended Columbia University, where he studied biochemistry. Before joining IOM, he worked with the Board on Agriculture and Natural Resources (BANR) of the National Research Council. While with BANR, he was involved with the committees that produced *Frontiers in Agricultural Research: Food, Health, Environment, and Communities; Nutrient Requirements of Dogs and Cats;* and *Air Emissions from Animal Feeding Operations: Current Knowledge, Future Needs.* At IOM, he has assisted on the reports *Veterans* *and Agent Orange: Update 2002; Medicolegal Death Investigation System: Workshop Summary; Learning From SARS: Preparing for the Next Disease Outbreak (Workshop Summary); Characterizing Exposure of Veterans to Agent Orange and Other Herbicides Used in Vietnam: Final Report;* and *Veterans and Agent Orange: Length of Presumptive Period for Association Between Exposure and Respiratory Cancer.*

Elizabeth J. Albrigo was a project assistant in the Institute of Medicine (IOM) Board on Health Promotion and Disease Prevention. She received her undergraduate degree in psychology from the Virginia Polytechnic Institute and State University. She is involved with the IOM Committee on Damp Indoor Spaces and Health. She also helped to facilitate the production of the reports *Veterans and Agent Orange: Update 2002; Veterans and Agent Orange: Herbicide/Dioxin Exposure and Acute Myelogenous Leukemia in the Children of Vietnam Veterans;* and Escherichia coli O157:H7 *in Ground Beef: Review of a Risk Assessment.*

Index

N

National Cooperative Inner-City Asthma
Study, 323
National Health and Nutrition Examination
Survey (NHANES II), 313
National Institute for Occupational Safety
and Health, 48
National Survey of Lead and Allergens in
Housing (NSLAH), 316
Nausea and related symptoms, and
exposure to damp indoor
environments, 246
Neuropsychiatric symptoms, 247–250
Neurospora, 57
Neurotoxic effects of indoor molds and
bacteria, 157–164
gliotoxin, 163
ochratoxin, 162–163
paralysis-inducing neurotoxins, 162
tremorgenic toxins, 161–162
trichothecenes, 163–164
Neurotoxic mycotoxins and effects, 158–
161
New York City Department of Health
(NYCDOH), 273–275, 287, 299,
324
NHANES II. *See* National Health and
Nutrition Examination Survey
NOAELS. *See* No-observed-adverse-effect
levels
No-observed-adverse-effect levels
(NOAELs), 126, 145
Nocardiopsis, 71
Nonculture methods, for assessing
microorganisms, 102–103
North American Air Duct Cleaners
Association, 296
NSLAH. *See* National Survey of Lead and
Allergens in Housing
NYCDOH. *See* New York City Department
of Health

O

Occupants of buildings
collecting histories of, 287–288
as sources of moisture, 41–42
Ochratoxin, 162–163
ODTS. *See* Organic dust toxic syndrome
Organic dust toxic syndrome (ODTS), 233

P

Paecilomyces, 57
Paralysis-inducing neurotoxins, 162
PCR. *See* Polymerase chain reaction
technologies
Peak expiratory flow (PEF), 96, 211
PEF. *See* Peak expiratory flow
Penicillium, 57, 66, 70–71, 128, 145, 157,
161–162, 200, 216, 225, 230–231,
236, 291
P. aurentiogriseum, 149–150, 169
P. bravicompactum, 54
P. chrysogenum, 149, 230, 292, 297
P. citreo-viride, 162
P. crustosum, 162
P. cyclopium, 169
P. expansum, 149
P. glabrum, 149
P. simplicissimum, 162
P. spinulosum, 135–136, 217
P. verruculosum var. *cyclopium,* 162,
169
P. viridicatum, 149, 169
Peptidoglycans, 68–69
Personal vs area sampling, 97–98
Phenylspirodrimanes, 147
Planning remediation activities, 295–301
decontaminating or removing damaged
materials as appropriate, 298–299
eliminating or limiting moisture
sources and drying the materials,
297–298
establishing appropriate containment
and worker and occupant
protection, 295–297
evaluating whether the space has been
successfully remediated, 299–300
reassembling the space to prevent
recurrence by controlling moisture
and nutrients, 301
Plumbing and wet rooms, 37–38
Polymerase chain reaction (PCR)
technologies, 103
Polyvinyl chloride (PVC) materials, 73–74
Potential for water and moisture sources
leading to excessive indoor
dampness, 3
Predictive exposure models, 109–110
Prevalence, 44–49
of dampness in buildings, 44–49